ROVING

MARS

STEVE SQUYRES

ROVING

MARS

Spirit, Opportunity, and the

Exploration of the Red Planet

HYPERION

NEW YORK

To Joe, Ray and Carl

ACKNOWLEDGMENTS

The most difficult part of writing this book was deciding what to leave out. The *Mars Exploration Rover* project was the work of literally thousands of dedicated, talented and passionate people. It's impossible to tell all of their stories in a book of this length, or even to hint at the full richness and complexity of their contributions. Indeed, many people without whom the mission could not have succeeded do not even appear in these pages. The inspired leadership of a Matt Wallace, the performance under pressure of a Glenn Reeves and the scientific insight of a John Grotzinger were matched thousands of times over, for many years, in stories equally deserving of telling. The few people here whose stories I do tell—the characters in this book—are also surrogates for uncounted others who did as much.

Many people gave generously of their time and effort to help me write this book. Mark Adler shared his detailed notes of some of the earliest days of the project. Leo Bister, Ralf Gellert, Steve Gorevan, Randy Lindemann, Rob Manning, Tom Myrick, Glenn Reeves, Adam Steltzner and Matt Wallace all sat for long interviews, telling their stories and patiently explaining the details of their work. John Beck gave me unlimited access to many hours of video footage that he shot during flight operations, helping enormously in getting the details of dialogue right. Many people on the project contributed snapshots for the photo inserts; their names are given in the photo credits.

Gentry Lee, Charles Elachi, Pete Theisinger, Barry Goldstein and Mark Adler all reviewed the entire manuscript, checking for technical errors and giving me good literary advice. Where mistakes remain the fault is mine, and where I have not followed their advice it has been at my own peril.

Peter Matson at Sterling Lord Literistic was an early believer in this book, and Gretchen Young at Hyperion came on board quickly and provided sound editorial guidance. Zareen Jaffery kept me on the straight and narrow, providing gentle prodding on the many occasions when I needed it. Pam Smith took on the Herculean task of pulling together our best attempt at a list of everyone who worked on the *MER* project.

The people who really have kept me sane through this story and its telling were my family. I owe my deepest thanks to my wife, Mary, and to our two daughters, Nicky and Katy. The *MER* project, for me, has lasted seventeen years, and there was not a moment in all that time when I did not know that I had my family's unqualified love and support. They made my contribution to the project possible, and I am more grateful to them than I can say. You've heard the expression that someone "married well"? I married well.

As I write this, we are now 445 days into our 90-day mission to Mars. The adventure continues, and each day *Spirit* and *Opportunity* bring us new discoveries. What I really wish I could do in these acknowledgments, with all of the depth and sincerity that they deserve, is individually thank every one of my friends, colleagues and comrades on the *MER* project who made these discoveries possible. The book itself must stand as my attempt to do so.

Ithaca, New York
April 4, 2005

CONTENTS

PROLOGUE....1

PART ONE
BEGINNINGS....7

1 MOSCOW, 1987 9
2 ATHENA 29
3 APEX 42
4 THE SPACECRAFT GRAVEYARD 54
5 ADLER'S INSPIRATION 73
6 THE DECISION 86

PART TWO
DEVELOPMENT....95

7 "YOU'RE BACKED INTO A
   HYPERDIMENSIONAL CORNER" 97
8 SHREDDING AND SQUIDDING......... 120
9 ATLO 142
10 THE CAPE 164
11 THE FUSE 179
12 LAUNCH 197

PART THREE
FLIGHT....219

13 FINAL APPROACH 221
14 GUSEV 237
15 "WE'VE GOT NOTHING" 267
16 EAGLE CRATER 288
17 ENDURANCE 322
18 POT OF GOLD 350

THE *MER* ROVER 379
GLOSSARY 381
APPENDIX........................ 385
INDEX............................ 415

PROLOGUE

When I was a kid, I loved maps. Still do. I grew up in the sixties, and in those days if you looked at a not-so-current atlas of the world, you could still find a few blank spots—places that were understood poorly enough by the mapmakers that they didn't know what to draw. I loved the idea of a map that wasn't done yet, with places on it still to be discovered. As a boy, I read everything I could get my hands on about exploration—Amundsen and Scott in the Antarctic, Beebe and Barton in the deep sea—tracing their adventures across my maps and dreaming about the exploring I'd do myself someday.

By the time I hit college, reality set in. There were no blank spots on the maps anymore. I was a student at Cornell University, in upstate New York. I loved climbing mountains and I had a knack for science, so I picked geology as a major, thinking that it might be a way to get paid to go climbing. After learning a little bit of geology I started to drift toward something involving exploration of the sea floor, since there were still some blank spots there. But it wasn't working for me. The geologists who have spent the last two centuries studying our planet, it turns out, have done a pretty darn good job of it. To me, geology felt like filling in details.

Then, early in the spring of my junior year, I was giving my girlfriend a tour of the Cornell campus. This was in 1977, just after NASA's *Viking* spacecraft had arrived at Mars, and while we were in the Space Sciences building she spotted a three-by-five card tacked to a bulletin board, announcing that a professor who was a member of the *Viking* science team was going to be teaching a graduate seminar course on Mars. What the hell, I thought, and I went.

I started by nearly getting kicked out of class. The first thing the professor asked when we were all in our seats was "Are there any undergraduates here?" One timid hand went up—mine—and he asked me to see him after the lecture. It was pretty obvious he was going to throw me out. What saved me was that one of my would-be classmates was a grad student from the geology department who knew both me and the prof, and who came over at the crucial juncture and vouched as best he could for my studious nature. The prof assented, though not without making it clear why he didn't like having undergrads in the course. "If I'm talking about temperatures on Mars," he said, "I don't want to have to stop and explain to the whole class what thermal diffusivity is." I nodded sagely, and ran back to my dorm room to look up thermal diffusivity.

Because the course was taught at the graduate level, we were expected to do some kind of original research project for our grade. A few weeks into the semester I figured I'd better start thinking about what I was going to do for my term paper, so I asked for a key to the "Mars Room," where all of the new pictures from the *Viking* orbiters were being kept. I found the Mars Room in Clark Hall, behind the Space Sciences building. It was a deserted and disorderly place, more like a warehouse than a scientific data archive. A few of the pictures that had been taken during the earliest part of the mission were in glossy blue three-ring binders, arranged in chronological order on gray-painted steel shelves. Most, though, were on long rolls of photographic paper, stacked on the floor or still in their shipping cartons. My idea had been to spend fifteen or twenty minutes flipping through pictures, hoping to find inspiration for a term paper topic. Instead, I was in that room for four hours, racing through the pictures, stunned. I understood almost nothing that I saw, of course, but that was the beauty of it. *Nobody* understood most of this stuff. In fact, only a handful of people in the world had even seen it yet. Sitting there cross-legged on the linoleum, I was exploring a new, distant and alien world. I walked out of that room knowing exactly what I wanted to do with the rest of my life.

The planet that I saw in those pictures is a beautiful, terrible, desolate place. It's cold: The average temperature on Mars is sixty degrees below zero centigrade. It's dry: If you could take all the water vapor in the martian atmosphere and freeze it out on the planet's surface, the layer of ice you would make would be barely a hundredth of a millimeter thick. The thin carbon dioxide atmosphere of Mars whips dust off the ground into storms that can darken the skies for months at a time. The planet that we see as a shining red point of light in the night sky of Earth is a barren, hostile world.

But it may not always have been that way. The pictures I was looking at, though I did not realize it then, showed evidence that water may once have flowed in abundance across the martian surface. There are dried-up riverbeds on Mars. There are dried-up lakebeds. There are features that tell of enormous floods that once poured across the martian surface. Most remarkably, there are small valleys, with branching tributaries, that wind sinuously through the martian highlands. These valleys must have been carved by streams that were so small that it's hard to imagine how they could form under the frigid conditions on Mars today. How can a trickle of water flow at sixty degrees below zero and not freeze? But there they are, in the pictures, demanding explanation.

Most of Mars's water-carved valleys are exceptionally old. It's tough to be sure, but many of them may date from the first billion years of Mars's 4.5-billion-year history. So despite the planet's forbidding climate today, its valleys are a clue that back in the earliest part of its history, Mars may have been a warmer, wetter and more Earth-like world.

And here's the thing: Four billion years ago is the very same time that, somehow, life first came into being on our own planet. We don't know how this miracle—the process of genesis—took place. But one thing that it surely required was liquid water. And if it happened here on Earth four billion years ago in environments that were warm and wet, then an obvious question arises: Could it also have happened on Mars?

Finding evidence that life arose independently on another planet would be one of the most profound discoveries that humans could ever make. If you only know that a miracle has happened once, then it may be a rare, or even singular, event. But if you can prove that it happened *twice in the same solar system*—recognizing that there may be countless solar systems out there—it means that, while no less wondrous a miracle, it may be a universal one.

So Mars is a world that can help us learn our place in the cosmos. If we go to Mars and find that it developed life, we've learned something fundamental about how common a phenomenon life may be. And if we go there and find that the conditions were once warm, wet and habitable, yet that somehow life *didn't* emerge, then we have learned something profound about the conditions that are required for life to develop.

And consider this: Suppose just for a moment that the miracle of genesis really did occur on Mars. On Earth, because of the intense geologic activity that our planet has suffered since its birth, the physical evidence of how that miracle took place is gone. That the miracle occurred is inarguable—we ourselves are part of the evidence. But tangible clues of how it actually happened are lost to us forever, because all the rocks from that earliest period of the Earth's history have been destroyed by later geologic activity. On Mars, though, they have not. Mars has been a more geologically quiescent world than Earth, and nearly half the martian surface is covered with rocks that are close to four billion years old. So if the miracle of genesis also took place on Mars, then evidence for *how* it happened may still be there, a story in the rocks waiting to be read.

The business of reading the story that rocks have to tell is the work of a geologist, and that was a subject I had learned something about. A geologist is like a detective at the scene of a crime. Something happened here long ago . . . what was it? Was it warm here? Was it wet? Was it the kind of environment that would have been suitable for life? The answers to questions like this can lie in clues that were left behind in the ancient rocks on the planet's surface.

Every rock preserves evidence of what conditions were like when it formed. When sediments are deposited, the coarse grains settle close to shore and the fine ones in deeper water. Look at the grain size in a sedimentary rock, then, and you learn something about where it was laid down. Ripples preserved in rocks can tell you about currents. Distinctive minerals like salts can tell you what was dissolved in the water. A good geologist can piece together clues like this to learn in detail what an ancient world was like. Geology on Mars, if there was a way to do it, would be something worth devoting a career to.

But the kind of Mars mission I wanted to do would be complicated. It wouldn't be enough just to fly by the planet, say, or to put something into orbit around it. Orbital spacecraft give you a great view, and they're the best way to get a really global look at another world. But the problem I wanted to attack wasn't a global one. The clues that I felt I needed were locked up in the rocks at a few special places on Mars. The only way to get at them, I was sure, was going to be the old-fashioned bang-it-with-a-hammer approach of a field geologist—albeit a robotic one—down on the martian surface.

And there was another thing about getting down onto the surface of Mars. Taking pictures from orbit didn't feel like real exploration to me. Lewis and Clark hadn't looked down on the Louisiana Territory from orbit. What I really wanted, when you got right down to it, was martian dirt on my own boots. And if I couldn't have that, I wanted the next best thing. I just didn't have any idea how to go about it.

Twenty-six years after my cathartic moment in the Mars Room, twin robotic explorers named *Spirit* and *Opportunity* were in final preparation for launch from Cape Canaveral in Florida. Built by a sprawling family of engineers and scientists, they were poised to carry the dreams of their creators to a planet where two out of every three spacecraft missions had ended in failure. Their mission was to study rocks on the surface of Mars, and to learn from those rocks whether or not the planet had ever had what it takes to support life.

That they were in Florida at all, however, was a small miracle.

PART ONE

BEGINNINGS

1 MOSCOW, 1987

T HE FIRST REAL JOB I ever had was at Ames Research Center, a
NASA facility in the heart of Silicon Valley. Ames was one of
NASA's first research laboratories, founded before the agency was
born, and for years it had been a player in the business of exploring the
solar system. Fresh out of grad school and freed from the shackles of finishing my Ph.D. thesis, I went off in a dozen directions at once at Ames,
several of them having to do with new ways to explore the planets. None
of them led anywhere; most space exploration concepts don't. But it was
a start.

After five years at NASA, I got an unexpected call from the astronomy department at Cornell. They had a faculty job open, and they were
wondering if I wanted it. I didn't, really. Ames was a good place, and I was
happy where I was. But I was recently married, and my wife, Mary, had
family in Ithaca, the small town that's home to Cornell. Ithaca is a leafy
and pleasant place, filled with earnest academics, left-leaning activists and
organic farmers. It was getting to be time for us to put down roots somewhere, and the countryside of upstate New York—especially with family
near at hand and, we hoped, a family of our own on the way soon—

seemed a better choice than the congestion of Silicon Valley. So I took the offer, and in the fall of 1986 we packed up and moved to Ithaca.

It was good to be back at Cornell. The astronomy department there was a strong one, with good researchers and sharp, aggressive grad students. But being there posed a problem. As long as I had been at Ames, I had a direct line to whatever NASA might be doing in the way of space missions. When I left California for the wilds of upstate New York, that line was cut.

Cornell isn't exactly an academic backwater. But if what you really want to do is space missions, a university isn't where you obviously want to be. At NASA, if you're clever and you're willing to put in a lot of hours, getting involved in some fashion in a mission to the planets isn't all that hard. Out in the remoter reaches of academia, though, it's a different story.

And flying a mission to Mars, ten years after the Mars Room, was still what I wanted to do.

Fortunately, I didn't need to learn how to build a spacecraft. That's the business of aerospace engineers, and there are plenty of people who know how to do it. But if I wanted to be a significant part of a Mars mission, at a minimum I was going to have to learn how to build a scientific instrument that could go into space. And even that was pretty much a mystery to me.

NASA doesn't do its missions into deep space all by itself. Most planetary spacecraft are NASA creations, but when it comes time to pick the scientific instruments that fly on their spacecraft, NASA casts the net more widely, with something called an Announcement of Opportunity. An AO is a document that NASA sends out to anybody who might be interested in sending scientific hardware into space, be they from universities, the government, industry or anyplace else. It describes the spacecraft that NASA intends to fly, and it describes in very broad terms the science that the spacecraft is expected to do. And then anybody—anybody at all—can send in a proposal.

Teams of scientists and engineers band together when an AO comes out, in marriages that are born out of anything from shrewd planning and friendship to raw necessity and ambition. Sometimes, scientists with a smart idea go out and find engineers that might be able to make their idea a reality. Other times, engineers with a slick new technology go out in search of scientists who can find something useful to do with it. Either way, once a team is formed, it's the scientists' job to decide what they want to learn, and what measurements they have to make in order to learn it. Then it's up to the engineers to try to come up with a detailed design for a scientific instrument—a camera, say, or a spectrometer—that can make that measurement and get the job done.

Problem is, scientists and engineers don't always work well together. Science and engineering are profoundly different disciplines, and their practitioners belong to two different tribes.

Scientists are seekers of truth. They're people who look at the world and wonder how it works. Scientists are people with hunches, and the good ones are people who know how to pursue those hunches to the correct conclusions, whatever they might be and however long it might take to find them. To a scientist, there's enormous appeal in an open-ended research project, where there's no telling where it might lead.

Engineers, on the other hand, are creators. They are tinkerers and inventors. To an engineer, the goal is to build a machine that works. And even better is to build a machine that works *right*. The best engineers can look at a problem and not just find a design for a machine that can solve it. They can find the *best* design that can solve it.

But engineering is a real-world pursuit, and in the real world you have to deal with realities like finite budgets and schedule deadlines. Engineers wrestle with these realities daily, often compromising their profound aesthetic sense of what makes a good design to arrive at one that is simply *good enough*—but that gets the job done on time and within budget.

The problem when engineers and scientists have to work together is that "good enough" is anathema to a scientist. There's no such thing as "good enough" when what you're after is the truth. So on every space project, there is a tension: the idealistic, impractical scientists against the stubborn, practical engineers. On the good projects, it's a creative tension that draws out the strengths of both disciplines. On the really bad ones, it's an acid that eats away at the collaboration until it's rotten.

There aren't a lot of things that can get scientists and engineers to pull together, but one of them is an Announcement of Opportunity. When NASA puts out an AO, teams form up to write proposals that respond to it. If the scientists on the team get their way, the design gets the job done. If the engineers on the team get their way, it gets the job done without breaking the bank, the schedule, or the laws of physics. And if everybody gets along, it's a big plus.

One way or another a proposal gets written, and off to NASA it goes. NASA's headquarters are in Washington, D.C., and that's where the decisions get made and the winning proposal gets selected. It's an intense, competitive process. The proposals take months to write, and preparing them can cost hundreds of thousands of dollars. Many teams compete, and only one wins. The theory is that the proposal that wins is the best one. But whether they're the best team or not, it's only the winners that get to go to Mars. There are a lot more losers than winners.

In 1987, just after my arrival at Cornell, the only kind of spacecraft instruments that I knew anything at all about were cameras, from my grad school days with the "imaging team" for the *Voyager* mission to Jupiter and Saturn. But *Voyager* had been built in the late seventies, using early seventies technology. By 1987, camera technology had moved far beyond the TV-like vidicon tubes that were at the heart of the *Voyager* cameras. Where it had gone, I understood only barely.

If I was going to write a proposal to send something to Mars, clearly what I was most likely to succeed with was a camera. But to do that, I needed to find a real camera expert—an engineer who could take whatever concept I came up with and turn it into something that could actually work.

As the semester began in the fall of '87, I unexpectedly received in the mail an elaborate envelope postmarked from Moscow. I knew no one in the Soviet Union, but this clearly was not a bulk mailing. I opened it to find an invitation to attend an international symposium on space exploration, commemorating the thirtieth anniversary of the launch of *Sputnik*. Where the symposium's organizers had come across my name I could not imagine.

Sputnik had been a very big deal; its orbital mission in 1957 had marked the beginning of the space age. More than that, *Sputnik* had provided the impetus for the birth of America as a space power. *Sputnik* carried no instrumentation, scientific or otherwise. Instead, it was a tool of propaganda, broadcasting a beeping tone that could be picked up almost anywhere on the globe with a simple radio set. *Sputnik*'s communist *beep-beep-beep* was heard across the United States, conjuring visions of Soviet spacecraft flying at will over the country, ready to drop bombs like cannonballs from the heavens. The American reaction to *Sputnik* was an outpouring of funds that led to the birth of NASA, *Apollo* and everything we know as the space program of today.

I sent in my RSVP.

The Americans who were invited to the symposium were an oddly assorted lot of former astronauts, aerospace engineers, business executives and space scientists. Our hosts flew us from Washington to Moscow on a lumbering Ilyushin jet that was the Soviet answer to the 747. Not exactly a model of efficiency, it had to set down to refuel in Newfoundland and again in Ireland, both overcast, cold and gray, before it got us to Moscow.

We touched down at Sheremetyevo, and our hosts eased us through passport control with none of the KGB-style intimidation that was usually encountered by American travelers to Moscow in those days. Bags gathered, we were whisked downtown in a motorcade, the blue flashing lights of our police escort reflecting off rain-slicked streets. The regal treatment ended as we were dumped abruptly at the Rossiya Hotel, a cavernous eyesore on Red Square.

The symposium was a hodgepodge, ranging from ordinary scientific and engineering talks to historical paeans to the glories of Soviet aerospace. The highlight was the banquet, held in a vast hall inside the Kremlin. Chandeliers sparkled overhead, and red cloths covered tables lined with heavy bottles of Georgian mineral water. My companion at dinner was an old man in a rumpled brown suit, wearing a small cluster of medals on his chest. His English was broken and I spoke no Russian, but I sat rapt through dinner as he described the elegant system he had designed to assure that the spacecraft of Yuri Gagarin, the first man to orbit the Earth, would be oriented properly when it began its plunge through the atmosphere at reentry. I was among people who had made spaceflight history.

As the banquet ended and I walked down the red-carpeted marble steps at the exit to the hall, I spotted someone I recognized. It was Alan Delamere, an engineer from Ball Aerospace in Colorado. Ball is a strange outfit. The same company that makes the famous Ball canning jars, it turns out, also makes scientific instruments and spacecraft. I knew that Alan was an expert in the developing field of charge-coupled devices, or CCDs—the light-detecting silicon chips that were used in modern spacecraft cameras (and that, today, are at the heart of every handheld digital camera). I didn't know him personally, though I'd admired work he had presented at a few conferences we had both attended during my NASA days. Elfin, wiry and built like the cross-country ski racer he was, Alan had a mop of white hair, a middle-class British accent and a pleasant smirk that made you feel like he knew something you didn't.

Emboldened by my conversation at dinner, I strolled over to him as casually as I could and introduced myself. We chatted about the meeting, and about our experiences at the banquet. Then, with what subtlety I could muster, I steered the conversation to the fact that I was looking for an engineering partner to help me develop a camera to go to Mars. I don't think I sounded much like I knew what I was talking about. But Alan had just been through the same banquet that I had, and I hoped he was in a similar frame of mind. To my surprise, he gave me his number and told me to give him a call at his office in Boulder once I was back in the States.

I stepped out of the banquet hall and wrapped my coat around me. The rain of the previous days had cleared, and a cold wind was whipping along the pavement. I looked up at the Kremlin towers as I walked slowly across Red Square to the Rossiya. I had absolutely no idea what I was getting myself into.

One thing that became clear very quickly was that I was going to have to put together a strong team if I was ever going to have any hope of getting real hardware onto a real spacecraft. With Alan onboard, the next guy I went after was Fred Huck from NASA's Langley Research Center, down in the Hampton Roads region of Virginia. Fred was a tall, aristocratic German who had been a driving force behind the *Viking* lander cameras back in 1975. He probably knew more about building cameras to go to the surface of Mars than anybody alive. I made a pilgrimage to see him on his own turf at Langley, and I told him about my hopes to send a new camera to Mars. To my joy and relief, he agreed to join forces with us—joy because he would bring so much to the team, and relief because if he was with us then he wouldn't be working with anybody else.

There was no mission out there for us to compete for yet, but something was bound to come along sooner or later. So we put a proposal in to NASA for a little bit of camera development money, with Fred as the principal investigator—the PI—because I was still too green to lead anything like a serious effort to develop space flight hardware. To my surprise, we got it: a whopping $100,000 a year for three years to build a prototype of a camera to go to Mars.

Just weeks after the money from NASA arrived, we all got together at Ball Aerospace in Boulder to work out the basics of what we wanted to build. I told everybody what I wanted the camera to be able to do, Alan figured out how to turn it into hardware and Fred kept everybody honest with his experience from *Viking*. It was my first taste of what later became one of my favorite parts of developing flight hardware: the blank sheet of paper. You start off with nothing but a bunch of smart colleagues and a vague idea of what you hope to accomplish, and then a concept starts to come together before your eyes. There's enormous potential at a moment like that, and a multitude of ways to go hopelessly wrong. Success or failure, maybe months or years down the road, can turn on the outcome of one decision you might make after half an hour of discussion. It's exhilarating and terrifying at the same time.

Trying to guess what future Mars missions might need, we decided to focus on panoramic imaging, since almost any kind of vehicle that might go to the martian surface would have to have some form of panoramic vision. Our idea was to build a true panoramic camera that could take a full 360-degree picture all in one shot. The design that we came up with was a variant on the "pushbroom" cameras that are used on many kinds of orbiting spacecraft. A pushbroom camera doesn't use film, and it doesn't use a two-dimensional CCD array that mimics film. Instead, it uses a simple one-dimensional linear CCD array, taking advantage of the spacecraft's motion across the ground to sweep the array along, pushbroom-style, building up an image as it goes. But whatever our spacecraft was, it wouldn't be orbiting above Mars; it would be sitting on the martian surface. So the way we conceived our camera was with the CCD linear array turned up on its end, oriented vertically and then slowly swept around the scene by an ultra-precise motor to build up one enormous image. Alan named it "Pancam."

By 1990, NASA had chosen their next big goal at Mars. It was something they were calling the *Mars Environmental Survey*, or *MESUR*. This thing was a real hummer: sixteen identical landers to be launched over a period of four years. The idea was to put them down at sixteen widely separated places on the planet, deploying a network of seismometers and weather stations, and also studying what would have been sixteen very diverse and interesting landing sites. Each lander would, of course, carry a camera. Sixteen cameras to build and send to sixteen different places on Mars . . . that sounded very cool.

MESUR was a complicated beast, though, and NASA was very worried that it would be tough to build that many landers, and make them all work, at an affordable cost. So before *MESUR* flew, they were planning to do a single test flight of one *MESUR* lander, in a mission they were calling *MESUR Pathfinder*. *MESUR Pathfinder* would be fundamentally an engineering mission, but word was out that there would be an AO to pick a camera for it. Whoever won the competition was expected to have the inside track for building the cameras for all of the *MESUR* landers, so it was a very tasty target.

NASA released the *MESUR Pathfinder* camera AO in the fall of 1992. The timing was perfect; the prototype camera we had built with our little development grant was now alive and taking pictures. We decided to propose a very slick color panoramic imager based on Pancam for *Pathfinder*. And for this proposal, after the experience I'd gained in building the first Pancam, I felt confident enough to be the PI.

The community of planetary scientists in the United States is pretty small. Everybody knows everybody else, and rumors spread fast. So when an AO comes out and teams form up, you can usually size up your competition quickly. Our competition sounded dangerous. Mike Malin was up to something, and that was trouble. I had known Mike for years. He is an intense and highly secretive guy. Mike is short and stocky, with closely cropped dark hair, a neatly trimmed beard, and close-set eyes behind round-rimmed wire-frame glasses. The guy is seriously smart, with a Ph.D. from Caltech, and he had recently won one of the coveted MacArthur Foundation "genius grants." Mike had been at NASA's Jet Propulsion Laboratory back when I'd first met him, and after that he had spent a few years as a professor in the geology department at Arizona State. But he'd left ASU when he won his genius grant, and had gone off to start his own company, Malin Space Science Systems, in San Diego. What he was doing there, I had no idea.

Mike and I had been competitors in almost everything I had ever done. We had both done Mars research, though he was a little bit older and a good deal more accomplished at it than I was. We had both done geologic field work in Antarctica. And now we were both competing to build cameras to go to Mars. Mike, though, was already the PI for the camera on NASA's *Mars Observer* spacecraft, an orbiter that was in development at the time, putting him far ahead of me on that score as well. So he definitely was somebody to worry about.

There were also rumors about a group at the University of Arizona that had teamed up with Martin Marietta, the big aerospace company. I was pretty sure we had the jump on everybody, because of the development money we had gotten several years before and the prototype Pancam we had already built. But Malin in particular had me worried.

In the spring of 1993 I took my first sabbatical leave from Cornell, and Mary and I left Ithaca for six months and moved the family to Colorado. We had two daughters by then: Nicky was a pretty blond preschooler and Katy a rambunctious redheaded toddler. We rented a little pink house that was within bicycling distance of Ball, and I spent every day in Alan's lab tuning the camera and getting ready for the site visit from NASA we knew would precede the selection. I bought books on electronics, motors and CCDs, trying to teach myself little bits and pieces of engineering. Every day I worked on that camera, and with time I came to know all its capabilities and quirks, and what every component in the electronics did.

One day in March I got the phone call from NASA Headquarters that we had been praying for: We had made it through the first round of the selection, and they were coming to Boulder for the site visit. They faxed us some questions that they wanted us to answer while they were with us. I pulled the fax off the machine and looked at the first question on a short list. It read "Given that the allocated volume envelope for the camera is x cm wide by y cm deep by z cm high, how do you plan to accommodate your design onto the lander?"

Huh?

I read the sentence three times before it began to sink in, and when it did a hard, sickening knot began to form in my stomach. I ran across the hall to Alan's office, pulled a copy of the AO off his bookshelf and flipped quickly to the critical page. NASA was right—we had gotten the volume envelope wrong! The AO had contained a diagram that showed the space on the *MESUR Pathfinder* lander that the camera was supposed to fit into. The diagram had been printed on the page sideways in the AO, and we had literally gotten confused about which way was up. They wanted a camera that was low and wide, and we had designed one that was tall and thin. With one terrible, boneheaded mistake we had thrown away five years of work. Or at least it seemed like we had, since something this bad should have disqualified us outright. You're not supposed to get second chances on these proposals; whatever you send NASA is what they judge you on, and that's it. But they were coming for a site visit anyway, so maybe we were still in the running.

We thrashed for a couple of days, cooking up some ugly last-minute fixes to the problem. We'd turn the camera over on its side to fit it onto the lander, and then flip it upright with some kind of mechanism once it was safely on Mars, or something like that. Everything we did was aimed at making the change seem as trivial as possible—the way we were going to spin it, this was just a small tweak to deal with an unanticipated contingency, nothing more. The delegation from NASA visited, led by Joe Boyce, a mid-level functionary from NASA Headquarters. We gave them our pitch, showed them the facilities, and even took their picture with our camera and gave them some nice prints to take home with them. No way Malin or Arizona could match that, I figured.

One evening a few weeks later, when we knew the final call from NASA was coming, Mary and the girls and I arrived at Alan's house for dinner. As I grabbed a beer from his refrigerator I noticed a bottle of champagne he had been keeping there, tucked way in the back, but I didn't say anything. Alan punched through the messages on his answering machine while I sipped my beer, until he found one from Boyce, who had said he wanted me to call him at his home in D.C. later that evening. A creepy feeling came over me. If we had won, would it be just Joe Boyce who would call? Wouldn't it be Wes Huntress, the NASA associate administrator who was responsible for making the final selection? It didn't seem right. The two-hour wait until the time Boyce had asked me to call him was excruciating.

The time finally came, and I dialed Boyce's number. He didn't try to break it to me easily. He offered no preamble, and no compliments on a job well done. We had lost, he said, and it was because of our screwup with the volume envelope.

"Well, why didn't you just tell us that at the time, instead of putting us through the whole site visit thing?" I asked him bitterly.

"We were all hoping that somehow we'd read your proposal wrong," he replied. "In retrospect we probably should have disqualified you off the bat."

Well, yeah, maybe you should have, Joe. "Who won?" I asked.

"Peter Smith, from Arizona."

Jeez. I felt like such a dope.

In the fall of 1993, NASA's $900-million *Mars Observer* spacecraft, with Malin's orbital camera onboard, went silent just three days before its arrival at Mars. Nobody was ever certain what killed it, though the most likely explanation was that a flaw in the propulsion system had allowed fuel and oxidizer to mix in a propellant line. If that had happened, the propellants would have reacted violently, burning through the line and venting into space. The spacecraft wouldn't have blown up in a ball of fire; it just would have tumbled out of control, slowly bleeding to death.

Something that really did blow up around the same time was the projected cost of the *MESUR* mission, and NASA killed that one off before it got anywhere close to the launch pad. *Pathfinder* somehow managed to survive both of these disasters, under the new name *Mars Pathfinder* and augmented with a neat little mini-rover called *Sojourner*. But with *MESUR* gone, what *Pathfinder* was supposed to be pathfinding for was anybody's guess. The rest of the Mars program crumbled.

Central to all of this was NASA's Jet Propulsion Laboratory, in Pasadena, California. Founded in the years following World War II, JPL had once actually specialized in jet propulsion. The lab was built out in an arroyo, safely distant from downtown Pasadena, as a place where Caltech rocket scientists could test their dangerous devices without alarming or endangering the citizenry. But JPL's focus had changed with time, and by the early sixties it had become the place where virtually all of NASA's robotic planetary spacecraft were developed. *Ranger, Surveyor, Mariner,* the *Viking* orbiters, *Voyager, Galileo* and many more had all been born at JPL. Now JPL was stuck with figuring out how to piece together a new Mars program after the loss of *Mars Observer* and the collapse of *MESUR.*

As big government organizations often do when they're in trouble, JPL responded by organizing a bunch of advisory committee meetings to try to help them decide what to do next. I went to all of them. And at one, to my utter surprise, Mike Malin quietly pulled me aside.

"Look," Mike said in a conspiratorial tone as we stood together in a hallway out of earshot of everyone else, "we've been busting our butts trying to beat each other for years. The result was that we both lost to Smith." I listened intently, wondering where this was going. "Whatever happens next," he said, "let's team up together. If we do it, I'll bet we can beat anybody who comes after us."

Coming from Mike this was astonishing. He had always seemed so secretive that I'd never considered him as a possible partner. But it was a great idea, and it made me feel ashamed that I hadn't thought of it first. We shook hands, and we agreed to bide our time and see what happened next.

Months passed, and a new Mars program gradually began to take shape. Its guiding principles would be the ones laid down by Dan Goldin, the recently appointed NASA administrator. Goldin felt that NASA had become hamstrung by its tendency to build ever-more-complex space-craft—*Battlestar Galactica* missions, he called them—costing many hundreds of millions of dollars each. *Mars Observer* was Exhibit A in his case against big missions, with too many eggs in that one $900-million basket. Goldin's new vision for exploration of the planets was "faster, better, cheaper" spacecraft that would swarm out across the solar system in unprecedented numbers. The idea was that it wouldn't be a big deal if you lost a mission or two, as long as they were cheap ones.

Mars Pathfinder already fit Goldin's faster-better-cheaper mold, and it would fly as planned in 1996. Also launching in 1996 would be a new orbiter called *Mars Global Surveyor* that would carry copies of some of the instruments that had been lost on *Mars Observer*. The next question was what would go in 1998. Mars and Earth line up in the right positions to get from one planet to the other once every twenty-six months, so missions tend to get launched during these every-other-year "launch opportunities." NASA had decided that they wanted to fly two missions in 1998, one of them an orbiter and the other a lander. And a new concept that arose out of all those committee meetings was to have the AO seek an integrated suite of instruments for the lander: the whole scientific payload together in one proposal, instead of a bunch of separate proposals for separate instruments.

This, to me, was something really different. What NASA was soliciting, for the first time ever, was every tool that would be needed to go after some big scientific problem on another planet, all in one package. The idea was to give the scientists who wrote the proposal the ability to create a payload that could work as a carefully crafted ensemble of instruments, maximizing the overall science return of the mission. It was an enormous technical challenge. But it also was close to being something that could enable one proposal team to mount a serious attack on a really big question—like whether or not Mars might ever have been capable of supporting life.

The advisory meeting where the idea of an integrated payload first took hold happened at the Hilton hotel in downtown Pasadena. Malin was there too, and we huddled together in a hallway outside one of the meeting rooms only moments after it had become clear this was the direction NASA was likely to go. This was an opportunity not to be missed, and we needed to pull a team together. Most important, we had to get the right people onto our team very quickly, before anybody else grabbed them. The same thing that had made us fierce competitors also made us good partners: We were both interested in the same kinds of science, and that made the choices simple. After five minutes of conversation we had locked in on what we needed.

Imaging was what we both knew best, so there were definitely going to be some serious cameras on our payload. Pancam would be there for sure, and also a nice downward-looking "descent camera" that Mike had developed to snap pictures as the lander plummeted toward the surface.

What else? We needed an infrared spectrometer. This lander wasn't going to be able to go anywhere once it touched down, so most of our measurements of what the stuff around the lander was made of would have to be made from a distance. If you look at rocks with your naked eye you can tell a little something about what they're made of, but not much. If you use infrared vision, though, at wavelengths longer than those visible to the human eye, you can tell rocks apart with much better accuracy. Most rock-forming minerals have their own distinctive infrared signature. Looking at rocks with infrared eyes enables you to read those signatures, which means that you can learn a lot about what a rock is made of without ever having to go over and touch it.

Looking at rocks from afar wasn't enough. We'd also need some kind of instrument package to look at stuff close-up—to get really detailed images and to measure composition in much greater detail. It'd also be nice to throw in some kind of meteorology package, in hopes of making our payload attractive to any atmospheric scientists who might review it. And finally, we'd have to go with instruments that were pretty far along in their development, since the schedule before the launch in 1998 was looking very tight (the "faster" part of faster-better-cheaper). So the ground rules were pretty clear. And everybody who was anybody in the Mars science game was at the meeting. So why not just do it, right then and there?

Who should we pick for the infrared? Phil Christensen from Arizona State was the PI for an instrument called the Thermal Emission Spectrometer, or TES, late of *Mars Observer* and now slated to fly again on the *Mars Global Surveyor* orbiter. He also had a concept for something he called "Mini-TES." The idea behind Mini-TES was that it would be a miniaturized version of Phil's TES instrument, shrunk down and adapted for use on the martian surface. It was a good concept, deriving a new instrument from one that was already both mature and accepted by the science community at large. On top of that, Phil was an old buddy of mine from my grad school days, someone with whom I had played many late-night hotel-room poker games. Mike liked him, too. So picking Phil was easy.

We needed something to look at the composition of the soil, and we settled quickly on a nifty little instrument that Jim Gooding at NASA's Johnson Space Center in Houston had been working on for years. The nice thing about this one was that all we had to do was lower it down to the surface somehow. After that it could do its work all by itself, using a little built-in scoop to pull soil grains into the instrument and analyze them.

Getting to rocks, though, was the real key to the science we were after, and that was a much tougher problem. There was no way we could count on the rocks that we wanted to look at—the ones that contained clues about what conditions were like on Mars long ago—being within easy reach of a lander. We needed some way either to get the rocks to our instruments or to get our instruments to the rocks. Problem was, NASA wasn't offering a lander that could go anywhere. The thing was just going to plop down and sit right wherever it first settled, forever. And if some wonderful rock was just out of reach, that was tough luck. NASA also wasn't offering enough money for us to build anything that could carry instruments or rocks around.

I knew of a way out, though, and I told Mike about it. Another old friend, Heinrich Wänke of the Max Planck Institute in Mainz, Germany, had been working for several years on a device he called the Nanokhod—*nano* for small, and *khod* after the old Soviet *Lunakhod* rover that had flown to the Moon in the early seventies. It was a tiny rover, about the size of a fat dictionary, that could creep for a few meters across the martian surface. It wasn't much to look at, but it carried a very nice little suite of three instruments. One was a microscope, to take close-up pictures of rocks. Another was called an Alpha Proton X-ray Spectrometer, or APXS, and it measured the elemental composition of rocks: how much silicon they contain, how much iron, how much magnesium and so forth. The APXS in the Nanokhod was the brainchild of Rudi Rieder, an eccentric genius who worked in Heinrich's lab in Mainz.

The third instrument on the Nanokhod was called a Mössbauer spectrometer, and it was built by Göstar Klingelhöfer, a physicist at the Technical University of Darmstadt, not far from Mainz. The Mössbauer technique can identify any minerals that contain iron. And Mars, which isn't called the red planet for nothing, promised to have plenty of rustlike iron-bearing minerals on its surface. Altogether, the suite of instruments in the Nanokhod was just right for what we wanted to do. And while the mobility it offered wasn't much, it was something. Best of all, the Nanokhod effectively came for free. NASA won't pay for foreign instruments to fly on their spacecraft; they expect the country that supplies the instrument to pick up the bill. And Heinrich had convinced the German government to do exactly that for the Nanokhod.

Finally, the obvious choice for meteorology was another friend: Ari-Matti Harri, from Helsinki. Ari-Matti had the best meteorology hardware around and, like Heinrich's Nanokhod, it would come for free, this time from the Finnish government.

So that was it. Everybody was at the meeting, and Mike and I hunted them all down in the corridors of the Pasadena Hilton, one by one. In half an hour the whole payload was together. We had no way then of knowing that in less than an hour we had formed the nucleus of a team that would work and struggle together for more than a decade.

At the outset, we didn't have a PI. We were just a bunch of hardware guys who had banded together to try to get our stuff to Mars. But somebody had to do it. It was going to be a NASA mission, so obviously the PI would have to come from the United States. That was just a political necessity, and it narrowed the choices down to Mike, Phil, Jim and me. Mike and Phil had both lost instruments on *Mars Observer*, so they were now up to their necks in *Mars Global Surveyor*, taking them out of the running. And Jim was just too nice a guy to crack the whip on a bunch of prima donnas like this. So I was it, by process of elimination.

After a while, we learned who the competition was: Ray Arvidson from Washington University in St. Louis, Larry Soderblom from the U.S. Geological Survey in Flagstaff, Arizona, and Dave Paige from UCLA. A very scary bunch. Arvidson was teamed with Lockheed Martin, the big aerospace company that NASA had already picked to build the lander. Paige was doing some kind of mission to the south pole of Mars. There was no telling what Soderblom was up to, but he had been one of the megastars of planetary exploration since the seventies, so he was somebody to worry about, too. There weren't going to be any weak proposals.

I teamed with Alan and Ball again to pull the whole thing together. The proposal was fun to write at the beginning, because it was another blank sheet of paper. The instruments had all been designed independently of one another, and we had to invent a whole bunch of things to get them to work together. We redesigned Pancam so it would fit with Phil's Mini-TES on a single device that would serve as both a mast for the camera and a periscope for the spectrometer. We designed a little crane that could get Jim Gooding's soil instrument down to the surface, and deploy the Nanokhod at the same time. And a lot more.

The final push, back out at Ball again in the summer of '95, was no fun at all. This was the hardest I had ever worked on a proposal. After the Pathfinder flameout it was time to win one. Every detail of this proposal was perfect, and you can bet we made damn sure that we got the volume envelope right.

The proposals went in, the review panel met and the rumors began to spread. NASA's proposal review process is supposed to be scrupulously confidential, but scientists are human, and word leaks out. This was the leakiest review panel I'd ever encountered, and the news, if news it was, was that we had come out on top in the science evaluation. All we could do was wait.

The final selection came in late October, while I was at a meeting at the headquarters of ESA, the European Space Agency, in Paris. It was gray and drizzly, and I was straight off the airplane and jet-lagged all to hell. The French wine at dinner went to my head, and I gave up the fight against the lag and went to bed at 9 P.M. Half an hour later the phone in my room rang, jarring me from a deep sleep.

"Steve? It's Joe Boyce."

Oh no no no no no. And he confirmed it: We had lost again. They picked Paige? A polar mission? I was woozy, and not thinking very hard before opening my mouth, and I tore into him. "What happened, Joe? I know we came in first in the panel review. What the hell happened?" I actually knew nothing; all I had was rumors. Boyce stonewalled, of course; it was all he could do when confronted with a tirade from a failed and furious would-be PI. After a few minutes of yelling at him I hung up the phone in frustration, almost instantly regretting my outburst. Any more sleep was impossible, and I spent the rest of the night walking miserably through the streets of Paris in the rain.

I never did get a straight story from NASA about what they hadn't liked about our proposal, but months later I did get some hints from an old friend who had been on the review panel. We had indeed done well in the science ranking, but we had made two serious errors. One was that we were proposing to look at rocks, which didn't seem very appealing on a lander that couldn't go anywhere. A polar lander mission like the one that Paige had proposed seemed a lot more attractive, and in many ways it was. Mars's polar deposits come in thin, flat layers, and all you have to do to study them after you land is dig straight down from wherever you are. So Paige's polar mission was well suited to the stationary lander that NASA was offering. Ours was not, even with the Nanokhod.

Our other big mistake was that we were proposing a tiny German rover that wasn't even as capable as the *Mars Pathfinder*'s little *Sojourner* mini-rover was going to be. It was all we could afford, of course. But it would seem like a step backward after *Pathfinder,* and worse, it would give the sexiest thing on the mission to a foreign partner. None of that made our payload very attractive to NASA. So we lost, and Paige's mission, called *Mars Polar Lander,* was born. After eight years we still had no ride to Mars.

2 ATHENA

A COUPLE OF MONTHS AFTER my miserable night in Paris, I went
to a meeting of the American Geophysical Union in San Francisco. The AGU is a huge scientific conference, held each December in San Francisco's vast Moscone convention center. As I stood
outside one of the dozens of meeting rooms scanning the list of talks for
something interesting to listen to, Ray Arvidson came up to me. Ray had
been one of our competitors for the '98 mission, but he was also a friend.
Round-faced and bearded, Ray had been one of the young hotshots on
the *Viking* lander mission back in the seventies. He and I had grown up
just a few miles apart from one another in South Jersey, and while I
hadn't known him in our youth, I had admired his work for a long time.
He is one of those rare people who consistently get more done than
everyone around them, and his ability to juggle dozens of complicated
tasks at once is the stuff of legend. Like Mike Malin, he had been a very
scary competitor.

Ray had something on his mind, so we found a quiet spot and sat
down to talk. It quickly turned into a replay of my conversation with
Malin a year before: Instead of beating up on each other, he said, next
time we should team up. That sounded fine to me, I said, though "next

time" looked at least a couple of years away. He also told me about a very cool instrument that his colleague Larry Haskin was developing at Washington University: a Raman spectrometer, an instrument that can give you the precise, grain-by-grain mineral composition of a rock by shining a laser at it and looking in detail at the light that was scattered back. It would fit in beautifully with the other instruments we already had.

When AGU was over, I headed across the Bay Bridge to Berkeley, where I had been invited to give a colloquium in their geology department. I tried to focus on giving a decent talk, but the conversation with Ray kept percolating in my brain. There was real potential here, but I couldn't see how to make it fit into the anemic Mars program NASA was planning.

I spent a couple of nights at Berkeley, and the geology department put me up at the university's faculty club. It was a nice place, all decorated for the Christmas holidays, and at dinner that evening there were monks from somewhere, in long brown robes, solemnly singing carols. I went to bed late, but I was restless, and I woke up at 5 A.M. As I lay there in bed, eyes open in the dark, I couldn't get the unfortunate state of NASA's faster-better-cheaper Mars program out of my head. Paige had won the '98 lander mission. The '01 mission was going to be more of the same—another lander. Most people seemed to think we had the inside track for that one, but it wasn't any better suited to what we wanted to do than the '98 lander had been.

In 2003, NASA was going to send three more copies of the same lander yet again, in a woefully diminished version of the old MESUR concept. There simply wasn't anything in a program like that to get excited about.

After 2003 things got hazy. There was talk of trying to bring some rock samples back from Mars in 2005, but I couldn't see how such an underfunded program could pay for it. And even if they could afford to get to Mars and back, there certainly was no way they could pay for whatever would be needed to really explore on the surface, or to collect a good set of samples. At best it would be a grab-and-go, and even that looked unlikely. All in all, the situation was grim.

Then, as I lay there turning it all around in my head, an idea came to me. Along with their Mars program, NASA had something else that they called the "Discovery" program. Discovery had a budget all its own, completely separate from the Mars program. And under Discovery, a PI could propose not just an instrument or a payload, but an entire mission: payload, spacecraft, rocket, the whole shebang. Any planet could be the target, even Mars, and you could propose whatever kind of mission you wanted to, as long as it fit within the budget cap. There was going to be a Discovery AO soon, with proposals due in the fall. Discovery could offer a completely different way to go at Mars, without all the restrictions that had made NASA's mainline Mars program so unappealing.

A Discovery mission to Mars would be another blank sheet of paper, and with it a whole new world suddenly opened up. What we really needed was mobility: the ability to put our payload on something that could move, that could really explore. So suppose we took the Lockheed Martin lander that was going to be used for Paige's mission in 1998, but revamped it so it could carry a big rover. We could use JPL's *Sojourner* de-sign from *Mars Pathfinder* as the starting point for the rover, but scale it up massively so that it could carry the whole payload and drive long distances from the lander. I ran quickly through the possibilities. We'd take the best pieces of our '98 proposal, but switch to an all-rock focus: drop the mete-orology package, drop the soil analyzer, and add the Raman. And to sweeten the deal, we could pick up some samples as we explored, so that another spacecraft could go there and grab them and bring them back, in 2005 or whenever NASA got around to it. In one shot we could do a killer Mars rover mission and get NASA positioned so they could pull off the sample return mission they really wanted.

This was too cool an idea, and there was no way I was going to get back to sleep now that I had it in my head. I needed both JPL and Lock-heed Martin onboard if I wanted to do this mission, and I had to lock them up before anybody else thought of it. I waited impatiently while the clock ticked and the sun rose to call Noel Hinners at Lockheed Martin and Charles Elachi at JPL—Noel first at 8 A.M. Denver time, and Charles an hour later in Pasadena. Noel was Lockheed's vice president in charge of their deep space missions, and Charles led JPL's space science division. I spun them each the sweetest-sounding tale that I could about the wonders of big rovers on Mars, and I had them both onboard before breakfast.

I called the first meeting of our new team at JPL in January of 1996. Arvidson was there, along with the nuclei of both his team and mine from our '98 proposals. Larry Soderblom, the megastar from Flagstaff, was with us now, too. This was the blankest piece of paper any of us had ever faced, and the meeting turned into an intellectual free-for-all. We had so many questions to answer that it was hard even to list them all. How much effort should we put into landing accurately? Precision landing would be really expensive, but it could get us to some very interesting places if we went for it. How mobile did the rover have to be? Ten meters a day? A hundred? Should we try to land with some spare fuel onboard and then hop the lander to someplace new once we could see where we had come down? That'd be quite a ride if we could pull it off.

And what about the payload? Did we want an APXS, like the one we had had in our '98 proposal, or some kind of more advanced X-ray experiment? How about some kind of little wet chemistry lab? Video capability would be cool, especially if we were going to hop and land again. Wouldn't that video be a kick to watch? It got pretty wild. There was a lot more creativity in the room than there was realism, but that's how things often are at the start of something new.

By March reality had taken hold, and we had settled on the basics. We'd start with the '98 lander design, but we'd completely clear off the deck to make as much room as we could for the rover. The rover itself would be as big as we could possibly make it without having to make expensive changes to the lander design. No precision landing, no hopping, and we'd shoot for a hundred meters a day. The payload would be Pancam, Mini-TES, Mössbauer, APXS, Raman, Microscopic Imager, all on the rover, plus a descent imager on the lander. The Mössbauer Raman and microscope would all be out on the end of a fancy five-jointed robotic arm. We'd also have a little scoop on the rover for picking up soil and pebble samples, and a detachable can to drop them into. The schedule for an '01 launch looked frighteningly tight, so we'd launch it all in 2002, using a flyby of Venus to get it to Mars. And I found somebody at JPL to help me manage the whole thing: Saterios "Sam" Dallas, a burly Greek with a bushy white mustache and a strong inclination to speak his mind.

With the mission concept starting to shape up, we needed a name for it. I hate acronyms, and I couldn't come up with a decent one for this mission that fit. (We had called our '98 proposal MACS, for Mars Ancient Climate Surveyor, and I had never liked it much.) Names for a lot of NASA missions have come from the classics over the years—*Mercury, Gemini, Apollo*—so I turned there. Ares was an obvious name from the Greek pantheon for a Mars mission, but Ares was such an obnoxious god that he seemed a poor choice. Next I came upon Athena. She was a warrior, too, like Ares, but a beloved one, and also the goddess of wisdom. That sounded pretty good. I popped off an e-mail to Sam, asking what he thought of *Athena* as a name. "I love Athena," he wrote back, "Athena is my wife." Greek proposal manager, Greek wife, Greek name. I took it as a good omen, and *Athena* was what our mission became.

The summer wore on, and the proposal came together. Getting it all down on paper was the usual grind, now woefully familiar. In August of '96, growing weary of it all, I vacationed with Mary and the girls at Mary's childhood summer getaway on a little island in Lake Huron. We spent a peaceful two weeks there, building a tree house and canoeing every day. It was good to be cut off from phones, from e-mail, and even from Mars for a while.

On the way from Michigan back to Ithaca, we stopped at Mary's grandparents' home in rural Ohio. I plugged in my laptop for the first time in a couple of weeks, downloaded my e-mail and stared at it, dumbfounded. Dave McKay had found fossils in a martian meteorite?!? Dave was a researcher at NASA's Johnson Space Center, and he was a major-league meteorite expert, so this was no off-the-wall nutcase. Still, it had to be a joke, or a hoax, or something, didn't it? But there was too much e-mail traffic in front of me for it to be anything but real.

And real it was. I flipped on the TV, and everywhere I looked it seemed there were Mars bugs. CNN had Dave live in a press conference, looking wide-eyed and a little alarmed by all the fuss he had created. His story seemed hard to believe at the outset, but as he laid out his case it sounded like there might be a chance he had actually found something. President Clinton came on next, declaring that the United States was going to get to the bottom of this question. Holy shit. It was obvious to me that the only way anybody was going to get to the bottom of this question was to send a rover to Mars to collect some rock samples. And here I was in Woodville, Ohio.

I went to the kitchen and dialed Norm Haynes and Donna Shirley, who led the Mars program office at JPL. We had kept our proposal pretty tightly under wraps, but Norm and Donna knew the basics of what we were up to.

"What the hell is going on, you guys?" I asked. "Have you heard anything from NASA Headquarters yet?"

"Nothing. . . . All this just happened yesterday."

"Do you have any idea what it means?"

"Not really," said Donna. "But *Athena* may suddenly be the best game in town."

It just might.

That same day I got an e-mail from somebody named Steve Gorevan, president of a little company in New York City called Honeybee Robotics. I didn't know much about Steve; we had met only once or twice. He was tall and congenial, with a huge and unruly nest of prematurely gray hair, and he seemed like a very smart engineer. Honeybee, though, I knew a bit more about. They were a really interesting company. They were tiny, with no more than twenty employees, but they had done some very intriguing things. Their specialty was custom mechanical gadgets, and they had done stuff ranging from a giant mechanized Coca-Cola sign in Times Square to a remotely operated robotic welding machine that could creep through the network of steam pipes that lies beneath the streets of Manhattan. Lately they had begun to branch out into mechanisms for spacecraft. Steve's note was terse. "I've heard you're working on a Mars rover mission," it said, "and I was wondering if there was anything we could do for you."

Well, yeah, just maybe you could. "Send me whatever you've got," I replied, wondering if he'd tip his hand or not. Things can get pretty weird when there's something as big as a whole Mars mission on the line, and you never know how much somebody will be willing to show you in such a competitive environment. What Steve had, it turned out, was five different concepts for how a robot might collect samples on a planetary surface. Four were useless for Mars, but the fifth was a very slick rock coring drill that they called the Mini-Corer. Scooping up dirt and pebbles had seemed okay before, but McKay's putative fossils had been inside a rock. With Mars Bug Fever spreading, it was clear we were going to have to be able to extract samples from martian rocks if we were going to get *Athena* selected for flight. Honeybee had what we needed, and I quickly added their Mini-Corer to the payload, and Steve Gorevan to the team.

By the fall of '96, it was starting to look like maybe the president had meant what he said. A big increase in NASA's Mars exploration budget was rumored, and NASA had sent very clear signals that they wanted to spend it on a rover sample return mission just like ours. But that sounded a lot better for us than it really was. McKay's Mars bugs, if bugs they were, had opened the floodgates, and there was no question that we had exactly the mission that NASA wanted. There also was no question that we had a huge jump on the rest of the world. But the whole idea behind proposing *Athena* as a Discovery mission had been to give NASA something that they couldn't afford in their uninspiring and underfunded mainline Mars program. If a big budget augmentation for a rover mission suddenly showed up, then NASA would make the rover the centerpiece of a revitalized Mars program, not do it as a Discovery mission. And if that happened, our Discovery proposal was doomed.

In September, I got a call from Lynn Lowry, an engineer in the Mars program office at JPL. "We've heard we're going to get a big budget upper for Mars next year," Lynn said, "and we're going to have to help NASA put out an AO to select the payload for a rover." Oh jeez, I thought, here we go . . . they're going to do a mission identical to *Athena*. And Lynn confirmed it. "We haven't been working on a rover big enough to do the job, though, or on a lander that can deliver a big rover. But we know that you have. So could you maybe send us some of the details of what you've been doing?"

I was speechless. We hadn't even finished our proposal yet, and already the Mars guys at JPL were writing its obituary and asking us for the remains so that they could use them to write an AO . . . giving us the chance to compete against the rest of the world to build the payload for a mission we had conceived and designed! I also was worried that Lynn was talking about a mission that would launch in 2001, not 2002, which would make the schedule appallingly tight. I told her we were too busy finishing our proposal to respond, but that I'd think about it once we were done.

And of course she called again, exactly one day after we had sent the proposal in. It was a very weird situation, but I took a deep breath and sent her the stuff. Not the whole works, of course; our greatest strength was our payload, and I had to keep that a secret. But I sent her everything else.

The weeks passed, and midway through their evaluation of the Discovery proposals NASA announced that they were indeed expecting to get their Mars augmentation from Congress. We were dead. I didn't bother trying to run down the rumors this time, or to wait by the phone for news. And when the news finally did come, NASA didn't even call. I read about it in the Pittsburgh airport, in an online press release that announced the Discovery winners. *Athena* was not among them.

So now it was the new 2001 mission, or nothing.

Just the thought of writing yet another proposal, so soon after the last one, drained and disheartened me. It was now 1997, more than eight years since the night in Moscow when I had first started down this road. The main focus of my career for all those years had been writing proposals to NASA, all of them the best I could do, and all of them failures. It was time to now begin a fourth, but the fun had gone out of it. The main thing driving me still was fear: the fear that if we didn't try again, and if we didn't win it this time, those years of work would have been in vain.

The first thing I needed if I had to put a payload proposal together for this new mission was a new proposal manager. JPL was going to build the rover that would carry the winning instruments, so it would help to have somebody from JPL manage the payload. Sam was a mission guy, but he wasn't a payload guy, and he bowed out. I asked around at JPL, and one of the names suggested to me was Barry Goldstein.

Barry seemed like a strange choice. He was the project engineer for Dave Paige's *Mars Polar Lander* payload, and rumor had it that Paige had eyes on 2001 as well. But sometimes raiding the enemy camp for talent is a good thing to do, and I went to see Barry.

Within minutes of walking into his office I knew I had found my man. It was love at first sight. Barry turned out to be a tough, fast-talking Jewish boy from Brooklyn. He was the son of a cigar store owner, educated at the University of Colorado, and he had an impressive combination of street-smart people skills and hard-won engineering talent. He had been at JPL since 1982, working his way through several of their toughest technical organizations. Everywhere I was weak—engineering, management, JPL politics—Barry was strong. And after being the engineering second-in-command on Paige's payload, Barry had the same attitude I did about wanting to win a big one. So now I had my proposal manager, and my payload manager, too, if we won.

The competition formed up fast, and the rumors flew. Dave Blake at Ames Research Center was doing some kind of X-ray diffraction thing, using an iffy-sounding explosive rock blaster to do his sampling. Paige had pulled together a bunch of his *Polar Lander* friends, with his deputy Candy Hansen, from JPL, as the PI this time. Those guys were good, so that proposal could be trouble. And then there was Chris McKay from Ames, who had jumped ship from the *Athena* team. Chris was apparently building a package around an amino acid analyzer, of all things, plus an infrared spectrometer and a camera. Amino acid analysis sounded a little goofy to me on the face of it, since there really wasn't any evidence that there were amino acids on Mars to be found. But it was hard to tell. Mars Bug Fever was spreading, and if the review panel went crazy with it then Chris could be trouble, too.

I had never liked the competitive part of proposal-writing much, and it was especially bad this time around. After the Discovery experience, this mission felt—rightly or wrongly—like it was "ours." Having to compete to put the payload on it, especially after all the failed proposals that had gone before, felt like somebody was trying to take something away that belonged to us. The fact that both Chris and Candy were good friends of mine made it even worse. I'd see them at meetings, or just on the sidewalk at JPL, and I'd go out of my way to avoid talking to them, or even coming face-to-face. This was a fight, and I had to stay focused on winning it. We could be friends again after it was over.

The AO came out, and it was a disaster. The budget augmentation from Congress hadn't been all that NASA and JPL had hoped for, and the total funding for the payload was a paltry $17 million—barely more than half what we had planned to spend for the payload on our *Athena* Discovery mission. Getting this thing into the bag was going to be painful.

The sampling scoop from the old *Athena* Discovery proposal was the first thing to go. After that went the Raman, which was our most expensive instrument. Instead of making it part of the main payload, we had to propose the Raman as an "above the line" option that NASA could select if they found some extra money somewhere. Next we cut the number of joints in the instrument arm from five down to a bare-bones minimum of three. Even with that, the cost was still way over $17 million. Something else big had to go overboard.

The obvious thing to kill off was Pancam. This hurt like hell, but I had to do it. The AO claimed that the rover was going to be outfitted with a bunch of ultra-cheap JPL-provided engineering cameras, to be used for stuff like navigation. If we replaced Pancam with copies of JPL's engineering cameras, slapping a new lens and some cheap color filters on the front of each one, it would save us a ton of money on paper. With this change we could get, barely, into the $17 million bag.

I took the idea to my old friends at Ball Aerospace, and things got unpleasant. With the cameras themselves gone, the only thing left for Ball to build would be the mast for the cameras and Mini-TES, which now amounted to little more than a couple of mirrors, a few motors and about a meter of pipe. After all the years we had worked together, I desperately wanted Ball onboard, and Alan Delamere especially. But Ball had a business decision to make, and there just wasn't enough interesting work in the mast to make it worthwhile. They passed, and the old Pancam that had gotten me started almost a decade before was no more. I decided to keep the Pancam name for the new cameras, though. After all the work that Alan had done, at least the name he'd given our camera might someday get to Mars. And I kept the name *Athena* for the payload, too, for Sam.

The proposal work went on. I had a few special places I hid out to do most of the writing—a shady porch on the side of our log house in the woods outside Ithaca, and some ornate Victorian-era library stacks at Cornell when I was on campus. When I was on the road, I'd take my laptop out to the beautiful old Huntington Gardens in Pasadena, and sit by a pond in a bamboo grove there for hours, watching the carp swim while I quietly did my best to pound the hell out of the opposition.

My assistant, Diane Bollen, and I spent the last week out in Pasadena, crashing at a bed and breakfast for a few hours a night, and otherwise living in our "war room" at JPL. A month before, a student had given me a bootleg CD with the complete works of the Beatles compressed onto it, and we listened to that constantly off my laptop—it was years before I could hear the Beatles and not think of that week. Barry was right there with us the whole way. We got the proposal printed, and we loaded the copies into a couple of big white cardboard boxes. Diane and I hand-carried them to D.C., and I personally dropped them off at NASA Headquarters, a day before the deadline.

Selection day was November 7, 1997. The rumor was that the phone calls would go out from NASA Headquarters around 2 P.M. Eastern. I couldn't work as the hours crept by, couldn't focus on anything. I just paced in my office, bouncing a red rubber ball off the wall and catching it, over and over.

At 1:59, the phone rang. I picked it up, and I recognized the voice. It wasn't Joe Boyce. It was Wes Huntress, NASA's associate administrator for space science. Oh my God. "Steve, this is the phone call I've been waiting years to make. Congratulations." I didn't know what to say; I'm not sure I could have gotten words out if I had. We really had won. "It was a slam dunk," he went on. "Even if I didn't like you I would have had to select you." They hadn't gone for the whole package, though. The Raman had cost too much, Wes said, but they were going to try to find some money to add it later. The new *Mars 2001* project office at JPL would contact me with details about my contract in a couple of days, and so forth and so on. I managed to maintain my composure enough for a few minutes of small talk, and that was it.

Diane had been down the hall when Wes's call had come in. She walked into my office, and I gave her the news and hugged her. Then it was time for the phone calls. Barry first, but there was no answer and I left a message for him to call me back. Sam was next, for old times' sake. He was in, so he got to be the first at JPL to hear, which was fitting. Barry called back while I was on the phone with Sam, and he told Diane not to say anything; he wanted to hear it from me. I gave him the news, and I could hear him ricocheting around his office at the other end of the line, pounding on his desk, yelping. We were a couple of happy guys.

It was time to head home. I left it to Diane to spread the news around the Space Sciences building. On Ellis Hollow Road I realized that Mary and the girls were in the car right in front of me, and I flashed my lights. They knew something had to be either very good or very bad if Daddy was coming home in the middle of the afternoon. We pulled into the driveway together, and when I got out of the car they knew what had happened from the look on my face, before I could even say it. I hugged Mary tight, and I realized that my whole life was about to change.

3 APEX

THE *MARS 2001* PROJECT began in early December, 1997, with a two-day gathering of the team at the Pasadena Hilton, the same place where Malin and I had first assembled our payload, four years before. At the same time that we'd been frantically writing our proposal, JPL had been trying to pull together a team of their own to lead the rover development once the mission got started for real. All of the competing payload teams had been "firewalled" off from this new rover team while the proposals were being written, so nobody could get an unfair advantage by talking to them. But now it was time for us to start figuring out how to forge a partnership with these guys.

The rover team was led by Jake Matijevic, who had been one of the original architects of the *Sojourner* mini-rover that flew so successfully on *Mars Pathfinder*. Jake and his team quickly let us know that there were a bunch of things they didn't like much about our payload. Back during the proposal review process, Jake's rover team had been asked to rate the four payloads for their "ease of accommodation," and they had ranked us third out of four. At the top of their list of things they didn't like was the big fat stovepipe of a mast that we needed for Mini-TES.

Mini-TES is an infrared spectrometer, and like all instruments it has some basic needs. One of the things Mini-TES needs is a short, fat, little built-in telescope to gather the infrared "light" that streams into it, and to feed it into the spectrometer. Another thing it needs is a nice warm, stable environment to live in. So instead of hanging the whole instrument out in the cold martian air on the top of the mast, we had decided to put it, telescope and all, down inside the rover body where we figured it'd always be nice and warm. The telescope would look up the mast, using mirrors at the top of the mast to peek out at the scene around it. Put simply, the mast was a periscope.

Mini-TES works by staring at one point in the scene long enough to get a good quality spectrum, and then moving its periscope mirror a little bit to step to the next point in the scene. So how long does it need to stare at each point? That, it turns out, depends on the diameter of the telescope. The bigger a telescope is, the more light it can gather . . . and the more light it can gather, the less time it has to stare at one spot before the spectrum is good enough to move on. We wanted Mini-TES to be fast, which meant we didn't want to have to stare at each spot for any longer than just a few seconds. And to get the length of each stare down to that few seconds, we needed to make the diameter of the telescope pretty big. Problem is, there's a price to be paid when you do that: The fatter the telescope, the fatter its periscope has to be. And the fatter the periscope is, the more it weighs and the bigger a shadow it'll cast on the solar array. So it was a delicate trade-off. We had finally settled on a telescope that was two and a half inches in diameter, which turned out to mean that we would have to stare at each spot for only four seconds before we'd have a good enough spectrum for that spot to move on to the next one. But that telescope diameter led to a very fat mast, which was giving Jake and his guys considerable heartburn.

We worked through this and a bunch of other issues, and we thought it was going pretty well, right up until the very end. Then, in the last hour of the second day, Donna Shirley from JPL's Mars program office swooped in and dropped a stink bomb on everybody, rover team and payload team alike.

"There's no way this rover of yours is going to work," Donna said with what sounded like utter conviction. "You're too big. You're too heavy. You're too complicated. And you're *way* too expensive. So what are you going to do to get back in the bag?" We all sat there silently and stared at each other. What the hell? But she was dead serious. Moreover, she hadn't even mentioned the schedule, which looked to everybody else in the room like the worst problem of all. Her little speech wasn't subtle, and it rocked us back on our heels. Just because Donna said we weren't going to make it didn't mean that we weren't going to make it, of course. But we obviously had some convincing to do.

We pressed ahead in the weeks that followed, making the decisions that stood between where we were and having a design we'd be ready to start building. We also began to get to know the other instruments that NASA had selected to go on the lander that would carry our rover. Loading the lander up with instruments wasn't something we had done on our Discovery proposal, but NASA was eager to squeeze everything they could out of this mission. So they had tricked out their new lander with what they called the HEDS—short for Human Exploration and Development of Space—instruments.

The HEDS instruments actually were pretty cool, for the most part, all aimed at helping to pave the way for someday sending astronauts to Mars. One would measure the dangerous radiation that an astronaut would experience on the martian surface, and another would study the soil for toxic properties. A third was aimed at learning whether or not it was possible to make rocket propellants using gases extracted from the martian atmosphere. All of them would go about their business after we had gotten off the lander and started to explore and collect samples.

And to top it all off, there was also an orbiter being readied to fly in 2001, being managed by the same project office at JPL that managed us. It was a very complicated project.

At the same time, back on the East Coast, NASA was working out their plans for 2003 and beyond. The first hint of what the future might hold came in the middle of March, when Wes Huntress called to tell me that in 2003 NASA planned to fly a rover and a payload that would be exact duplicates of the ones we were building for 2001. Just that abruptly, we had not one but two rides to Mars.

So that was the face of NASA's new Mars program. The reality, though, was that it was all façade, with nothing behind it. What Wes didn't realize—because nobody at JPL had worked up the nerve to tell him yet—was that his program was busted. At JPL they had known for months that they were in too deep a hole financially to pull our mission off in 2001. That was why Donna had come down on us so hard at our kickoff gathering. The JPL guys had kept pretty quiet about it, hoping that by cutting a corner here and shaving a little off the design there they could get everything back into the bag. But the closer they looked at how they were actually going to do everything they had promised NASA, the more obvious it had become that they couldn't. By the end of March things had gotten so bad that they couldn't keep the news from the East Coast any longer. Their best estimate was that they were at least $90 million in the hole.

A conference call was scheduled, with most of the key *Mars 2001* project staff on the line, to break the bad news to Wes. I can only imagine what it must have felt like from his end, to hear suddenly that the program he had been sold was a financial impossibility. What JPL got back from Wes, once it had all sunk in, was a simple ultimatum. JPL had made a commitment, and he intended to hold them to it.

"First, you can't have any more money," Wes said. "Second, you have to fly an orbiter with at least a gamma-ray spectrometer on it, a lander with at least one of the HEDS instruments on it, and a rover with the whole *Athena* payload on it, including the Raman. Period." Wes's unspoken implication was that if the Mars guys at JPL couldn't do what they were told, he would find somebody else who could. I was happy to see him looking out for our rover, but it was obvious that something was going to have to give.

It didn't take long. The same work that had revealed how badly the budget was busted was starting to show that the schedule was busted, too. Only a few weeks after our phone call with Wes, Jake came to a painful conclusion: JPL's promise to have the rover ready for a 2001 launch could not be kept, no matter how much money NASA might be willing to throw at the problem. There just was no way they could get the rover done by then. I got the news in a short, painful phone call from Barry. And as simple as that, it was clear who was going to get yanked off the '01 mission to get the costs back in the bag. It was us. Wes called me the same day, and assured me that we'd still fly with everything intact in 2003. But the '01 rover was gone.

So that was it—we had gotten ourselves kicked off our mission before it was even six months old. I sent an e-mail out to my team, breaking the news as gently as I could and trying to make it clear that this was just a slip to 2003, not a cancellation. And I kept myself going with the belief that what Wes had said about 2003 was really true. But so soon after the euphoria of selection, it was hard to look at this as anything other than a devastating setback.

Without the rover, I couldn't see much point to what was left of the 2001 lander mission. It still had the HEDS experiments onboard, but to me those alone didn't add up to a mission that was worth a quarter of a billion bucks. I figured that NASA would just kill the thing outright, but they didn't. They had promised Congress and the White House that something with a NASA logo on it would go to Mars in 2001, and they intended to keep that promise, no matter what.

There were only two options that would fit the budget, and they were both pretty lame. One was to fly just the HEDS stuff. The other was a hybrid that would take a fraction of the HEDS stuff and pair it with our Pancam and Mini-TES, bolted down to the deck of the lander. On paper the HEDS-only option cost $7 million less than the one that included our hardware. And with money tight and science pretty much out the window anyway, the HEDS-only mission looked like the leading option to everybody on the West Coast.

It wasn't up to JPL to decide what would fly, though, it was up to NASA Headquarters. And, as usual, NASA invited in a panel of scientific experts to help them choose. On June 1 and 2, 1998, the experts gathered in a big conference room at JPL and our fate was debated. It was a very unpleasant business.

The management team from the beleaguered *Mars 2001* project got up and told the experts that they'd prefer to fly just the HEDS stuff, because it was the least expensive option they had in front of them. After that, I got up and argued for flying Pancam and Mini-TES, hoping that my lack of enthusiasm for the whole unsavory enterprise didn't show.

It was a strange and difficult performance to give. NASA's panel of experts was made up of friends and colleagues of mine, and the situation demanded scientific candor and honesty. But it was obvious to everyone in the room that the real problem wasn't science, it was money, and my honest opinion was that the science we would get out of the mission simply wasn't worth the money it would cost. On paper, we still had a ride to Mars for the whole rover in 2003, which was the mission I really wanted to fly. If I had really believed that any money that might be saved in 2001 would be applied to 2003—and if I'd had any confidence that the Mars program wouldn't turn inside out yet again before then—I would have fallen on my sword and told the experts to fly just the HEDS stuff in 2001, save the money and be done with it.

Unfortunately, it wasn't that simple. Eleven years had now passed since I had started trying to send hardware to Mars, and in all that time I hadn't seen a single plan for Mars exploration survive for more than about eighteen months before there was some kind of cataclysm. I had a promise in my pocket from NASA that they'd fly our rover for real in 2003, but history said that such promises were not to be trusted, no matter how sincere the intentions. It was hard to convince myself that the mission I had in hand, lame though it was, should be given up. So I sucked it up and argued with all the passion I could muster to fly Pancam and Mini-TES on the lander in 2001.

It was a pretty good performance, I thought, with lots of facts, lots of figures and every scientific argument I could pull on short notice from our failed Mars '98 proposal—which, after all, had been the one that had Pancam and Mini-TES mounted on the deck of a lander.

I didn't fool anybody, and neither did the *Mars 2001* project team. The experts saw right through everything, and by the end of the first day of the meeting they had concluded that both options were awful. The answer they were headed for was to tell NASA just to cancel the whole mess and move on to 2003.

It didn't happen. Wes had sent a deputy to the meeting to keep an eye on things, and at the end of the first day the deputy laid down an edict: A lander had to go to Mars in 2001, like it or not, and it had to include Mini-TES.

This was a bit of a jaw-dropper, especially this far into the meeting; nobody had mentioned at the outset that Mini-TES had to be on the payload. But as I thought about it, it made sense, at least politically. The Space Science part of NASA was paying the bills for this mission, but the HEDS experiments were from the astronaut side of NASA. If Space Science was going to have to pay for this thing, Space Science wanted to be damn sure that some of their instruments were onboard.

The Mini-TES pronouncement effectively ended all discussion. It took the experts a painful evening of digesting NASA's directives over beer and Chinese food, but they rolled over. The next morning they threw up their hands, accepted what they'd been told and recommended that NASA fly the option that included Pancam, Mini-TES and some of the HEDS stuff. It sort of made you wonder why they had been gathered together in the first place.

As soon as it was clear that Pancam and Mini-TES would fly I saw an opening, and I jumped into it. What about our Mössbauer spectrometer, too? It was being paid for by the German government, so it would hardly cost NASA anything to fly it. We could mount the thing on the lander deck, looking at a magnet. The magnet would gather dust from the martian atmosphere, and we could use the Mössbauer to figure out what the dust was made of. It sounded good and it sounded cheap, at least the way I spun the tale, and the experts went for it. NASA did, too, right there on the spot. I walked away from the meeting with three of our instruments on the payload.

We were back in business, but it didn't feel right. The rover mission that we had proposed for was something that I had felt really good about. Instead, the mission we were getting was almost embarrassing to be part of. Worse, I had compromised my principles to get on it, justifying it to myself with a "bird in the hand" argument. My flight back to Ithaca was an uncomfortable one.

Within weeks, there was a new twist. Months before, NASA administrator Dan Goldin had directed $60 million that had originally been intended to pay for the HEDS experiments into the floundering Space Station program instead, leaving the Mars program holding the bag. Goldin is an interesting character. He had come to NASA almost a decade before from the aerospace giant TRW, and since then he had shaken up the agency in a very big way. Goldin is a bold, visionary leader, full of wild and occasionally brilliant ideas. He is also an aggressive manager who isn't above getting results through threats and intimidation if that's what he feels it takes. And while he was good at charming Congress most of the time, this particular maneuver backfired on him. Congress keeps pretty good track of such things, and they didn't like what Goldin had done with the $60 million. They directed the money back into the Mars program.

What NASA could have done with this money, of course, was give the *Mars 2001* project enough money to do their job right. Or they could have plowed it into getting ready for the much more complicated rover mission that was coming up in 2003. Instead, still smarting from the harsh words of their panel of experts, they decided to use it to jazz up the '01 mission. First, they'd add back the full set of HEDS experiments. On top

ey'd also add the spare copy of *Pathfinder*'s now-famous *So-
ni-rover*, which was named *Marie Curie*. *Marie Curie* would be
mounted on the lander deck, and we'd use a robotic arm on the lander to
lift her up and plunk her down on the martian surface, where we could
watch her run around next to the lander with Pancam.

Flying *Marie Curie* was not without risk. *Marie Curie* had borne the
brunt of much of the testing for *Pathfinder*, and it wasn't clear that she was
in particularly good shape to go to Mars. But NASA wanted to do it, and
all of a sudden it meant we had yet another instrument to fly. *Sojourner*
had carried an APXS instrument that was the forerunner of our APXS,
which meant that an APXS was also the only useful thing that *Marie Curie*
could carry without an expensive overhaul. There had been an APXS in
our proposal, which NASA figured to mean that any APXS on *Marie
Curie* would be our responsibility, too. So now we had a four-instrument
payload to fly on the '01 lander: Pancam and Mini-TES on the lander
deck, Mössbauer staring at a magnet, and APXS on the *Marie Curie* rover.

To my dismay, the guys at Lockheed Martin, where the lander was being built, started calling this whole mess Athena Junior. I quickly changed
that to APEX, short for Athena Precursor Experiment, the idea being to
remind the world that this thing was *not Athena* . . . it was just something
that would lead to *Athena* two years later. But APEX was about to become
the main focus of our lives.

The mission lurched forward, with the focus slowly shifting from
money and politics to how to solve the technical issues involved in getting
real flight hardware to Mars.

We fixed up Pancam, dumping the cheap engineering cameras and
designing a new high-quality camera body and a really slick little color
filter wheel to go on the front of it. It wasn't the old Pancam, but it was
something that'd take pictures we could be proud of if we got it to Mars.

Mini-TES went from conceptual drawings to a detailed design. Step
by step our uncluttered and elegant original concept slowly filled in with

detail and complexity until it was so dense that there was hardly room for another wire or screw.

The biggest change was that we found a better way to do the Mössbauer experiment. The Mössbauer-on-a-magnet thing had been a nice trick for getting our foot in the door, but it really was a bad way of doing business. It would be much better to put the Mössbauer somewhere on the lander's robotic arm, where we could at least get it down to the martian surface and use it to measure the composition of the soil and maybe a rock or two. Barry and I fought hard to get it on the arm and we won, but it wasn't pretty. The sensor ended up bolted to the forearm of the arm, growing out of it like a big wart. It wouldn't give us access to the hundreds of acres of martian terrain that a big rover could get it to, but at least we'd be able to get it onto a patch of dirt about the size of a doormat.

With the Mössbauer out on the arm, the design of the APEX payload was complete. We locked down the design and through the fall of 1998 and into early 1999 began on the grind-it-out process of building the hardware and getting it ready for flight.

Around the same time, the new Mars program for 2003 and 2005 began to take shape. We'd use a rover to collect samples, but how would those samples get back to Earth? The breakthrough, if that was what it was, came when an engineer at JPL came up with a wild new idea for what was called a Mars Ascent Vehicle, based on a classified Navy program that his father had worked on all the way back in 1958, right after *Sputnik*. It seemed like an elegant concept: a tiny two-stage solid rocket that could boost a spacecraft the size of a coconut into orbit around Mars. It was called the Mini-MAV.

When you put the whole scheme together it went something like this: In 2003, a big three-legged lander would carry our rover and a Mini-MAV to Mars, with the rover astride the Mini-MAV like Slim Pickens on the bomb in the scene at the end of *Dr. Strangelove*. We'd land on Mars, drive down some ramps, pick up rock samples, drive back *up* the ramps, stuff the samples into the coconut-sized spacecraft at the top of the Mini-MAV, drive back down the ramps again, and hide behind a rock so we wouldn't get toasted. Then the Mini-MAV would be cranked up to a vertical position and fired, smoking the lander but launching the coconut into orbit around Mars. In 2005, we'd do the same thing again: another rover, another Mini-MAV and another coconut full of rocks blasted into orbit around Mars. And then, also in 2005, a big orbiter built by CNES, the French space agency, would be launched to Mars, go into orbit, somehow find the two precious coconuts in the vastness of space around the planet, stuff them into a canister and rocket the canister back to Earth, where it would land with a thump in the desert of Utah in 2008.

I kid you not.

By early 1999 the reality of just how complex this whole thing had become was starting to sink in. Wildly optimistic cost estimates were showing essentially no budget margin, which was no place to be at the start of a project that was more complicated than anything since *Apollo*. Stack that budget situation on top of the massive technical challenges, and the situation looked bleak.

There were several ways things could play out. One was that we would chip away at the mission bit by bit until whatever was left looked like it would just barely fit. But that would have given us an ugly, risky mission that did bad science. Another was for NASA to step in with enough new money to let us do it right. But they had made it abundantly clear that they had no intention of doing that, and I don't think they had the money to do it if they'd wanted to. The only other possibility seemed to be that the whole Mars program would fall apart yet again.

I spent the last week of February at a NASA meeting at Cape Canaveral. To get to the Cape from the Orlando airport, you take Florida Route 528 straight east. The landscape of central Florida is about as flat as land can be, but just before you hit the coast there's a causeway that crosses the Banana River. I reached it at about ten o'clock in the evening, and as my car climbed up and over the bridge, the lights of the launch pads came into view. I pulled over to the side of the road, killed the engine and got out. I could see the shuttle pads glowing at the far north end, up on Kennedy Space Center, and then the other pads stretching to the south across Cape Canaveral Air Force Station . . . the Titans, the Atlases and, at the south end, the Delta pads of Launch Complex 17. Those were the ones that we'd use, if we ever got that far. For ten minutes I just leaned back against the hood of my car, looking at them all. Most of the great events in space exploration had started right there on that coast, and the sense of history in that place was overpowering. For us, it was the promised land. I wondered if we'd ever get there.

4 THE SPACECRAFT GRAVEYARD

IN LATE SEPTEMBER, THE first of the two spacecraft that had been launched to Mars in 1998 was on final approach to the planet. This one was called *Mars Climate Orbiter,* and its main job was to study the martian atmosphere. To us, *MCO* was a martian comsat—something that would pass over our lander twice daily once we got down onto the surface, relaying data back to us when we needed it. *Mars Global Surveyor* was in place around Mars now, and doing good science, but it would be a mediocre relay satellite at best. So *MCO* really mattered. The night it arrived at Mars, a bunch of us gathered in the Space Sciences building at Cornell to watch the action on NASA TV. We didn't have any hardware onboard the thing, but it still felt like part of the family. And we really cared about getting that relay capability in place.

There isn't much to see on TV during an orbit insertion burn . . . just a bunch of engineers sitting at computer consoles, plus a few squiggly lines plotted on a screen. But the plots on the screen that night told a life-or-death story that was unfolding two hundred million kilometers away. As the spacecraft went sailing past Mars, its big main engine would be commanded to fire, slowing it down and easing it into a long, looping elliptical orbit around the planet. The radio signal would cut off as the spacecraft passed out of sight behind Mars, and we'd learn whether or not the burn had worked properly by watching exactly when the signal reappeared on the other side. Too soon would mean that the burn had terminated prematurely, leaving the spacecraft traveling too fast and, maybe, on a trajectory to fly by the planet. Right on time, though, would mean a good burn, and a good orbit.

The signal disappeared on schedule. We waited for it to come back, watching for the bright green points that represented good data to reappear on the black telemetry plot. The time for reacquisition of the spacecraft came and went, but nothing happened. There was no signal at all. Engineers stared at their screens while we stared at ours, silent and uncomprehending. Something had obviously gone wrong, and we went home that night expecting the worst.

Within a day, JPL put out a terse press release saying only that some kind of "navigation error" had caused the spacecraft to pass just 57 kilometers above the martian surface, instead of the planned 120 kilometers. If they were right, this was a colossal mistake, and one that would have sent the spacecraft deep into the planet's atmosphere. What JPL was saying was that somehow they had managed to fly a perfectly healthy spacecraft too close to Mars, burning it up.

It took less than a week to figure out what had gone wrong, and it was devastating. One part of the complicated business of navigating in deep space is keeping track of various "small forces" that operate on a spacecraft. These aren't big things like the gravitational attraction of the sun and the planets. Instead, they're tiny things like pressure from the solar wind and the little jolts you get when you fire your attitude control thrusters. Lockheed Martin had built the spacecraft and knew it best, so they had the responsibility for computing these small forces and letting JPL know what they were. Then, the JPL navigators used the numbers that Lockheed Martin gave them to help predict and interpret the spacecraft's trajectory.

Lockheed Martin, like many aerospace companies in the United States, still uses English units like inches and pounds for a lot of their work. It's an archaic holdover, but it can be hard to change your ways when you have tens of thousands of employees and hundreds of millions of lines of computer code to look after. So when Lockheed computed the small forces on the spacecraft, they worked them out in pounds, which is the English unit of force. And then they gave those numbers to JPL. Problem was, the JPL navigation team thought that the numbers they were getting were in newtons, which is the metric unit of force. And that meant that all the small forces used by the JPL navigators in their calculations were wrong, by about a factor of four.

Small forces have only a minute affect on a spacecraft's trajectory, so the error was insidious. As the navigators watched *MCO* approach Mars, it seemed that it might have been drifting ever so slightly from its expected path. The drift was so subtle, though, that they couldn't be sure what they were seeing until the spacecraft had gotten very close to the planet. And once they began to realize just how bad the problem really was, the time left before the orbit insertion burn was measured in hours instead of days. It was too late to do anything, and the *MCO* team had watched their spacecraft disappear behind Mars fearing that it wouldn't come out the other side alive. You care about these spacecraft when you work on them, almost like you'd care about a living person. Losing their creation like that tore some of the *MCO* team up in ways that they didn't recover from for a very long time.

If you wanted to find fault, the immediate fault was Lockheed's. The paperwork said they were supposed to give JPL the small forces in metric units, and they hadn't. The real failure, though, ran much deeper. Somehow, the Mars program had become so screwed up that nobody had caught a high-school mistake like mixing up English and metric units.

Next up was the arrival of *Mars Polar Lander*, Dave Paige's mission, at Mars. Even though we had lost the '98 competition to Dave, Mike Malin had succeeded in proposing his little descent camera separately, going on the lander as an augmentation to Paige's payload. Mike had invited me to be on his team, and I traveled out to Malin Space Science Systems, Mike's place in San Diego, for the landing. My notes tell the story:

Landing day:

10:43 PST: We're in wait mode. About five hours ago they did the final trajectory correction burn . . . a last-minute propulsive kick to nudge the landing target away from a scary-looking crater. That went as hoped for, by all accounts. Right now the spacecraft is off Earth-point and silent while it uses its star tracker to give the onboard attitude control system a final fix. We should get it back in six minutes.

11:02: The Deep Space Network tracking station in Madrid has two-way lock on the spacecraft. The spacecraft is playing back old data from its memory now, but once there's fresh stuff coming down they'll be polling all the subsystems for the final pre-entry status check.

11:19: Polling subsystems . . . Propulsion is go, Thermal is go, everyone else is calling in nominal. Things are looking good.

11:26: The pyro event to pressurize the propulsion system is about to happen. That's what killed *Mars Observer* back in '92, so this is a big one for the *MO* survivors in the room.

11:27: Pyro system initiated, enabled for firing.

11:28: Pyro has fired, but we don't see any change in the pressure data yet. Sweating it out a little . . .

11:30: Pressures are up. Yowza.

11:56: Heaters and valves are all doing what they're supposed to. We're going to lose signal in about seven minutes as the spacecraft makes its final turn to the entry attitude. This is all as we're seeing it on Earth, of course, with the signal dribbling back to us at the speed of light. Out at Mars, the spacecraft actually hit the top of the atmosphere about a minute ago.

12:00: Guidance system initialized. About to start the turn to entry attitude. Autopilot on.

12:03: All stations reporting loss of signal. Here we go. Systems to Flight Operations Manager: "Happy landings, Sam."

So now we wait. It's a strange feeling. . . . Earth receive time for the first signal from the surface is at best half an hour from now, but whatever is going to happen has already happened. If everything went right, we're already on the surface with the solar arrays deployed. It sorta messes with your concept of what a seemingly simple word like *now* means.

12:14: The navigation team just got their final solution for the spacecraft's trajectory. The entry flight path angle was just 0.12 degrees off nominal . . . right down the pipe. Their guess at the most likely landing point looks very bland and featureless in the best orbital images. First signal should be at 12:39 if everything's gone perfectly. The nav guys are shaking hands all around; their job is done.

12:38: DSS-14, the big 70-meter Deep Space Network dish at Goldstone, out in the Mojave, is going to signal acquisition.

12:39: Nothing yet.

12:41: Nothing. Tense faces in a quiet room.

12:45: Terse chatter on the net from Goldstone. . . . All their spectrum analyzers are running, but they see nothing yet. They're going to a wider frequency range in case things have drifted a little bit. Here in San Diego there's some talk now about what would have happened if maybe the lander had POR'ed—done a power-on reset, a reboot—on touchdown.

12:53: DSS-14 may have a signal—41 kilohertz off nominal. Oh boy . . .

12:55: Nope . . . "negative acquisition." They're still looking at it, but it seems to have been wishful thinking.

12:59: Yeah, that blip six minutes ago was interference from some other source. . . . They're saying now that it didn't look anything like a spacecraft signal.

13:03: We're not going to get it on the first try here. The medium gain antenna on this lander just points at a fixed location in the martian sky during this communication pass, rather than tracking where it thinks the Earth is, so if we were going to see something we would have seen it by now.

16:00, and the Cornell contingent is just back from the driving range next door. I sure hope we get a signal back from that spacecraft, because the golf pro thing definitely looks like it's not gonna work out.

18:30: Canberra's got Mars in the sky now, and they're radiating commands to the spacecraft, Aussie accents over the net. These

commands should make the spacecraft initiate a search for Earth
that'll last from 20:08 to 22:40, and it should be our best chance
of hearing something. If things are still quiet by late tonight, it'll
be time to start worrying.

20:08: Okay, here we go. This may be our best shot at this
thing . . . an hour and a half of the lander's antenna wandering
around the martian sky, looking for home.

21:07: We're more than halfway through—according to the clock,
the spacecraft should have finished 25 points now out of a
40-point scan.

21:28: 35 of 40 now. Nothing.

21:33: Done. We didn't get it.

22:30: Things aren't looking real good, but I'm not sure how much
there is to worry about yet. From what I've been hearing, Paige's
team was fiddling with the plans for what their instruments were
supposed to do right after landing as recently as just a few days
ago. This lander is kinda twitchy, and if you give it a bad command it's real easy to send it into "safe mode," where the thing
gets confused, and then just hunkers down, does nothing, and
waits for Earth to solve the problem. So there's a more-than-
decent chance that we're in safe mode just because of some
harmless problem in a command sequence. It'll be a long
24 hours, but there's still a pretty decent chance we're alive.

Day 2:

20:30: Waiting again. If we went into safe mode after touchdown, and if the antenna is pointed at the Earth, this is when we should hear from it. Nothing yet.

20:40: Still nothing. They're widening the frequency window.

20:44: Looks like we're probably not going to get it. That's not all that surprising. . . . It's not all that likely it'd be right on Earth point.

22:29: They're radiating a command load now to try another direct-to-Earth pass tomorrow night. Meanwhile, the next decent shot at hearing something is 10:50 tomorrow morning, which is when we'll get the first relay data through *Mars Global Surveyor* . . . *if* we didn't go into safe mode on landing. And on and on it goes . . .

Day 3:

10:40: Relay time. This one isn't too likely, I'm afraid. If we landed successfully, and if we didn't go into safe mode right after landing, and if there's a very serious problem with the direct-to-Earth antenna or radio, and if the relay system is fine, then now is when we'll hear from it. But that's a pretty long string of *if*s. I still think our last good shot is tonight, but we'll see.

11:10: Nope. All we got was housekeeping telemetry from *MGS* itself. Not a surprise. Next shot is in about ten hours. Hell, I've got time to go to Mexico and back. . . .

21:51: This is it. Either we get it in the next 61 minutes or, most likely, we don't ever get it.

22:02: Ten minutes in . . . nothing yet.

22:07: Ten points into the 40-point scan.

22:25: Halfway through . . . the odds are starting to head down a little. . . .

22:32: 26 points down . . .

22:45: 35 down. This is starting to look bad.

22:50: Damn. We're now down to the stage where if it's alive it's had one or more major hardware failures . . . failures so bad it'll never work right even if we do get it back.

There's not much else to say here, except that I don't ever want to go through this again.

Stacked on top of the *MCO* embarrassment, the *Mars Polar Lander* failure was a sledgehammer blow to JPL, to NASA and to the Mars program. News articles came out calling 1999 NASA's worst year since 1986, the year that the *Challenger* and her crew were lost. Jokes from Jay Leno and David Letterman took their toll. The pain was worst at JPL, where people take a passionate pride in the laboratory's deep space missions. *Mars Pathfinder* in 1997 had brought JPL public adulation; now, two years later, they learned that the price of failure at Mars could be public ridicule. To me, comparing *MPL* and *MCO* to *Challenger* was ludicrous. No lives had been lost when those two little Mars-bound spacecraft had disappeared, and the cost of *MPL* and *MCO* together was a miniscule fraction of the cost of a shuttle. But this was Mars, and NASA began to realize that missions to Mars can attract attention, including very negative attention, that is far out of proportion to their cost.

Unlike *MCO*, it was hard to tell what exactly had killed *MPL*. NASA convened a failure review board, and the usual sort of investigation was conducted. The board found almost two dozen potentially fatal flaws in the spacecraft's design, but there was no way of pinning down for sure which one of them, if any, had been the one that had done it in. Because money was tight, the *MPL* team hadn't given their spacecraft the ability to talk to Earth as it descended through the martian atmosphere. After that final turn to the entry attitude, there was no way for them to hear from *MPL* again unless it made it safely to the surface. So the review board had to work in the dark.

Of all the possible flaws, there was one that was so obviously bad that the board concluded that it probably would have been fatal if something else hadn't killed *MPL* off first. Like our lander, *MPL* entered the martian atmosphere with its three legs tightly stowed, descending toward the surface using rocket motors for control. What the review board learned was that when the lander's legs popped into place 1,500 meters above the ground, they may have triggered a switch on the legs by accident—one that had been put into the design to sense when the lander had touched down.

If the leg switch triggered, that event would have flipped a bit in the lander computer, called the Indicator State, changing it from a zero to a one. Flipping the bit wouldn't do anything . . . yet. The lander would continue to descend, using the signal from its landing radar to throttle its engines. Once it was forty meters above the ground, though, the spacecraft would be too close to the surface for the radar to work accurately enough any more. From that point onward, it was supposed to just descend at a slow and constant rate, shutting down its motors when the computer sensed that touchdown had occurred.

As soon as the radar was turned off, forty meters from touchdown, the computer would have started looking at the Indicator State bit, checking to see if it had gone from zero to one yet, indicating touchdown. And, according to this theory, the first time the computer looked, the bit already would have been set to one—by accident—back when the legs had deployed. Thinking that it was safely on the surface, the computer would have cut the engines off, sending the lander into free-fall from a height of forty meters. It would have been traveling almost eighty kilometers an hour when it hit the ground.

If this was the fatal mistake, then one line of software to set the bit back to zero after leg deployment would have saved the mission. But nobody had known that that line of software was necessary. The *MPL* team had violated one of the most basic principles of space flight: "Test as you fly, fly as you test." They had, of course, run a complete test of the process of entry, descent and landing, including deploying the landing legs, before they had launched their spacecraft. But the first time they ran the test, the leg switches had been miswired and didn't produce a signal at all. Once this mistake was found, the engineers knew they had to run the test again. But their schedule was getting tight, so they ran only part of it. They reran the simulation of the actual touchdown, proving that the switches were now wired correctly. But they neglected to precede it with another test of leg deployment, and leg deployment was where the fatal flaw lurked. In flight, of course, touchdown would always be preceded by the hidden danger of leg deployment. But in the rerun of the crucial test, it was not. "Test as you fly" had been violated, and the fatal flaw had been concealed.

It went a lot deeper than that, though. Like the *MCO* loss, the real reason for what happened with *MPL* was much more fundamental than a bad test procedure or a missing line of software. Dan Goldin's vision may have been for faster, better, cheaper spacecraft, but small, cheap spacecraft were not something that came naturally to JPL. To force Goldin's vision on them, NASA had put a tough set of restrictions on the Mars program. The cost of the program was capped. This meant that no overruns were allowed, and if a project got into financial trouble along the way then it was up to that project to get their costs back under control. Two missions had to go to Mars every twenty-six months whether they made scientific sense or not, and one of each pair had to go to the surface. And no spacecraft could be so big that it required anything bigger than the smallest version of the Delta II rocket to launch it. It was a pretty tough set of rules, and those rules shackled JPL.

Like any big technological project, a mission to Mars has four basic elements: cost, schedule, performance and risk. With all their restrictions, NASA had dictated to JPL exactly what the cost, the schedule and the performance of every Mars mission had to be. What that meant was that when things went wrong—when time or money got tight, as they always do—there was nothing JPL could do but let the risk increase. Everything else was nonnegotiable. NASA's rules meant that cutting corners and taking chances were JPL's only management tools.

It would be wrong to put all the blame on NASA Headquarters, though. There were people at JPL who knew the risks that were being taken, but none of them were willing to stand up and say that there simply wasn't enough money to do it safely. Nobody likes telling his boss that he can't do his job, especially if he's afraid that the job might be given to somebody else. So JPL swallowed hard and hoped they'd get away with it.

And we scientists were to blame, too. Yes, we had complained when NASA first had said that they wanted to fly cheap and risky missions. But once the deal went down, we didn't exactly walk away from what many of us feared could be a doomed enterprise. Nope . . . instead we fell all over ourselves fighting for a chance to fly anyway, me included. There was plenty of blame to go around.

With the loss of *MCO* and *MPL,* reality took hold at NASA Headquarters. They had just lost two Mars missions in very embarrassing ways, and the missions they had lost were both a hell of a lot simpler than the *Mars Sample Return* extravaganza that was on the books for 2003 and 2005. Something had to change.

The guy who got stuck with the whole mess was Ed Weiler, the new head of Space Science at NASA Headquarters. A twenty-year NASA veteran, Ed had made a steady rise through the Headquarters ranks. One of the formative experiences of his career had been serving as chief scientist for the Hubble Space Telescope when its optical flaw was discovered. He took the fire then for a problem that hadn't been his fault, and it gave him a taste of public and Congressional scrutiny in the face of failure that he never wanted to repeat. But here he was, barely into his new job, under fire again for something he hadn't done. Ed is compact and solidly built, with blond hair going to gray, a perpetual tan, and a blunt, plain-spoken manner. You might love him or you might hate him, but you always knew where you stood.

Ed did the logical thing under the circumstances: he killed the *Mars Sample Return* mission. There was never a press release to kill it off, no public announcement. I didn't even get a phone call. Instead, Ed simply made it known that the Mars program was about to undergo a profound change. The pace of the program would slow dramatically, he said. No more corners would be cut just to save money. There would be no sample return mission for the foreseeable future. You couldn't find fault with his reasoning, but NASA's new position meant the end of our rover. We had tied ourselves to sample return back when Mars Bug Fever hit, and when sample return died, our rover died with it.

The sense of loss among everyone on my team when we realized it was all over was profound. The *Athena* rover had become real to us, and losing it before we could even begin to build it felt worse in a way than losing it in flight ever would have been. We never even got a chance. I asked Barry to take all the rover drawings down off the walls in our offices at JPL; I couldn't stand to look at them anymore.

So our rover was gone. The only question left was whether or not we could salvage something from APEX and the *Mars 2001* mission. It didn't look good. From our standpoint, it hardly mattered which of *MPL*'s many flaws had done it in. The '01 lander that our APEX payload was supposed to fly on was essentially a knockoff of *MPL,* and there was no avoiding the conclusion that there probably were fatal flaws lurking in our spacecraft, too. With only a year until the 2001 launch, it was hard to believe that we had much of a chance of fixing them all.

Then one morning at the end of January, 2000, my phone rang.

"You're not going to believe this." Barry wasn't much for preliminaries on the phone.

"What?"

"They think they heard a signal."

"From *Polar Lander*?"

"Yep."

This was bizarre. Fact was, nobody *really* knew what had gone wrong back in December. Even the tripped-the-leg-switch theory was nothing more than an educated guess, since you could never be certain that the switch had actually tripped when the legs deployed. If it hadn't tripped, maybe the thing actually would have landed safely. Clearly *MPL* hadn't been able to talk directly to Earth after it landed, and clearly it wasn't able to talk to *Mars Global Surveyor* via its relay antenna. So something had definitely gone wrong. But what if the relay problem was the orbiter's fault, instead of the lander's? Nobody expected *MGS* to be much of a Comsat. There was a slim chance that *MPL* was alive on the surface with nothing worse than a busted direct-to-Earth antenna, and that we hadn't heard from it simply because something had gone wrong with the relay link aboard *MGS*.

To test this theory, JPL had quietly sent commands to *Polar Lander* on December 18 and again on January 4, long after everyone else in the world had concluded that the spacecraft was dead. The commands they sent it told *MPL* to blast out a full-strength relay signal, as if *MGS* were overhead. But then, instead of listening with *MGS*, they listened with a giant radio telescope at Stanford University. And amazingly, they heard a faint whisper of a signal. Even more remarkably, the whisper seemed to have the exact characteristics that a signal from *MPL* ought to have. It was astounding.

JPL kept this very quiet, not wanting news to leak out to the press before they knew what was really going on. The next step would be to command *MPL* to turn its relay radio on and off in a distinctive sequence, and to see if that sequence could be seen at the Stanford antenna. It was like teaching the spacecraft to send smoke signals.

On February 4, Barry called again. "To quote Dr. Frankenstein, it's alive!" They had done a successful test a couple of days before, in which the antenna at Stanford saw the relay transmitter on *MPL* turn on within a second of the time it had been commanded to turn on. Now the *MPL* project faced a race against time to figure out what had happened to their lander before the martian polar winter came on and killed it for good. *MPL* wouldn't be able to get any science done just by sending smoke signals, but maybe they could get a hint of whatever had gone wrong so we could try to fix it on our 2001 lander.

What all of this really meant for *Mars 2001* and APEX, though, was anybody's guess. In the Mars program office at JPL they had already written off the '01 lander mission. In fact, they were busy figuring out how to spend all the money they'd save from the anticipated cancellation of our lander to bail out whatever program they'd have left once things finally began to settle down. But in the *Mars 2001* project office, also at JPL, it was a very different story. The collective wisdom there was that a live *MPL* on the surface of Mars meant that they could fly after all. All they had to do was fix up a few little potentially fatal design flaws. It looked like maybe we were back in business with APEX.

The problem was that APEX still wasn't what I wanted to do. It never had been.

Times of turmoil can be pivotal: moments when a small idea, properly timed, can fundamentally change the course of what is to come. And we obviously needed a new idea if we were going to change the course of the program at this juncture. As soon as Weiler killed off *Mars Sample Return*, it was obvious to Barry and me that an *Athena* rover would be a great mission all by itself for launch in 2003, just as we had first conceived it back in 1995. Problem was, we had no way of getting it to Mars. Every mission concept we had ever thought about, going all the way back to

1995, had assumed that we'd use a three-legged lander, like the one that had just been lost. And landers like that were grounded until further notice, the apparent resurrection of *MPL* notwithstanding.

The breakthrough came to Barry on a weekend in the middle of February, revealed to me in a wee-hours-of-the-morning e-mail. Suppose that instead of delivering our rover to Mars on a three-legged lander, Barry suggested, we came up with a new landing system that was based on what JPL had done for *Mars Pathfinder*? *Pathfinder* had demonstrated a radically different way of landing on Mars, using inflated gas bags, like the airbags in a car, to cushion the lander's impact at touchdown. It wouldn't look like the *Pathfinder* system, of course, since our rover was too big to fit inside the *Pathfinder* landing structure. But the basic idea was the same. This mission would do all the science we had planned almost five years before, with the only difference being the way we'd get the rover down onto the surface. It was a beautiful concept; I was embarrassed for not having thought of it myself. We called it Athena-in-Bags.

How to sell it? We had to play up its strengths. One was that it was the only idea out there that used a technique for landing on Mars that we knew actually worked. Given everything that had gone wrong in the past year, the next mission to the surface of Mars *had* to succeed. With the success of *Pathfinder*, airbag landing was now a known quantity. If we used airbags to deliver our rover, *Mars Pathfinder* would have been a true pathfinder, instead of the evolutionary dead end it had seemed to be back when three-legged landers became all the rage.

Another big plus of Athena-in-Bags was that it would use instruments that were already built. We had been working hard on APEX, and Pancam, Mini-TES, the Mössbauer spectrometer and the APXS were now built, tested and ready to go to Mars. If we flew them in 2003, NASA could say legitimately that some of their investment on the *Mars 2001* lander hadn't been wasted.

It was also good for JPL. In fact, it was *really* good for JPL. The big money on these projects tends to get spent wherever the spacecraft is

built. On *Mars 2001*, most of the money had been spent at Lockheed
Martin in Denver. But it was JPL who knew how to build rovers, and JPL
who knew how to build airbag landers. Athena-in-Bags would be almost
a pure JPL in-house job if it happened, and that would feed a lot of hun-
gry mouths at the lab. This argument obviously wasn't going to get us
anywhere with NASA Headquarters, but it would sure sound good to the
management in Pasadena.

Just as Barry and I started to figure out how to make Athena-in-Bags
happen in 2003, the *Polar Lander* story got even weirder. To confirm the
Stanford results, JPL had asked two other radio telescopes, in Holland and
the U.K., to take a look, too. The folks in Holland thought they had seen
something, and that they'd seen the signal turn on when it was expected to.
But they also saw almost the same thing when their antenna wasn't pointed
at Mars at all, which made it look more than a little bit fishy. And the team
at the famous Jodrell Bank Observatory in the U.K. saw absolutely nothing.

With no convincing detection from across the Atlantic, scrutiny
turned to the Stanford results. Stanford had been using a search technique
that was tuned to pick up signals with exactly the characteristics they were
looking for, and if you push that kind of thing too far you can see things
that simply aren't there. (The phrase scientists use for this phenomenon is
"I wouldn't have seen it if I hadn't believed it.") And indeed, within a
week of the ambiguous results from Europe, the Stanford team reluctantly
concluded that they had been spoofed, victims of their own optimism.
There had been no signal. *MPL* really was dead, and the *Mars 2001* lander
mission, APEX and the last traces of everything we had worked for were
dead with it.

So what was going to happen? JPL was all over the problem, convening
review boards and advisory committees with abandon. Somehow, it was all
supposed to come together in a three-day meeting in Pasadena at the end of
February. This thing was huge, involving a distinguished-sounding team of
scientists and engineers that JPL had asked to help them fix the Mars program, a separate and enormous independent science advisory committee

and the usual hangers-on that show up whenever something of this magnitude is going on. It filled JPL's biggest conference room.

The first two days were spent on scientific yammering that didn't amount to much. But on the evening before the third day, a dinner was held for a small subgroup at the Athenaeum, Caltech's elegant faculty club, hosted by Charles Elachi. Charles had become a major player in the Mars game over the previous couple of years. He was the head of JPL's Space Science Directorate, which meant that he now owned all the Mars problems at the laboratory. Lebanese by birth and educated in Grenoble and later at Caltech, Charles was an expert in space-based radar who had worked his way up through the JPL ranks, much as Weiler had at NASA Headquarters. And while his credentials as a scientist were superb, throughout his career he had also shown uncommon political instincts and skills as a salesman. Charles was in full-up salesmanship mode as he scribbled on a transparency what he planned to present to the group the next day.

Twelve of us sat around the dinner table in front of what was left of our desserts, staring at the image Charles projected onto a rough stucco wall. In 2001, all his transparency had on it was the '01 orbiter. After that, he had four options. They were:

- Fix the *Mars 2001* lander and fly it in 2003 with our APEX payload.

- Fix the *Mars 2001* lander and fly it to the south pole of Mars in 2002 on a Venus flyby trajectory, with Paige's *MPL* payload.

- Fly a totally new mission in 2003 that would send half a dozen very small landers to the surface, like the old *MESUR* concept.

- Fly the *Athena* rover in 2003, using a *Pathfinder*-like airbag system to deliver it.

I did the best I could not to choke on my coffee when Charles wrote his fourth option down. He had heard about it from Barry, and he had liked it. I glanced nervously around the table to gauge the reactions, but there seemed to be general assent that it belonged on the list with the others. My God, I thought, could this actually happen? In fact, only three of Charles's possibilities were viable; there was no way the small landers could be ready by 2003. And the Venus flyby thing looked like a long shot too, since that spacecraft had never been designed to go that close to the Sun. If Charles's options went forward, we had at least a one-in-three shot, maybe better, of actually flying the mission we wanted to fly. And when Charles presented it to the whole group the next day, they went for it, too.

It was starting to look like maybe we had a chance.

5 ADLER'S INSPIRATION

W ITHIN TWO WEEKS, A quick run of the numbers showed that Barry's Athena-in-Bags idea wouldn't work for 2003. There were only two major problems with it, but they were the really bad kind: cost and schedule. Getting our rover down onto the surface with airbags would require a completely new lander and airbag design, and the conclusion of everyone who looked at it hard was that there just wasn't enough time in the schedule or money in the budget to design and build a new lander and airbag system for launch in June 2003. (Just how prescient this conclusion turned out to be became apparent many months later, under very different circumstances.) Disappointed, we filed it away as something to think about for 2005 if the chance ever came.

Other options for 2003 sank from sight, too. JPL dropped the Venus-flyby reflight of *MPL,* and also the *MESUR*-like mission with all the little landers, both because they seemed too risky. But a new idea popped up in their place. Nicknamed the Gavsat, after Tom Gavin, the gruff, crew-cut JPL old-timer who had dreamed it up, it was a big orbiter with a new generation of cameras and spectrometers hung on it. It wasn't sexy, but it was safe.

Then, around the beginning of April, a brash young JPL engineer named Mark Adler came into the picture. Mark is tall, strong-limbed and quick in both thought and speech. When he's around, there's no doubt that he's nearby, laughing easily and often, and cracking his knuckles loudly enough to be heard across a crowded room. Favoring gaudy shirts and rarely without sunglasses hanging from his neck, Mark is a product of sunny climes: born in Miami, raised in south Florida and trained with a Ph.D. in physics, from Caltech. He scuba dives and flies light aircraft in his spare time, and his willingness to confront risks intelligently in his daily life also comes through in his work. He had once helped NASA to assess the risk of flying the plutonium-laden *Cassini* spacecraft close by the Earth in a gravity-assist maneuver, receiving a medal from the agency for his efforts.

When Mark looked at the Mars situation, it looked bad. The Gavsat was safe and boring. The *Mars 2001* lander was dangerous and boring. JPL's credibility was at its nadir. Nobody asked Mark to solve the problem; it was just typical of the way that people work at JPL that he decided on his own to do it.

Mark realized the same thing that everybody else did: What the Mars program really needed was an exciting mission that could land on the planet, and do it safely. But his breakthrough was to recall that JPL actually did have a safe way of landing on Mars, though everyone seemed to have forgotten about it. It was the *Pathfinder* landing system, exactly as it had been built four years before. There was even spare hardware still lying around.

In essence, Mark turned Barry's Athena-in-Bags idea around backward. If you didn't have time to build a new lander to deliver the existing rover design, could you maybe instead cram a new rover into JPL's existing lander design? The lander was a lot more mature than the rover was; in fact, the thing had already been to Mars once, and it had worked. Maybe you wouldn't be able to fly quite as capable a rover, and maybe it couldn't carry the whole rover payload. But suppose you could make it fit?

Adler got together with Rob Manning, who had been one of the young stars on the original *Pathfinder* team. Manning thought that maybe it was doable. They signed on three other *Pathfinder* veterans, Jake Matijevic, Joy Crisp and Howard Eisen, and very quietly they got down to business.

By mid-April, they had roughed out a concept. It looked like the rover would have to fold up in a very crazy-looking way to fit inside the lander, but it wasn't a clear mechanical impossibility. The whole Athena payload wouldn't fit, but they were pretty sure they could get at least Pancam, Mini-TES, APXS, Mössbauer and the Microscopic Imager onboard. For the time being, they called it *Athena-Pathfinder.*

Adler and his accomplices weren't the only ones at the lab with wild ideas, though, and there were a lot of mission concepts floating around. Desperate to come up with something that would be safe enough to fly in 2003, JPL had put one of their most experienced hands, John Casani, onto the job of picking through all the various mission concepts and evaluating how risky they were. Mark pitched the new mission, which he had now renamed *Mars Geologist Pathfinder,* to Casani's group on April 26. They liked it. In fact, they liked it enough to proclaim it to be the safest way of getting something useful to the surface of Mars in 2003, a fixed-up version of the *Mars 2001* lander included. Mark's idea was out in the open, and it was beginning to gain some traction.

Time was growing short. Three years is a frighteningly small amount of time in which to prepare a mission to Mars; if anything at all was going to fly in 2003, a decision was going to have to be made very soon. To figure out what the laboratory's preference ought to be, JPL director Ed Stone called a meeting of his executive council, all the high-and-mighties at the lab. Their decision, after a gratifyingly short debate, was that their preference was for *Mars Geologist Pathfinder.*

It made a lot of sense when you thought about it. Much as it would've been nice to think so, the council's decision had little to do with JPL loving us or our science. Instead, it came down to three very practical things. First, of all the options, it was the only one that would be built entirely at JPL. To put it bluntly, MGP would vector several hundred million dollars to the Jet Propulsion Laboratory, and that's not chump change. Second, *MGP* really was less risky than flying the *Mars 2001* lander, and JPL was terrified of the next Mars landing not working. In fact, they were acting as if the future of the lab depended on it. It probably did. The third and final reason was closely related: If your future truly does depend on the success of a mission, you do not contract that mission out. That's just common sense. You line up the very best people you have, and you do it yourself.

So JPL loved our rover now.

As always, it wasn't up to JPL to decide what would happen next, it was up to NASA Headquarters. And to bring things to closure, NASA set up a meeting in Pasadena for May 3 and 4 at which all the possible options—all the ones in work at JPL and still others that were favored back east—would be put on the table. When you added up all the ideas and the many ways in which they could be combined, it came to over a hundred possibilities. It was a wild situation; in fact, it would have been a fun race to bet on if betting on it had been optional. Somehow, the participants at this meeting were supposed to consider all the possibilities and then narrow the field down to just a handful. Outside scientists were pointedly not invited—it was a pure NASA-JPL show, and cost and safety were going to rule the day. I put every spy I could into the room, and I had them dashing out to phone me back in Ithaca whenever they had the chance.

Adler and company pitched the new rover mission to the group, under yet another name, *Mars Mobile Pathfinder* this time. The options were hashed and rehashed, and by noon on the second day, Thursday, just three missions had survived. They were the Gavsat, the *Mars 2001* lander and *Mars Mobile Pathfinder*.

Ed Weiler was at the meeting, and he listened to the proceedings impassively. By Thursday afternoon, though, it was time for him to make his feelings known. The three surviving options were reasonable, he said, but he had his druthers. "I can't imagine flying the 2001 lander, and I'm relieved you didn't ask me to do it. And we could fly the orbiter, but it doesn't seem to me like the medicine that the program needs." Instead, he said he was leaning toward flying *Mars Mobile Pathfinder*.

After this pronouncement, Weiler and his lieutenants spent the rest of their time at JPL roaming around the lab talking to rank-and-file Mars program engineers. One got the sense that they didn't entirely believe everything that the JPL managers had to say about what was really going on, and that they wanted to take the pulse down among the people who do the real work. What they got was an impassioned plea to get off the dime and fly *something*. Time was passing, and there were a lot of wounds to be healed at JPL. Only a fresh challenge crowned with eventual success at Mars would do it.

At 7:30 the next morning, in anticipation of a NASA announcement, JPL Director Ed Stone called a meeting in his office to start forming the *Mars Mobile Pathfinder* project team.

The project manager would be Pete Theisinger. Portly, gray-haired and rumpled, with a salt-and-pepper mustache, Pete looks like your uncle the elevator repairman. The minute he opens his mouth, though, it becomes obvious very quickly that you are dealing with a gifted, natural leader, and with one of the clearest thinkers at JPL.

The flight system manager would be Richard Cook. Richard was a veteran of both *Pathfinder* and *MPL*, and he had been cool under fire as JPL's main public spokesman during the unfolding disaster of the *MPL* landing. Richard was young and aggressive, and was widely considered to be the engineer with the brightest future at JPL, especially if he could pull off something as tough as leading the development of a new spacecraft that was as complicated as this one would be. Undeniably brilliant, Richard also had a reputation for having a healthy disdain for scientists, whom he seemed to consider undisciplined, unruly and a bit of a pain in the ass. Barry was chosen as Richard's deputy, giving me a trusted friend and confidant in the flight system office when I needed one. It was a powerhouse team.

Then, on May 13, NASA did something that proved that they'd made a decision. They put out a press release. The release was brief and blunt, and it announced that the choices for the 2003 mission to Mars were now down to just two. One was the Gavsat, renamed the *Mars Science Orbiter*, or *MSO*. The other one was us.

It was stunning. We were back to our roots . . . a big rover to do pure scientific exploration on Mars. The difference this time, though, was that now we were one of just two candidates for flight and maybe the leading one, with every resource that JPL had at their disposal solidly behind us. It was an astonishing reversal of fortune.

It is hard to overstate how much this mission meant to JPL. *Mars Mobile Pathfinder* would be both a cash cow and a chance to restore their ruined reputation. If they could make it happen, *MMP* would be a mission of redemption. They were going to go after it harder than anything I'd ever seen them do.

For me, it was hard to believe that this was happening—that we really were at the point where our chances of doing a rover mission were back to fifty-fifty or even better. We had been getting bad news for so many months that my brain didn't seem capable of processing good news. For a team that had been flat-line dead just months before, this new chance at resurrection felt like nothing short of a miracle. We had to find a way to make it work.

The day the press release came out, I was in the Spacecraft Assembly Facility at JPL with a bunch of the hardware guys from the science team, testing Pancam and Mini-TES together for the very first time. It had been a three-day marathon of eighteen-hour sessions, and by the third day we were so whipped we could barely think, let alone manage to go out for food. So we called for Chinese takeout, and after the food arrived and was sloppily consumed on the lab desks, I reached into the bag.

"Okay, guys, this is gonna be it . . . the official *Athena* Science Team fortune-cookie fortune." I cracked the cookie open. It read: *"You will soon be the center of attention."*

So now our job was clear. If we could beat the orbiter mission we'd get our trip to Mars. NASA gave us four million bucks and two months to figure out what we were doing, and they did the same thing for *MSO*. The whole thing would be decided in a two-day shootout at NASA Headquarters in July. We got to work.

The landing system would be the easy part, since it was already done. *Mars Pathfinder* had pioneered a new way of getting hardware onto the surface of Mars. Unlike *Viking* back in 1976, *Pathfinder* had used no orbiter to linger first above the planet, waiting to find the right time and place to drop the lander to the surface. Instead, *Pathfinder* had plunged straight in from a deep space trajectory, and we would, too. *Pathfinder*'s entry into the atmosphere was done pretty much the same way it's been done since the sixties, with the spacecraft tucked inside a two-part "aeroshell" consisting of a forward heat shield and a cone-shaped aft cover called the backshell. As it entered the martian atmosphere, the *Pathfinder* aeroshell looked just like an *Apollo* space capsule, and its job of protecting precious cargo while bleeding off hypersonic velocity through friction with the atmosphere was substantially the same.

Once it fell deeper into the atmosphere, though, the *Pathfinder* entry system started to fly like nothing ever seen before. At Mach 2, twice the speed of sound, it threw out a parachute to slow it further. The heat shield, no longer needed, dropped away. Still falling at hundreds of miles an hour through the thin martian atmosphere, the lander was lowered down from the backshell on a long cord, called a bridle, to put it a safe distance from three solid rocket motors in the backshell that were poised to fire. A radar system turned on to sense the distance of the spacecraft above the ground.

Next, about eight seconds before impact with the surface, a cluster of twenty-four fat, spherical airbags inflated explosively around the vehicle.

The airbags were *Pathfinder*'s signature engineering achievement. Conceptually similar to the airbags that protect passengers in a violent automobile accident, their job was to absorb the shock of impact with the martian surface. Then, a split second before touchdown, the solid rocket motors in the backshell ignited, bringing the vehicle's descent to a halt about ten meters above the surface. The bridle was cut, and the lander fell away. The *Pathfinder* lander, cocooned within its airbags, had bounced and rolled for almost a kilometer across the martian surface before finally coming to rest.

Once at rest, vents opened, deflating the airbags. Cords that were spun like webs through the bags were reeled in, collapsing them flush against the lander's sides. With the airbags retracted, the *Pathfinder* lander was shaped like a tetrahedron—a pyramid that, cast like a die, could have come to rest on any one of its four faces. All the lander's equipment, including the camera and the tiny *Sojourner* rover, lay hidden inside this pyramid. Still executing a preprogrammed series of events, three of the four sides of the pyramid opened out like the petals of a flower. If the pyramid had come to rest on its heavily weighted "base petal," this flowering would simply reveal the hardware concealed within. But if it had settled on one of its side petals, the process of opening the petals would flip the lander upright. With its petals open and its life-giving solar cells now exposed to the sky, the lander was safe and ready to begin its work.

End to end, the *Pathfinder* landing system would give us one hell of a ride. Using it, we'd go from our first 12,000-mph contact with upper fringes of the martian atmosphere to bouncing on the surface in six terrifying minutes.

So that was it. We'd simply use the *Mars Pathfinder* backshell, heat shield, parachute, airbags and lander structure, exactly as they'd been built four years before. End of story.

The science payload issues that we faced weren't trivial, but things looked pretty good there, too. We figured we could fly the Pancams, the

Mini-TES and the Mössbauer that we already had, though we'd have to prove they could survive the jolt that an airbag landing would give them.

The Microscopic Imager would have to be new, but it looked pretty easy. The rover would need a bunch of new engineering cameras anyway, and the idea was that they'd be loosely based on our existing Pancam design. We'd use this same new camera body for the MI as well, just screwing a different close-up lens onto the front of it. We looked real hard at ways to get some color filters onto the thing, since a color microscope would have been much cooler than a black-and-white one. But room was so tight at the front end of the rover that we just couldn't make the filters fit. So black-and-white it was.

The APXS situation was complicated, since the APXS that we had built for APEX really worked only with a *Sojourner/Marie Curie*–style mini-rover. But I knew that Rudi Rieder, the mastermind of the APXS lab in Germany who had played a big role in building the *Sojourner* and *Marie Curie* instruments, had a new APXS design that he had developed for a European comet mission called *Rosetta*. It was a much better instrument, and I thought it would work beautifully on our new rover. I twisted Rudi's arm to build us a copy, and he agreed to.

So that was the core of the payload, and it was in good shape. Where it got hairy was the Mini-Corer and the Raman, since those were both new and complicated things compared to everything else.

We didn't need the Mini-Corer, since nobody expected us to collect any samples on this mission. But we needed something. Our rover was supposed to be a robot field geologist. When you see field geologists on Earth, they've got their boots, they've got their backpacks and, always, they've got big rock hammers. Rocks that sit around on a planet's surface get exposed to the elements—sunlight, moisture, wind—and if they sit around long enough they can become weathered. As weathering takes place, the surface of a rock can become dirty, brown and crumbly. When a rock first forms, its composition and its texture preserve evidence of the conditions under which it formed. But the composition and texture of a rock can be changed by the weathering process, and the evidence of conditions when the rock formed can be altered or destroyed. Geologists carry rock hammers so that they can break weathered rocks open, exposing the pristine evidence within.

Nobody knew if martian rocks were weathered or not, but we didn't want to take a chance on it. The Mini-Corer had been our way of getting inside rocks, but now it was gone. We needed something to replace it. A hammer would be way too dangerous; smart geologists wear safety goggles when they're swinging their rock hammers, and they do it for a good reason. Instead, we needed some kind of a grinder . . . something that we could press or rub against a rock and have it grind away the outer layers. Not even knowing what it was yet, I dubbed it the RAT, for "rock abrasion tool," and I asked Honeybee Robotics, the developers of the Mini-Corer, to come up with something. It was a cute acronym; we could figure out what it really meant later.

The painful part of the whole deal was the Raman. This instrument was a complicated sucker, with fiber optics that would run up the rover's arm and a lot of optics and electronics inside the rover body. When we looked hard at the problem, it was clear that a rover that was large enough to accommodate the Raman would just be too damn big to fit inside the *Pathfinder* landing system. So it couldn't fly.

It ached to lose the Raman. The instrument was maybe our strongest tool for learning what minerals were in the rocks, and it was our only way of looking for organic molecules if any happened to be there. And it wasn't just losing the Raman science that hurt. The Raman was Larry Haskin's baby. Larry was nearing retirement, and he had been working on his instrument for years. If we didn't get it to Mars for him in 2003, it wasn't clear that it would ever get there. It broke my heart to have to tell him that his instrument couldn't fly.

Where things got really complicated was the rover itself, where Adler's idea meant that almost everything had to change. The *Pathfinder* system was a wonderful way to land on Mars, but it was definitely not rover-shaped on the inside. Using the *Pathfinder* lander would mean that we'd have to find a way to fold our rover up into a tetrahedron, and then do the same trick again in reverse once we landed. The JPL mechanical engineers thought they had found a way, but it looked scary as hell, with more joints, folds, motors, hinges, springs, gears and latches than anybody had ever seen on a planetary spacecraft. An awful lot of moving parts were going to have to work just right for this thing even to be ready to begin doing its job on Mars.

The rover electronics all had to change, too, because now they had a new job. In Adler's concept, the rover possessed the only computer on the spacecraft, which meant that the rover was the brains of the operation all the way to Mars. On top of its old duties, then, the rover computer now also would have to handle everything during the cruise to Mars, and, especially, during the high-speed and mission-critical events of landing.

And of course we changed the name once again, this time to *Mars Geological Rover*.

In the middle of all this, NASA announced something strange and important and new. Pictures taken by Mike Malin's camera on the *Mars Global Surveyor* orbiter had revealed small gullies on Mars that seemed to have been carved by liquid water, perhaps in geologically recent times. It was a new and fresh discovery that would take years for the scientific community to digest. But a knee-jerk reaction by NASA could be fatal for us. The gullies were close to the poles, where sunlight is faint, and there was no way that our solar-powered rover could work there. The orbiter, on the other hand, could take new and sharper pictures of them. If NASA reacted quickly to this discovery and decided that the new mission had to go after the gullies, which suddenly were the scientific puzzle of the moment on Mars, we were dead.

To decide between us and *MSO*, NASA put together a group of twelve wise men—senior scientists and engineers drawn from academia, industry and the government. Both project teams were summoned to appear before them on July 13 and 14, 2000, in MIC-7, the big conference room on the seventh floor of NASA Headquarters in D.C. It was a head-to-head shootout, with hundreds of millions of dollars on the line and only one winner. It was a very tense affair.

We started with the science, and on Thursday afternoon NASA gave each team an hour and a half to present its case. Rich Zurek, the project scientist for the orbiter mission, won the toss and went first. His presentation was scary. Rich was good, their payload was good, and there really wasn't a serious weak spot anywhere. I got up next and gave it my best shot for *MGR*, hoping for a draw.

Friday morning came the engineering discussions. The best we could really hope for here was a draw, too, since it was obvious to anybody with half a brain that our mission was riskier than theirs. Pete Theisinger gave the pitch, and he did fine. But the orbiter pitch was solid, too, and maybe even stronger than ours was. It was looking like we weren't about to convince anybody, including ourselves, that our mission was better than theirs on either the science or the engineering.

But we held two trump cards, or at least we thought we did.

The card that we played aggressively came from simple celestial mechanics. The year 2003, it turned out, would provide the best opportunity to send a solar-powered rover to Mars for the next eighteen years. The trip required less rocket fuel than it would in most years, and once the rover arrived, Mars would be comparatively close to the sun, which was good for solar power, and also close to Earth, which was good for communications. All this came about through no virtue of ours; it was simply a matter of the fortuitous way in which the planets lined up that year. If NASA ever wanted to do a mission like ours, now was the time, and we made *real* sure that everybody in the room understood that.

Our other trump card was one that everybody knew we had, though we never mentioned it. Our mission was just plain sexier than theirs.

We wrapped up our presentations, and shook hands all around. The twelve wise men closed their door and went into executive session. I headed for the airport.

6 THE DECISION

NASA HAD TOLD US to expect a decision in less than a week. It didn't turn out that way.

On Tuesday and Wednesday after the shootout, there was a Mars planning workshop at the Lunar and Planetary Institute, a sleek, modern research and conference facility near Johnson Space Center in Houston. The point of the workshop was to bring scientists together to talk about what NASA should do at Mars from 2005 onward. And indeed most of the presentations and panel discussions in the meeting rooms were about that topic.

In the hallways, though, it was another story. The subject on everyone's minds was what was going to fly in 2003. The only two people who actually knew something—Scott Hubbard and Carl Pilcher from NASA Headquarters—were poker-faced. Scott had been brought in from NASA Ames some months before to fix the Mars program after the *MCO/MPL* debacle, and Carl ran the rest of NASA's planetary program. With 250 people at the workshop, 248 of whom had not the slightest clue what was going on, the silence of Hubbard and Pilcher made for a lot of speculation. Rumors abounded. The rover's got it locked up. The orbiter's got it locked up. They're deadlocked and they can't reach a conclusion. My favorite was that they had decided to fly the rover, but without any wheels! And everybody was trying to handicap the decision based on what seemed to be sentiment among the workshop participants, as if that somehow mattered.

Of course, all the speculation was irrelevant. By midday on Tuesday the deed had been done.

The decision meeting with Ed Weiler was supposed to have taken place on Monday, but Ed had airplane adventures on his way back from a vacation in California, and he wound up stuck and grumpy in a Denver airport hotel. On Tuesday, though, the deal went down. At about noon, Hubbard and Pilcher disappeared from sight, and they didn't reappear until three hours later. Where they had been was at a teleconference with Weiler, JPL Director Ed Stone and Firouz Naderi, who was Hubbard's new counterpart running the Mars program office at JPL. And at that telecon the decision had been made.

As soon as Scott and Carl reappeared, I began to get good vibes. Scott came to sit in on all the talks about our payload, and then he pulled me out into the hall and started to ask some basic questions about our science. The answers to his questions were the kind of things you'd need to know if you were going to explain the rover mission to your bosses up the line. What I desperately wanted to do, of course, was just flat-out ask him what had happened, and whether we were finally going to Mars or not. But I couldn't bring myself to do it, and he wouldn't have been able to give me a direct answer if I had. Hard as it was, it was better to wait.

Wednesday was more of the same, with lots of speculation, lots of whispering in the hallways and not a hint from Hubbard or Pilcher.

Then, late Wednesday afternoon, things took a strange turn. I caught Hubbard talking on a hallway phone with an anxious look on his face, and he waved me over. He finished his conversation, hung up and then, to my amazement, let it all out. "Weiler picked the rover, so you're going to Mars if we can get it past Goldin." The rush of emotion that should have followed news like that was flattened instantly by the look on his face and the tone in his voice. "But there's a problem," Scott continued. "Can you get together with me and Carl for a few minutes?"

Well yeah, I said, I think I can probably fit that into my schedule. So he found Carl, and the three of us huddled on a couch in the middle of the institute's big "Great Room," pretending that nobody noticed that we were talking. It turned out that earlier that day Weiler had run into Goldin in the NASA Headquarters cafeteria. Goldin had asked Weiler if the Mars decision had been made yet, and Ed had said yeah, it's going to be the rover. And Goldin didn't get it. They had just had a press conference two weeks ago to announce that they'd discovered recent water activity on Mars, and now they were going to have another press conference to announce that they *weren't* going to explore the gullies? Goldin had been expecting the orbiter to be selected, and he let Ed know that he was going to be asking some very penetrating science questions when he got briefed formally on Friday.

All of which sent the Space Science guys at NASA Headquarters into a tizzy. Suddenly they were faced with really having to tell our science story to Goldin, and none of them knew quite how to do it.

So what should we do? The thing to do, we concluded, was for me to get my butt on an airplane immediately, and spend Thursday at NASA Headquarters with Jim Garvin, Hubbard's chief scientist, going through the story with him so he could pitch it to Goldin.

Into my rental car I hopped, leaving a vapor trail out of the LPI parking lot. I managed to catch Ray Arvidson before I sped away, and then called Dick Morris, a Mars soil expert on the team from Johnson Space Center, from my car. I asked them both to put together their own thoughts on how our mission would go after questions related to water and life on Mars. By midnight I was holed up in a Holiday Inn near National Airport, whipping together PowerPoint charts. Arvidson's stuff came in by e-mail at about 1 A.M., and Morris's at 4 A.M., bless their hearts. By eight in the morning I had a killer story to tell. I got a couple hours of sleep, and I was in Garvin's office at Headquarters by 11:00. We spent a few hours going over it, I cabbed back out to National, and I was back in Ithaca in time for dinner.

And that was it. All that was left was to sit by the phone on Friday and wait for the call from Hubbard that would let us know if Goldin had gone for it or not. The meeting in Goldin's office was at 10 A.M., and I figured it would last an hour or two. And then we'd know one way or another.

Or so I thought.

At 11:30 the phone rang, and it was Scott. A very rattled-sounding Scott, as it turned out. Plus Jim Garvin, plus Carl Pilcher, all on a speakerphone. Well now, this is interesting, I thought.

"Hi Steve," said Scott. "We just got out of the meeting with Goldin. I have to ask you something."

"Yeah, go ahead."

"Can you build two?"

Huh? "Two what?"

"Two payloads."

It still wasn't sinking in. "Why would you want two payloads?"

"For two rovers."

Say what?!? I sat silent and thunderstruck as Scott ran me quickly through what had happened. Goldin had bought the science story, and he was convinced that Weiler had made the right decision by picking the rover. But, naturally enough, he was worried about the risk. So he asked one simple question. One simple, insightful question that nobody— nobody at all—had thought to ask before.

Could we fly two of them?

Two rockets, two landers, two rovers, two of everything. Launch them both in 2003. Land them both and operate them both in 2004. Full redundancy, the way NASA used to do it in the old days.

This was classic Goldin. There are a lot of bad things you can say about the guy. He has terrible people skills. He can be mercurial, abrasive and disheartening to work for. But he is a brilliant strategic thinker. Flying two of everything instantly solved the biggest problems facing our mission. Risky landing? No room in the rover for redundancy? Fine: Double up everything. Not sure if it's better to land where it's the topography or the chemistry of the martian surface that provides evidence for water? Fine: Try both.

None of us saw this coming. Not me, not Scott Hubbard, not Firouz Naderi, not Pete Theisinger. Of course, it would have been suicide for us to have gone into the shootout saying we had to fly two. Two rovers would have jacked our costs way up compared to *MSO*, and asking for two would have been a tacit admission that our mission was too risky. But somewhere in there we could have dropped a subtle hint that flying two might be a good idea, and we sure as hell could have figured out what a second rover would cost so we'd know the answer if somebody asked.

Instead, Crazy Dan caught us all flatfooted. Scott and Carl and Jim needed to know if I saw any problems with building a second flight payload. No problems at all, I said, just send money. The next phone call came from Barry and Mark Adler, who had been given forty-five minutes to come up with a cost estimate for another complete spacecraft. I answered a few questions and off they went. And sure enough, forty-five minutes later, they called back. Their answer was $665 million, compared to the $440 million we had figured it would cost to do one. Fifty percent increase, more or less. They gave the number to NASA, and we all sat back and held our breath.

All we got was silence. NASA Public Affairs, which had earlier announced a press conference at which the decision was to be revealed, put out a terse statement. The decision about what to fly in 2003 had turned out to be "more complex than anticipated," it said. The answer would be delayed by "one to two weeks."

The reason for the delay, obviously, was money. Adding a second rover would cost almost a quarter of a billion, and you don't just round up that kind of money with a couple of phone calls, even if you're the NASA administrator.

There was no easy place to get it. One obvious approach would be to go to the White House and Congress and ask for more. That didn't sound real likely to succeed, though, administration comments about their support for Mars exploration notwithstanding. The other would be to carve it straight out of the NASA budget. NASA's budget was about $14 billion a year, and Barry and Mark's estimate of the cost of adding a second rover in 2003, spreading it out over three years, would be about one half of one percent of that. Given the fallout from the *MCO/MPL* fiasco, Goldin could easily conclude that taking it out of his own hide would be a pretty cheap insurance policy. But it would be painful for somebody.

They tried the White House first. On Sunday afternoon, I got an e-mail from Scott. NASA wanted us to produce some video animation of the mission, showing a second landing and a second rover. "The sexier the better," wrote Scott, "to be shown at the highest levels of government." Uh huh, good luck. I didn't have much hope, but we sent him the video.

Meanwhile, I was getting inquiries from the news media. Andrew Lawler from *Science* magazine called on Monday morning. He had nothing, but he smelled a story. Andrew is a shrewd guy. He worked out of D.C., and he knew all the players, both at NASA and at the White House. When NASA puts out a press release one day saying a decision worth half a billion dollars is about to be announced, and then puts another out the next day saying that everything has been postponed, somebody like Andrew is going to know that something's up. I stonewalled, but I didn't like the sound of it.

On Tuesday morning Andrew called back. He had the story, and he was going to run it. Moreover, he told me something that I hadn't known: The White House had refused to provide any more money for the second rover. Fortunately, *Science* is a weekly, not a daily, and Andrew's piece wouldn't come out until Friday. A NASA press release on Thursday could beat him to it. But if Andrew had the story, who else also had it? I sent an e-mail to Hubbard letting him know that the news was about to break, in three days or sooner.

On Tuesday afternoon, JPL confirmed to me that the White House had said no. That left NASA with three choices: They could fly two rovers and come up with the money themselves. They could fly one rover and take all the risks associated with that. Or they could bail out and just fly the orbiter. I didn't sleep very well Tuesday night.

Very early Wednesday morning, before I left home for work, Scott sent me an e-mail asking me to call him as soon as possible. This was it, it had to be. I took the phone out onto the same porch where I had written so much of the proposal three years before, settled myself and dialed. Scott picked up on the first ring.

"Hi Scott, it's Steve."

"Ah, good, you got my message. I wanted to tell you myself. You're going to Mars."

There were no qualifiers this time, thank God, no conditions that would have to be met before it became real. Goldin had decided that NASA was going to fund the mission, no matter what. We had done it. Publicly NASA's position was going to be that no decision had been made yet on one vs. two, since they were still figuring out exactly where the money would come from. But internally they had already decided that they would find the money somehow.

In the afternoon, Scott called me again. A press release would come out at 3:00 Thursday. NASA wanted to break the story before Lawler did it for them.

At 2:30 on Thursday, Barry called. Firouz Naderi had just announced the news to all the Mars troops at JPL, which made it official. The new name for the project was *Mars Exploration Rover*, or *MER*. I sent an e-mail to my whole team, and then I phoned Steve Gorevan at Honeybee. Keeping a business as small as Honeybee afloat is tough, and Steve had let me know, without saying it in so many words, that his company had a lot riding on NASA's decision. Five minutes later, the entire company—all fifteen of them—called me back on their speakerphone, sounding like they had just won the lottery. They popped a champagne cork for me so I could hear it over the phone.

It was time to get back to work.

NASA's decision to fly *MER* was like our original selection three years before in some ways, but it felt completely different. I had been very naïve in 1997. Back then, it seemed as if we were just going to rocket off to Mars and have a glorious adventure. I didn't foresee then how many things could go wrong in such a short time, or how close to death our mission would come.

This time I had no illusions. NASA's decision process had dragged on for so long that our schedule was almost impossible before we started. Adding another rover had solved some of our biggest problems, but it had created a bunch of new ones. With the schedule we'd have to face, it wasn't clear we could even make it to the launch pad, let alone get things to work at Mars. We had made a whole bunch of promises during the shootout. Delivering on them now—twice—was going to be a struggle.

And there was one more frightening thing. With everything that had happened, a big chunk of the credibility of the nation's space program was now riding on our mission. Nobody had intended it to be that way, but there it was, whether we liked it or not. The magnitude of what that might mean over the next three years was just starting to sink in.

PART TWO

DEVELOPMENT

7 "YOU'RE BACKED INTO A HYPERDIMENSIONAL CORNER"

P ETE THEISINGER AND I didn't hit it off. I barely knew the guy. Pete had worked on JPL flight projects for years, and he'd seen a lot of PIs come and go. From what I could tell, some of his past experiences with scientists had been bad ones. And goodness knows I'd had my problems with other project managers. I frankly didn't trust him, and it was pretty obvious before long that he didn't trust me either.

My trouble with Pete began only a few weeks after the official start of the project, in a dispute over the RAT. I wanted the RAT to be under my control, because I didn't trust the project to do it right. Pete wanted the RAT to be under the JPL's control, because he didn't trust me to do it on time and on budget. I fought the battle and I won it, but it left a very bad taste. A month into the most challenging thing either of us had ever tried, and the project manager and the PI couldn't work together.

It came to a head late one Friday when I walked into Pete's office at JPL after nearly everybody else had gone home. I'm not sure what I'd hoped to accomplish by going in there, but before long we were going at each other loudly, each with our list of grievances. They were all pretty stupid ones, I began to realize as the argument wore on, because we hadn't worked together long enough to have any good grievances yet. It was mostly just baggage we both carried around from past missions we'd done with other people.

After an hour or more of arguing, it was Pete who made the breakthrough. He simply looked straight at me and said that if I wouldn't act like some of the pushy scientists he'd had to deal with in the past (and he named names), he wouldn't act like some of the stubborn engineers I'd had to deal with (he named names there, too). If I was willing, he said, we'd find a middle ground.

Hearing that from Pete stunned me. I'd never had a JPL engineer speak to me plainly and honestly about the science-engineering conflict before. I told him I'd give him a chance, and he said he'd do the same for me. Within weeks of that cathartic talk I began to realize that he really did want the mission to succeed, in the true scientific sense, almost as badly as I did. A trust between us began to grow, and it kept growing as the months went on.

Pete and I had to find a way to work together for a whole bunch of reasons, but the most important was the schedule pressure we had come under when NASA selected us. As Pete put it, from the day NASA said go, "Our top three problems are schedule, schedule, and schedule. There's nothing in fourth place." His "nothing in fourth place" was hyperbole, but it was hard to argue with the other three.

The cause of the dilemma was the launch window. You can't just head off to Mars any time you get the urge. The planets are always moving in their orbits, and in order to get from one to the other efficiently, Mars and the Earth have to be in just the proper alignment when you launch. The launch window is the short interval of time in which the geometric conditions are right to leave one planet and head for the other.

Launch windows to Mars are infrequent, coming once every twenty-six months. The twenty-six-month rhythm is a fact of nature, driven by the unchanging orbital periods of the two planets. Our window would come around in the summer of 2003, when we'd have three weeks to get each rover off the ground. If we didn't fly then, we wouldn't fly at all.

The summer of 2003 wasn't far away, and that was our problem. With the months of turmoil that had followed the loss of *Climate Orbiter* and *Polar Lander,* we hadn't gotten NASA's blessing to start work on our rovers until just thirty-four months before they had to be on the launch pad. And conventional wisdom said that thirty-four months was not enough time to do a complicated Mars mission. In order to do *MER* at a sane, safe pace, with acceptable margin for error, all of JPL's past experience said that we needed a good six or eight months more than we'd been given.

But there was nowhere we could go to get those six or eight months. NASA could give us more money if we got into trouble, at least in principle. But the launch window was where it was, and NASA couldn't print more pages for the calendar. From the very start, we knew we were facing a brutal, arduous race to get our spacecraft done in time.

The motions of the planets in their orbits weren't the only aspects of celestial mechanics that waited to torment us. The rotation of the planets about their axes was going to be a problem, too.

Earth takes twenty-four hours to spin on its axis; that's why our day is twenty-four hours long. But Mars spins a little more slowly than Earth does. In contrast to the Earth's twenty-four-hour day, the martian day is twenty-four hours and thirty-nine minutes long. We adopted the word *sol* for a martian day, following the lead of the teams that had worked on *Viking* and *Mars Pathfinder.* If we managed to get the rovers to Mars, that thirty-nine-minute difference between the length of a day and the length of a sol was going to create havoc with our lives.

Our rovers would be solar powered. They would work during the daytime, when power was plentiful and temperatures were relatively warm. During the bitterly cold martian night they would sleep, conserving energy for the next day. But the rovers wouldn't care if it was daytime or nighttime in Pasadena, or in Ithaca, or indeed anywhere on Earth. They wouldn't even know. Each would know only about daytime and

nighttime at its landing site on Mars. And because we would send instructions to the rovers only once per sol, those of us back on Earth who were
entrusted with their care and feeding would have to live and work on
their schedules.

I tried to imagine what rover operations would be like. It would be
possible to plan activities for each sol only after we'd learned what had
happened on Mars the sol before. So if the planning process started at
8:00 A.M. Pacific time today, then tomorrow it would have to start at 8:39
A.M., slipping thirty-nine minutes later to account for the longer duration
of a sol. The next day it would be 9:18 A.M. Then 9:57. Then 10:36.
Three weeks later, it would be in the middle of the night, sliding inex-
orably onward around the dial of an Earth-time clock. The only way you
could wake up at the same time every day would be if your alarm clock
kept Mars time instead of Earth time. All of us who wanted to work
flight operations would have to shift our sleep schedules constantly, keeping pace with the planet where we worked instead of the planet where we
lived.

And it got worse the more you thought about it. We would have two
rovers, each landing at a different place on Mars. So we'd have to split the
team in half, one for each rover. All of us would live and work on Mars
time, but in two different martian time zones. And if you were working
on one rover and you had to switch to the other, you'd get martian jet lag.
It was going to be confusing and exhausting at best, and maybe dangerous. Would sleep deprivation affect our judgment, leading us to a fatal
mistake at some critical juncture?

The most pressing question about sols was simply how many of them
our rovers were expected to operate. The thing that would kill us, we figured, was the buildup of dust on the solar arrays. *Mars Pathfinder* had seen a
continuous buildup of dust on its lander's solar arrays, steadily decreasing
the electrical power from the arrays as the mission progressed. It stood to
reason that we'd see the same thing. We planned to put our main electron-

ics in an insulated enclosure deep inside the rover that we called the Warm Electronics Box, or WEB. The name sounded cozy, but the WEB would be warm only if we had enough electrical power to keep it that way. Sooner or later there'd be so much dust on the arrays that we wouldn't have enough power to keep the WEB warm overnight. When that happened, the electronics inside would get too cold, and they'd die. With too much dust on the solar arrays, the rovers would simply freeze to death.

We looked at all kinds of ways to get dust off the solar arrays, of course. We looked at windshield wipers. We looked at compressed gas that could blow the dust away. My favorite was the idea of covering the solar arrays with clear plastic on rollers, and when the plastic got dirty we'd just roll the rollers and bring in clean plastic.

But none of these schemes worked. The problem with all of them was that they were too complicated and too heavy. The windshield wipers that would keep the arrays clean would have to be bigger and heavier than the entire instrument arm and all the instruments on it. There was a limit to how much stuff we could send to Mars, imposed by how powerful our rockets were and how good our landing system was. Our design was close to the limit already, so there was no way we could add big heavy gadgets to keep the arrays clean without dropping something even more important. All we could do was make the arrays as big as we could get them, and live with the results.

When Mark Adler had first cooked the mission up, the lifetime he'd been willing to advertise was thirty sols. What that meant was that even with the worst rate of dust buildup that Mark could imagine, he was confident that a rover would survive thirty sols before it died.

But thirty sols wasn't enough to get anybody excited about the mission. We just couldn't get enough science done in that short a time. If a month on Mars had been all we'd had to offer, we would've had little chance of beating the orbiter and getting selected by NASA to fly. In fact, if the mission were that short, even I wasn't sure it was worth doing. So

Mark and his accomplices had worked very hard to stretch out the mission lifetime beyond a meager month.

The key to it was margin. If there's one thing that's sacred to engineers, margin is it. Margin is the difference between what you think your system can do, and what you know it has to do. You always want the thing you build to have more capability on paper than it really needs in practice. If you think, theoretically, that you can survive a fall from a third-story window, you're going to feel a lot more comfortable agreeing to jump from the second floor than to jump from the third or the fourth. You want some margin for survival.

Applying what he thought was prudent margin, Mark had been willing to sign up to only 30 sols of operations at first. But his team had worked hard for weeks, increasing their estimates of how big they could make the solar arrays and decreasing their estimates of how much power was needed to keep the rover alive. By the time we'd taken the mission to NASA Headquarters, there was enough margin in the design to make us comfortable signing up for ninety sols.

So ninety sols was it. It became one of our Level One Requirements.

When NASA puts a project together, the Level One Requirements (always written with capital letters, so you'll never forget how important they are) are the agency's fundamental statement of what the mission is supposed to do. If you do not meet your Level One Requirements, then You Have Failed. To meet our Level One Requirements, we had to launch two rovers to Mars in the summer of 2003, land them both safely, and operate them both on the surface for ninety sols. We also had to drive at least one of them six hundred meters, a figure we'd arrived at after much contentious negotiation, and much consideration of margin. Two rovers, ninety sols, and six hundred meters became numbers that drove every decision we made.

There was another critical number, and it was $688 million. It wasn't in the Level Ones, but that was the budget we had agreed with NASA that we would not exceed. There was margin in that number, too, enough for us to get the job done even if a lot of things went wrong.

But the number that most haunted me and Pete wasn't any of those. Instead, it was the one you got if you took the cost of the mission and divided it by the number of sols we expected to operate on the surface. It came to about $4 million a sol. That was four million dollars each and every twenty-four hours and thirty-nine minutes . . . *if* both rovers made it to the martian surface. Whatever we built, and whatever we managed to do with it, the result was going to have to be worth at least that much.

Most of the Level One Requirements were numbers I could live with. But ninety sols wasn't one of them. I wanted more.

Surviving for ninety sols didn't mean that we'd get to do science for ninety sols. On *Mars Pathfinder,* nearly a third of the sols hadn't worked out as planned for one reason or another. Maybe our luck would be better, but our mission was a lot more complicated than *Pathfinder* had been. If we survived for ninety sols, history said we'd probably have only sixty sols or so where things went right.

There would be other hits. It might take us ten sols just to get off the lander and start doing useful science, dropping the total to fifty sols. And NASA had said that we had to drive one of the rovers at least six hundred meters. At thirty meters a sol, which was our best guess for what we could do, we'd burn up another twenty sols just meeting the distance requirement. Add it all up, and a ninety-sol mission might leave only thirty sols that'd be devoted to scientific measurements. For me, that wasn't nearly enough.

There was only one way of getting more sols: We needed bigger solar arrays. Strangely enough, the things that stood in the way of getting them were the Level One Requirements.

The Level One Requirements said we had to live for ninety sols. That number was burned into our brains; it was part of how we defined Mission Success (another favorite capital-letter phrase). And while it'd be nice if we lived longer than ninety sols, according to the Level Ones it wasn't necessary.

Therein lay the problem. A favorite adage among engineers is that "better is the enemy of good enough." Our mission was so hard, and our schedule was so tight, that it was difficult to convince anybody to make anything better than it needed to be. Putting extra effort into some part of the design always meant that something else would be shortchanged. So prudent engineering practice was to make each part of the design— even something as important as the size of the solar arrays—only as good as necessary to meet the requirements with adequate margin. No better.

I understood this, but it chafed terribly. My dream of more than a decade was finally coming true, and I couldn't bear the thought that our adventure on Mars would be cut short because we weren't willing to try hard enough at something. My instinct was to fight for bigger arrays, but my cathartic conversation with Pete nagged at me. Was I reverting to the pushy, impractical scientist I'd promised I wouldn't be?

Luckily, I had an ally on the engineering team. His name was Randy Lindemann.

I knew Randy from time he'd spent on the old *Athena* rover. He's a big guy, well over six feet, with dark hair, penetrating eyes and an easy smile and laugh. Like many of the senior engineers at JPL, Randy grew up during the *Apollo* program, and as a kid he'd dreamed of building spaceships someday. He'd come to JPL in the late eighties, and he worked his way up through their mechanical division. When *MER* began, he was named the mechanical lead for the rover.

What made Randy special was his passion for science. It wasn't just that he wanted to build a rover that could do good science on Mars. A lot of the engineers felt that way, I was coming to realize, with Pete setting the tone. The thing that set Randy apart was that he was genuinely curious about Mars himself. This was a guy, I learned, who unwound in the evening by reading planetary geology books. He was the perfect person to solve the solar array problem.

What it really came down to was how many "strings" he could squeeze into the design.

Our arrays would be built up from individual solar cells, each about the size of a credit card. The cells would be glued onto the arrays, wired together in groups of about sixteen cells each. Each group was called a string. How much power we'd have, and how long our rovers would stay alive, would depend on the number of strings we had. I'd looked hard at the numbers, and the best I could figure was that I wanted something close to thirty-six strings. With that many strings, we might last for four, five, or even six months on Mars, instead of just three, if things broke our way.

The lander that the rover had to fit into was shaped like a three-sided pyramid. The rover had to fold up inside that pyramid. So the design had a central solar array on the rover deck, surrounded by three foldout solar panels. The central array would be triangular, just like the lander. The foldouts, each a trapezoid with its long side hinged along one side of the central triangular array, would unfold after the lander had opened up on the surface. With the panels fully deployed, the overall shape of the array would be a hexagon. Randy figured it would be easy to get twenty-five or twenty-six strings onto that hexagon, and he thought there was a chance he could push it up to thirty or so. But my target of thirty-six strings was hopelessly out of reach.

Randy wanted to get me more strings. The hexagonal array with three foldouts was the simplest design that would fit into the lander, so the only way to get more strings was to add more foldouts. Randy got his team together and they began to work the problem. It was like doing origami in reverse, trying to find the simplest way to make a small pyramid unfold into a big, flat sheet.

I tracked their progress closely as the weeks went by, from both Ithaca and JPL. They weren't doing well. One afternoon in the main *MER* conference room at JPL, as I was taking a break between the day's seemingly endless meetings, Richard Cook walked in with a disgusted look on his face and dropped some drawings of some of the latest solar-array designs on the table in front of me. He didn't say a thing, but he didn't have to. Even I could see that the designs were unworkable. Some had three or even four levels of foldouts, with bizarre shapes and terrifying complexity. If this was what it was going to take, I was never going to get bigger arrays.

Randy kept trying. Every few days, it seemed, he'd send around some new scheme, some new way of doing the foldouts. Each one was shot down in turn.

At some point, the sensible thing for Randy to have done was quit. The hexagonal array was still good enough. Nobody up the chain of command ordered him to stop working on it, but every time a new design came around, the collective response from most of the engineers on the project had been "Oh God, the solar arrays again? Call it a day, Randy. Finish it."

And then one day everything changed.

Engineers pride themselves on knowing the right design when they see it. They live for the "ah hah" moment when everything falls into place. If you're too close to a problem, though, if you've lived with it for too long, sometimes it's hard to know when you've stumbled onto the right solution. Each new design is a little like a child to an engineer, each one makes sense to them. Randy and his team had been lobbing out one design after another, each one a change from something they'd tried before. Their latest attempt was a pared-down version of some multi-multi-foldout scheme that had been shot down earlier. They threw it out to the team, just like all the rest, and to their surprise, this time they got back a unanimous chorus: "That's it, don't change it!" Randy wasn't sure why, but as soon as the rest of us saw the design, we knew right away. This was our rover.

It was beautiful. The rear panel of the hexagon was unchanged from the original design. But this time the other two panels had been hinged down the middle, each along a line perpendicular to their first foldout, creating secondary panels that would give us the additional surface area we needed. The result was a swept-wing design, blunt at the front and flaring toward the rear in two aerodynamic-looking winglets.

Our rovers didn't have to look aerodynamic. Their job was to creep across the martian surface, not soar above it. And they didn't have to be beautiful. Form follows function in a good engineering design, and we didn't have the luxury of making the rovers pretty just so they'd be pretty.

But somehow, after weeks of trial and error, Randy and his team had accomplished the ideal. They had found a design that was both functional and beautiful. The swept-wing solar array looked like nothing that had ever been created before. It looked so good it just had to be right. And the calculations said that it might be able to hold as many as thirty-six strings.

When I talked to Randy about it a while later, he didn't have a good explanation for how they'd come up with it.

"There's an intrinsic reaction to most designs," he said, "when you feel in your gut that something makes sense. There's a proper design for a human face; it's the composite of all the faces you've ever seen. If the nose were over on one side, you'd say, 'That's wrong.' It's not that the person wouldn't be able to breathe; it just wouldn't look right. But this was different. Nobody's ever seen a MER rover before. So the right design wasn't obvious to us from past experience. We didn't know when we saw it that this was the right one."

Still, when everybody else had seen it, we'd all known. It just looked so damn good. "So did you intentionally try to make it beautiful," I asked, "just to shut everybody up?"

"No!" he answered, looking surprised at my question. "We never said, 'Forget engineering; let's just design a cool-looking rover.' Everything we did was functional and logical. But in the end, the design won because people liked it."

He thought about it a little more, and shrugged. "Maybe they did like it because it had a certain aesthetic appeal."

Yep . . . maybe we did. That, plus the thirty-six strings it offered.

So the rover evolved, level upon level of detail added to the design.

But as the work went forward, a fear began to gnaw at us. The whole concept behind our mission, the thing that had convinced NASA to pick us, was that we could land safely using the *Pathfinder* landing system, just as it had been built six years before.

Slowly, though, our rover had begun to grow. It was a subtle thing, not wildly alarming at first. In fact, the solar arrays were part of it. As the engineers began to refine the design of whatever part of the rover they were responsible for, they'd improve their best guess of how big and heavy their finished product would be. And with each little change, the guesses went mostly up. Each change by itself was small, but together they added up. The rover was growing.

A growing rover was a bad thing. If the rover got bigger, it'd get harder to squeeze it into the lander. If the lander had to grow to accommodate it, then the lander would also get heavier. There was a limit to how much mass the parachute and the airbags could safely land, and if the lander and rover together exceeded that limit we might have to change those, too. Any changes would take time, which was something we didn't have enough of. Growth of the rover was the kind of thing that could ripple through the whole mission.

The problem was insidious, and it crept up on us over months. Slowly, incrementally, almost without realizing it, we were slipping away from the base-it-on-*Pathfinder* concept we had sold to NASA.

The first guy to see it coming was Richard Cook. As the manager of the whole flight system, Richard saw the big picture better than anybody. And Richard realized we were growing too much. To get the problem under control, Richard decreed that we had to take fifteen kilograms, or about thirty pounds, out of the design.

Fifteen kilograms didn't sound like much in the grand scheme of things. The spacecraft would total more than a thousand kilograms as it sat poised atop the rocket in Florida. But we didn't have any frivolous items on board, anything we could toss off without some kind of impact. Whatever we went after, it was going to hurt.

We went after the little things first, like a backpacker sawing the handle off a toothbrush to save weight. One thing that died early was the Suncam, an upward-looking camera that'd be on the rover deck, staring at the martian sky. The original idea had been to use the Suncam for navigation, reckoning which way was north from where the sun was. But Pancam could do that job if it had to, so we tossed the Suncam overboard to save three tenths of a kilogram.

Little things like the Suncam helped, but we saw quickly that we weren't going to nickel-and-dime our way out of the problem. Bigger things had to go, too. The list of possibilities was frightening.

The list included the high-gain antenna. The HGA was the dishlike thing on the rover deck that'd let the rover talk directly to Earth. Maybe we could get rid of it, we thought, because maybe we'd be able to get all our data down by relaying it through the two Mars orbiters, *Mars Global Surveyor* and *Mars Odyssey*, which was the new name for what was left of the 2001 mission. But it was a frightening prospect. If the orbiters died, we'd be finished.

So maybe we could drop the UHF antenna, which was for communicating with the orbiters, instead. Maybe we could do everything with just the HGA. But that was scary, too. We were counting on the UHF link through the orbiters to get most of our pictures back from Mars. And if the HGA ever failed, the mission would be over.

To my horror, parts of the payload popped up on the list. Maybe we could drop the Microscopic Imager, some of the engineers suggested, and just be satisfied with Pancam. Maybe we could drop the RAT, and just look at the surfaces of rocks. It was a measure of how desperate we had become that we were seriously considering tearing our payload apart to get out of trouble.

In the end, none of these cuts happened. They just weren't big enough. The RAT and the MI together totaled just a kilogram, so the savings they'd provide wasn't worth the pain we'd suffer by losing them. We had to go after something bigger.

There were only two things big enough to save us the mass we needed. One was the egress ramps, and the other was Randy's beautiful solar-array winglets.

The egress ramps hadn't been in the original design. Our rovers were so big that we'd figured that driving them off the landers would be easy. We'd just take a look around once we landed, pick the best route, and go monster trucking onto the martian surface. But we quickly realized that the lander deck could be as much as a foot and a half off the ground if we landed on a rock or if there were airbags bunched up under it in a funny way. And when we tried driving a test rover off a step that high, the thing flipped over.

So we'd added some ramps. They'd be made of heavy fabric, the same stuff the airbags were made of. They'd fold up like the wings of a bat between adjacent lander petals, opening out and snapping into place when the lander opened. Together, the three of them weighed enough to solve our mass problem, if we were willing to take the risk of dumping them. But it was a big risk.

However we solved the problem, we were going to have to do it before our critical design review. CDR was the most important in the seemingly endless series of reviews that NASA and JPL management had put in front of us. JPL had convened a group of "graybeard" engineers who had done complicated missions before and who could give us an outside perspective on how to solve our problems. On top of that, NASA Headquarters had put together their own Independent Review Team, a separate group that would report directly back to Washington on how we were doing. That group, in particular, had teeth. At CDR, we'd have to get up in front of both groups and tell them what our mass situation was. It wouldn't be okay to be busted.

The decision on what to throw off the spacecraft came on July 19, 2001, two weeks before the CDR, in a crowded gathering in one of our cramped little conference rooms at JPL. I flew across the country for it and then back, ten hours on airplanes for a one-hour discussion. This was one that really mattered, and I wanted everyone to see my face, not just hear a voice from a speakerphone. Every senior engineer on the project was there, with me the lone scientist at the table, feeling a little like a voice in the wilderness.

The outcome was obvious from the start. If we dropped the egress ramps, there was a chance we'd flip the rover upside down on Mars, an embarrassing end to everything we'd worked for. Dropping the winglets would give us a shorter mission, but it wouldn't give us a humiliating failure.

We went around the room, one at a time. The way Pete ran things on our project, everyone got their say. One by one, each person who spoke said we should drop the winglets. I waited impatiently for my turn to come.

"There's a freight train rolling down the tracks here," I said. "And as the guy who's about to get hit by it, I want to say that I think we're doing the wrong thing. The legacy of this mission is going to be what we accomplish on Mars. And what we'll accomplish on Mars is in direct proportion to how big our solar arrays are.

"I'm not saying we should get rid of the egress ramps," I went on. "We need those to get six wheels in the dirt. But if we drop the winglets now, I think we'll be sorry once we get to Mars. I'd rather keep both the winglets and the ramps, and press on as best we can. Maybe we can make it work."

My pleas went unheeded. Everything we knew said that the arrays would be good enough for ninety sols even if we dropped the winglets. And the mass situation was just too scary. The consensus, with one dissenting voice, was to take the winglets off the rover. The decision made, I flew back home dejected.

Four days later, I got an e-mail from Joy Crisp, the project scientist at JPL. The subject line at the top of her message was "Bad news." Now what? I opened it.

"The solar-array vendor has come back with their prognosis," Joy's e-mail said. I knew that they'd been looking at the design carefully, confirming how many strings it'd really give us. "They think they can fit 27 strings on the no-winglet version of the rover, and 34 strings on the with-winglet version."

Twenty-seven strings! All along we'd been certain that without the winglets we could get thirty strings. Twenty-seven would kill us.

I had a moment of despair, but then it dawned on me that this "bad news" actually might be the best news I'd gotten in months. With only twenty-seven strings, we'd never meet our Level One Requirement of ninety sols. That meant that if we dropped the winglets, we'd have to go back to NASA Headquarters and try to convince them to change the definition of Mission Success. And when the news spread through the project, Pete and Richard did the right thing. They put the winglets back on, and they kept the egress ramps, too. We'd go into the critical design review with a mass situation that was busted. As Richard put it, "We're going on risk."

We got beat up badly at the CDR. We put on the best show we could, and everybody on both review boards sounded impressed and sympathetic. But the review exposed to the world what we already knew: that the story we had sold to NASA had been a lie, albeit an unwitting one. With our mass over the line, we were going to have to redesign the whole works: the lander, the airbags and the parachute. And with all that new work to do, our costs would grow inevitably. Our schedule, which had been bad from the start, was now broken.

Every space project has many dimensions, including mass, budget, schedule, electrical power, data return and so on. A skillful team can trade one for another to a remarkable degree. If your budget's busted but you have lots of mass margin, you can replace a lightweight and expensive component with a heavy and cheap one and save some money. But those tricks work only if you've got room to maneuver somewhere. We were hemmed in on all sides. Gentry Lee, a board member who's both a brilliant spacecraft system engineer and a charismatic science fiction writer with a knack for words, summarized our situation perfectly: "You're backed into a hyperdimensional corner."

Steve Gorevan, from Honeybee Robotics, lived in an apartment in lower Manhattan, just blocks from Honeybee. Honeybee's building was on Elizabeth Street, just south of Houston, in a vibrant neighborhood on the border between Little Italy and Chinatown. Steve loved the place. The building had once been a Con Ed substation, with massive walls and floors that would be able to support the heaviest equipment Honeybee might ever need. There were rumors that Thomas Edison himself had had a hand in designing it. Honeybee was Steve's creation, a company he'd helped build up from nothing, and it was a source of enormous pride for him that they were playing a crucial role in a mission to Mars.

Steve is sensible about exercise, and even though he lived just blocks from Honeybee, his morning routine was to ride his bike to work by a circuitous route that took him through Hudson River Park from midtown all the way south to the Battery, then north through the plaza at the World Trade Center and up to Honeybee.

A couple of weeks after CDR, on the morning of September 11, 2001, Steve was on his bike, six hundred yards north of the Trade Center, when he heard jet engines directly over his head. Head down, concentrating on the traffic, he didn't look up, but something about it sounded very wrong. This wasn't a normal approach into LaGuardia or Newark. Instead, the plane was low and accelerating, engines pitched high and screaming. An instant later there was a terrible concussion. Steve braked hard, and pulled over. Dismounting, he turned and looked to the south. To his astonishment and horror, smoke and flames were erupting from high on the Trade Center's north tower.

Steve's first thought was that it hadn't been an accident. Somebody had been gunning those engines. He stood in the street next to his bike, surrounded by dozens of others, many also out for their morning exercise.

Together they looked helplessly at the tower, transfixed, with no idea what to say or do.

After what felt like a very long time, Steve's mind began working again. If this wasn't an accident, then his city was under attack. Suddenly worried about his children, and about Honeybee, Steve remounted his bike and pedaled up Elizabeth. He pulled up to Honeybee, out of breath, and locked his bike to a signpost outside the building.

By the time he stepped into Honeybee's main offices on the second floor, a second plane had struck the south tower of the Trade Center. Both towers were in flames. A coordinated attack on the city was under-way. Who was doing it, or how bad the destruction would be before it ended, was impossible to tell.

The scene inside the building mirrored the one in the streets. Most of Steve's employees were up on the roof, watching with growing anguish as the smoke poured from the towers, and as people trapped by the flames on the highest floors lost hope and jumped to their deaths. Some of the Honeybees walked the hallways, trying desperately to get news on their cell phones of their kids, their spouses. Others clustered around a televi-sion, hoping to learn more than what they could see with such horrifying clarity from the roof.

And then, one at a time, the towers fell.

Honeybee stands almost a mile northeast of Ground Zero, and with the wind blowing to the southwest they experienced no immediate phys-ical effects from the towers' collapse. Within minutes, though, a strange migration began, as thousands of people began streaming northward on foot up Elizabeth Street. Some were staggering, a few were bloodied, and most were covered, ghostlike, in gray soot. It went on for most of the day. Many of the Honeybees lingered in the building long into the after-noon, stunned into inactivity. For some, it was the worst day of their lives.

Within two weeks, Honeybee was functioning again. But there was a nameless, horrible stench in the air that lingered from the collapse of the towers, and a deep trauma that permeated everyone in the company. It was going to be a long time before they recovered.

About the time that Honeybee started working again, I got a phone call from Steve Kondos. Steve was the engineer at JPL who managed the Honeybee contract, and he'd been hit with a compelling idea. Was there some way, he asked me, that we could get some metal from the wreckage of the World Trade Center, and fly it to Mars?

I thought it was brilliant. Like people around the world, everyone on our team had felt helpless when the towers fell, and in the weeks that followed. Even at Honeybee, in the heart of lower Manhattan, the relentless pressure of getting a NASA mission to the launch pad made it impossible for anybody to pitch in at Ground Zero. We all felt like we had to do something, but we were locked into a Mars mission we couldn't step out of, even briefly.

So suppose we could put a simple memorial onto the surface of Mars? It wouldn't bring anyone back, and it wouldn't help put the city on its feet again. But somehow it felt right. The point, of course, was to honor the memory of the thousands who had died. But another reason I wanted to do it was to begin the healing process at Honeybee. When the 'bees heard the idea, they embraced it.

The first step was to get the right material. On the face of it, this sounded easy. Firemen from local precincts were working at Ground Zero around the clock, and there was wreckage from the Trade Center at every firehouse in Manhattan.

But that wasn't the right way to do it. As Steve Gorevan told me on the phone, "We're touching something sacred here. It's not enough to walk over to a plastic bucket in a firehouse and take something. We have to have a pedigree, some official transfer of material." It was so ironic; endless truckloads of debris from the towers were flowing by Honeybee every day. But if we were going to do this, we couldn't build a memorial by going Dumpster diving, or by grabbing something off the back of a truck. We had to do it right.

Steve had an idea. A friend of his, John Sexton, was the dean of the NYU Law School, and well known within the administration of New York's mayor, Rudy Giuliani. Working through Sexton, Steve made contacts at a series of increasingly high levels within the overburdened city administration. At every step he struggled to convince people that we were legit, that we really were sending something to Mars. But Sexton's help opened the doors we needed. At 1:00 in the morning on November 30, the idea was taken to Mayor Giuliani.

The next day, Steve was sitting in his office when his secretary stuck her head into his office.

"There's somebody downstairs who says he wants to see you," she said.

He walked down to street level, and was met at the door by a crew-cut man he didn't know. The man had a badge from the mayor's office on his lapel and a small cardboard box in his hands.

"I'm told that you need these," he said solemnly, and he handed the box to Steve. Steve thanked him and gingerly carried the box upstairs.

He called Tom Myrick to his office. Tom was one of the lead engineers at Honeybee, and the mechanical genius behind the RAT. When Tom walked in, the box was on the floor. Steve motioned to it without saying a word.

Tom knelt and opened the box. He knew what he was looking at. He would have known if he had been blind. The same acrid, plastic-metallic burnt smell that had hung over Manhattan for weeks after the attack emanated from the box. Inside was a handwritten note from Richard Sheirer, who ran the office of emergency management at Ground Zero. The note said simply, "Here is debris from Tower 1 and Tower 2." And beneath the note lay a steel bracket, two large bolts, and a twisted plate of aluminum, about a quarter of an inch thick and five by six inches in size.

Tom's eyes fell immediately on the aluminum plate. His RAT design had a lot of aluminum parts in it, including external "cable shields" that were there to protect the electrical cables that ran to the motors from damage if the RAT bumped against a rock. Tom quickly saw that the metal from the mangled aluminum could be cut into cable shields. The place to do the work would be the machine shop in Round Rock, Texas, outside of Austin, where many of the other RAT parts were being built.

To make cable shields, the machinists in Texas would need to work with a clean, regularly shaped piece of metal, not the contorted thing that Tom had in front of him. So Tom had to start by banging the twisted aluminum into shape between two steel plates with a hammer. It was a hard, violent job, and a painful one for Tom, but it had to be done. The emotion nearly overcame him as he did it.

A few days later Tom flew the squared-off aluminum to Austin himself, and he stayed with it the entire day as the machinists worked it into four perfect cable shields, two for flight and two spares. The Texans must have thought hard about why Honeybee's top mechanical engineer was giving so much personal attention to such a strange and simple task, but if they suspected what they were working on, they never said so.

Tom flew the cable shields back to New York, put an American flag decal on each, and pumped them down in a vacuum chamber to make sure that the decals wouldn't peel or blister in space. They didn't. The shields emerged from the chamber looking clean, perfect and beautiful. If we could get the rovers safely to Mars, a recovered and reborn piece of the World Trade Center would be there with each of them.

When the report on the CDR finally came out from NASA's Independent Review Team in early October, it was scathing.

The IRT was a hard group to like. They tended to be quiet in our reviews, lurking in the back of the room and letting the JPL board ask the penetrating questions. Their job was to catch our mistakes before they happened, and to report back to Washington if they thought we were likely to screw up. Having an independent group like the IRT was obviously a good idea if you were a manager at NASA Headquarters and you wanted unbiased input on how a project was going. But being on the receiving end of such severe criticism could be demoralizing.

The real surprise in the IRT report was what they'd had to say about our budget. As soon as we'd realized we were going to have to redesign everything, we'd known we were going to bust our $688-million ceiling. So we had confessed at the review that we were probably going to need another $60 million or so before the mission was over. But the IRT didn't buy it. Instead, they'd told NASA Headquarters that they thought we were going to run over by anywhere between $84 million and $158 million, an average twice as high as the number we'd come up with.

It couldn't possibly be that much. It just couldn't. But when the news got to NASA Headquarters—and it got there fast—very loud alarm bells began to ring. If somehow we overran by as much as the IRT had predicted we would, there would be a severe impact on Ed Weiler's space science program. The money would have to come from somewhere. So the response we got from back east was the obvious one: Why not pull the plug on the second rover, and get the budget back under control by going back to one?

We were horrified at the thought of losing a rover. Two rovers had caught us by surprise back when we'd first heard about it, but by now the idea of flying two was etched deeply into the psyche of our project. So we spent all that fall trying to cut our costs instead. We were spending more than a million dollars a day, weekends and holidays included. We were growing tired and demoralized. We cut back everywhere we could, but with each new redesign the numbers got worse. On December 7, 2001, we got hauled into NASA Headquarters to explain why we shouldn't give up a rover.

It was an ugly discussion. Along with the budget problems, it was clear that NASA was becoming worried about stress and fatigue on the team. If they took a rover away, they said, maybe it would make our jobs easier. But in the end they relented and let us keep going. Weiler bumped our budget up by the amount we said we needed, taking the cost of the mission to $746 million. But the new money came with a stern warning. If we overran again, we were going to lose a rover.

8 SHREDDING AND SQUIDDING

THE REAL KEY TO making the mission work, more than anything else, was going to be landing. As long as we made it to the launch pad on time, the single biggest obstacle between us and success was going to be getting our hardware from out in space down onto the martian surface safely. We called it Entry, Descent, and Landing, or EDL.

Not all parts of EDL carried the same risk. The fiery, scary-looking entry, burning a hole through the martian sky at twenty-five times the speed of sound, was actually the easy part. Spacecraft have been riding heat shields through atmospheres for decades, and we were convinced that the *E* part of EDL was going to be pretty straightforward. It was the last bit, riding the chute down and bouncing on the surface, where the big risks were.

A lot of it came down to where we chose to land the things. Mars is a big planet, with as much surface area as all the continents on Earth put together. Somewhere in that vast expanse we had to pick two tiny places to put our rovers. I wanted the sites to offer good science, of course. I wanted them to be places that might once have been able to support life, which meant that they had to be places where there once might have been water. But more than anything else, they had to be safe. Having good science at your landing site is great, but you don't get any science at all if you don't land in one piece. So safety dominated the site selection process.

There were a lot of places we couldn't go. Our rovers were solar powered, so we had to land near the equator, where solar energy would be plentiful. Our vehicles used parachutes to slow down, so we had to land where the altitude was low. At high elevations on Mars there simply wouldn't be enough air to slow our chutes down before we hit the ground. And the landing site had to be smooth and flat, so our airbags wouldn't ricochet off steep slopes or roll out of control for miles. All together, there were about 155 places on the planet where we *might* be able to land, each of them defined by an ellipse about one hundred kilometers long and a few tens of kilometers wide.

Once the initial 155 sites were located, we invited scientists from all over the world to come and argue about them. Some sites fell off the list because they weren't safe enough when we looked more closely. Others fell off the list because there simply wasn't a good scientific reason to land there. After a lot of talk, over many months, it came down to three candidates: Melas Chasma, Meridiani Planum, and Gusev Crater.

Melas was the glamor site on the list, the one that everybody was jazzed about. The site was down on the floor of the Valles Marineris, the gigantic canyon system that stretches for thousands of kilometers along the martian equator. There was some evidence that water had once filled the canyons, although people were still arguing about that. But nobody argued that there wouldn't be one hell of a view when we opened our eyes and took a first look at those towering canyon walls.

Melas had problems, though. Parts of the landing ellipse had very tasty-looking layered rocks in them, but other parts were covered with enormous fields of sand dunes. There was no way to pinpoint-target the good stuff; our landing system wasn't accurate enough to do that. All we knew was that we'd come down *somewhere* in that hundred-kilometer ellipse. And if thirty or forty percent of the ellipse was covered with dune fields, that meant that there was a good chance that one of our rovers would see nothing but sand the whole mission.

The death blow for Melas, though, came from winds. Strong winds could kill our rovers. When we fired our three rocket assisted deceleration motors, our RAD motors, a split second before impact, we figured that they'd pretty much zero out our vertical velocity. A fall from ten meters, the height we expected the RADs to stop us at, was something we thought our airbags could deal with. But if a wind was blowing, the whole shebang—airbags, parachute and all—would also be dragged sideways by the wind, at whatever speed the wind was blowing. Horizontal and vertical velocity both matter at impact, and if our horizontal velocity was too high the impact would shred our airbags, no matter how low the vertical velocity was. So winds were very bad.

We had no way of measuring the winds on Mars, of course, at Melas or anywhere else. The best we could do was bring in a bunch of experts in the esoteric field of forecasting martian weather, and ask them to give us their best advice. We did that, and our experts told us, in very certain terms, that all the spectacular topography at Melas could lead to some powerful winds there. Steep slopes on Earth can generate their own winds, and the same was likely to be true, they said, on Mars. If we went to Melas, there was a good chance that we'd destroy the airbags, and the rover inside them, on impact.

That left Meridiani and Gusev.

Meridiani was the safest site we had, in one of the smoothest, flattest and least windy places on Mars. In fact, it was so smooth and flat that a lot of the scientists were worried that there'd be nothing much there to look at.

But there was one thing that made Meridiani tantalizing, and that was hematite. Phil Christensen's TES instrument on *Mars Global Surveyor,* the instrument that was the orbital big brother of Mini-TES, had spotted hematite at Meridiani from orbit. Hematite is an iron-bearing mineral, an iron oxide. It's found in many places on Earth, and in most of those places it was formed by the action of liquid water. Hematite can form in deep basins of water, as it has in iron ore deposits in the upper midwestern United States. It can be deposited in fractures when hot water percolates through rocks, or as coatings when a little bit of moisture dampens the outer surface of a rock. So the hematite at Meridiani was like a mineralogical beacon, visible from space, that said, "Water may have been here."

Problem was, there are other ways to make hematite that don't require water. Under the right conditions, the mineral magnetite, which is common in some volcanic lavas, can transform to hematite without water being involved. So Meridiani was a gamble. It was a tantalizing place, but there was a decent chance that water had never been there at all.

Gusev, on the other hand, was a place where it was obvious there had once been water. Gusev is a big crater, about 160 kilometers in diameter, in the southern highlands of Mars. The thing that sets Gusev apart is that there's a gigantic dry river channel, hundreds of miles long, that empties into it. There's no water in the channel now, and there hasn't been for billions of years. But the thing *had* to have been carved by water. So Gusev is a big hole in the ground with a giant dry river bed flowing into it. At some time long ago, Gusev must have held a lake.

Unfortunately, Gusev was a gamble, too. The floor of Gusev Crater, which once must have been covered with sediments deposited in the lake, is billions of years old. A lot can happen in billions of years. Volcanic lava, or dust blown by the wind, could have buried the lakebed sediments, putting them out of reach of our little rover. There was no way to be sure from the orbital pictures whether or not this had happened. If it had, then our best chance would be to drive to one of the hundreds of little craters that pockmark the floor of Gusev, hoping that it had dug through the cover of boring stuff and down into the good stuff underneath. The gamble at Gusev wasn't that water had never been there. It was that we might get down on the ground and find that all the evidence for water had been buried.

So Meridiani and Gusev were the favored sites, each with its own attractions and each with its own risks. But only Meridiani was in the bag, safety-wise. Our martian weather experts thought that Gusev, like Melas, might be prone to high winds. The Gusev winds weren't expected to be as bad as the ones at Melas, so they weren't an obvious showstopper. But the numbers were discouraging, and with all the uncertainty about how our parachute and airbags would perform with a heavy (and growing) lander, there was a good chance that Gusev wouldn't be safe enough. So we put together a team of meteorologists and geologists to pick the best "wind safe" site they could find on the planet: another place with winds almost as benign as those at Meridiani.

They found one, in a place called Elysium Planitia. It was a *terrible* site, and I hated it as soon as I saw it. Elysium is smooth, and it is flat, but unlike Meridiani there isn't even a decent hint that water had ever been there. It is just a smooth spot on Mars, nothing more. I didn't see any way that we could do our science if we were forced to land at Elysium. Somehow, the EDL guys had to come up with a design that could get us down safely at Gusev.

There wasn't much I could do to help, but to provide some incentive, I sent an e-mail to Wayne Lee, the chief engineer for the Entry, Descent and Landing team:

> Be it known that if Gusev Crater is selected as one of the two *MER* landing sites, the PI shall treat the EDL team to as much beer as they can reasonably consume in one evening, at the watering hole of the team's choice.

It was an offer I prayed I'd have to pay up on.

Airbag testing began, in the world's largest vacuum chamber in Sandusky, Ohio. The idea for the airbags, of course, had been to use the original *Mars Pathfinder* bag design, exactly as the bags were built for that mission. But with the growth in the mass of our vehicle, there were growing concerns that the *Pathfinder* bags might not do the job. So Wayne and the rest of the EDL team took a set of *Pathfinder* bags to Ohio, put a fake MER lander inside them, pumped the chamber down to martian atmospheric pressure and let the bags fly against a rock-studded test target at the bottom of the chamber.

To their horror, the result was a catastrophic failure. A rock ripped completely through all the outer "abrasion layers" of the bags, ruptured the inner bladder and deflated the bags instantly. The *Pathfinder* bags were far too weak to land our fat new vehicle. With that one failed test, our technical problems catapulted to the same level of desperation as our cost and schedule problems.

Over a period of months, Wayne's team added new features to the airbag design, including beefed-up abrasion layers and double inner bladders. New bags were built for us, sewn on Singer sewing machines by some hard-working ladies in Dover, Delaware. The changes made the bags stronger, but of course they made them heavier, too, eating further into our rapidly vanishing mass margin. Dozens of tests in Ohio revealed incremental improvements in the bags' performance, and eventually the bags looked like they might be good enough to land us safely in the benign, low-wind conditions at Meridiani.

Gusev, though, was another matter. Unless we could find a way of dealing with the predicted Gusev winds, one of our rovers was headed for Elysium.

It wasn't an airbag problem anymore, it was a system problem. The airbags were as good as they ever were going to get. The only way to solve the problem was going to be to make some fundamental change to the whole EDL system that—somehow—would mean that the airbags we had were good enough.

Our lead system engineer, the guy who had the whole design in his head, was Rob Manning. Rob is burly and bearded, a former ski instructor who'd made the big jump from Whitman College, in Walla Walla, Washington, to Caltech, and then to JPL. Rob's a jazz trumpeter in his spare time, playing weddings, sushi bars, just about anyplace he can get a gig. Rob had been one of the guys on the little team that Mark Adler had put together when he first came up with the idea of stuffing an *Athena* rover into a *Pathfinder* airbag system. More than a year later, he was still trying to get it to work.

In October of 2001, Rob, Wayne Lee and a bunch of the other EDL guys were reviewing airbag test results. Recent tests had ripped bags repeatedly when the horizontal velocity at touchdown was high, and no reasonable amount of reinforcement seemed likely to solve the problem. What we really needed, Rob realized, was some way to know when the wind was carrying us sideways, and to jolt us back in the opposite direction.

We actually had hardware that could give us a jolt. Along with the big RAD motors that fired downward from the backshell an instant before impact, we also had three smaller rocket motors, called the Transverse Impulse Rocket System, or TIRS, that pointed sideways, sticking out from the backshell at 120-degree angles to one another. The TIRS motors could be fired in any combination to jerk us this way or that before touchdown, the idea being that they could yank us back upright if the vehicle sensed that it was tipping over before impact.

What we didn't have, though, was any way to know if we were moving sideways in the wind. Realizing that the system was tipping over was easy—you could do that with a gyroscope, and we had a couple of those. But sensing a steady horizontal velocity in a wind was different. It's like being in a hot air balloon: with your eyes closed you can't sense any motion at all, even if the wind is carrying you across the ground at high speed.

As Rob went through the latest depressing airbag data, he realized that what we really needed was a horizontal velocity sensor. We needed something that could sense our motion over the ground, letting us use that information to fire one or two of the TIRS motors if we were going too fast, counteracting that sideways velocity.

Mars landers had used horizontal velocity sensors before. *Mars Polar*

Lander had had one, and *Viking* before it. Any lander that uses throttled rocket motors and legs to land has to be able to cut its horizontal velocity to zero before touchdown, so it won't turn over. What both *Viking* and *Polar Lander* had used was a "multibeam radar," a radar that sends pulses to the ground in multiple directions, using all the returns that come back to sense both altitude and velocity across the ground. But a multibeam radar would weigh something like five kilograms, and we didn't have enough mass margin left to afford it. Rob's problem wasn't just finding a horizontal velocity sensor that would work. It was also finding one that we could fit onto our embarrassingly overweight system.

Rob wrapped up the airbag test review and left the building, walking dejectedly across the JPL campus to his next meeting. Coming the other way, on the opposite side of the street, he spotted Miguel San Martín, a spacecraft attitude control expert. Miguel, Rob knew, was about to leave town for a while, flying to Buenos Aires to visit his family.

"Hey Miguel, c'mere," shouted Rob. "You got a second?"

"I've got a problem, Miguel," Rob began after Miguel crossed the street, "and I need your advice." Rob explained the airbag problems, and why he needed some way to figure out the horizontal velocity before touchdown. Velocity sensing is an attitude control problem, and he hoped that Miguel would know of a piece of hardware that could do it. "So I need a horizontal velocity sensor, Miguel. Can you do a quick search around, before you leave, see if there's anything out there?"

Miguel thought for a moment.

"No," he said, with absolute finality.

No? thought Rob. Just like that?

"No," said Miguel again in his strong Argentine accent. "Just give me two pictures. Two pictures."

Rob thought for a long moment.

"Shit, of course!"

Miguel had seen the answer. It was just like being in a hot air balloon. If your eyes are closed, you can't sense your motion across the ground.

But if you open your eyes and look down, it's easy. If we had a camera that was pointed at the ground while we were still airborne, we could take two pictures with it, look at the pictures to figure out our velocity, and fire the TIRS motors to counteract it.

Rob quickly ran through the possibilities in his head. It was very late to be adding another camera to the design, but he knew a way to do it. The key was the Suncam, which we'd thrown off the rover a few months before to save a lousy three hundred grams. The electronics inside the rover had been designed to handle inputs from ten cameras: two Pancams, two navigation cameras, or Navcams, four "hazard avoidance" cameras, or Hazcams, the Microscopic Imager, and the Suncam. With the Suncam gone, it would have made sense to go from ten camera ports in the electronics to nine. But that was just one more piece of work for an overworked team, not worth the effort. So we still had ten camera ports in the electronics, even though there were only nine cameras left on the vehicle. And that meant that Rob could add a camera if he wanted to. If it solved the Gusev wind problem, it'd be well worth a few hundred grams.

Adding the camera wasn't the whole job, though. It wasn't even the hard part. The hard part was what to do with Miguel's "two pictures" once the camera had taken them. There was no way, during the high-speed rush of EDL, to transmit those pictures to Earth, work out the velocity, and send a TIRS firing solution up to Mars. EDL was so fast that even with signals traveling at the speed of light, the lander would be a pile of debris on the surface by the time the signal got there.

No, the pictures would have to be analyzed on board the speeding vehicle, the computer given only seconds to look at the images, figure out which way the vehicle was moving, and decide what to do. Computers can be taught to analyze images, as long as the task is simple enough, so it wasn't out of the question. But it was a big burden to heap onto a computer that was already going to be pretty busy with the process of landing itself on Mars.

Rob also immediately realized that he now had a new problem. A descent camera was a radical idea, a huge perturbation to the EDL design, and it wasn't going to go over well with Pete, JPL management, or NASA Headquarters unless it was very likely to work. So Rob decided to go underground. He'd pull together a small, trusted group of experts, and spend some time trying to smoke out any showstopper problems before anybody else did. Only then, once he had an idea that he knew could stand up to scrutiny, would he go public with it.

Rob went to Richard Cook and explained his plan for setting up a tiny clandestine project-within-a-project to look at a descent camera. "I'm just forming a little group, nothing to worry about, it's just a few meetings," Rob said with his most disarming smile. "It's not going to cost much money."

Richard, of course, would have been well within his rights to cut Rob off right there. But Richard had worked with Rob on *Pathfinder*, and he knew his friend's talents well. If Rob thought he was on to something, there was a chance it would be money well spent, no matter how goofy it sounded at first.

"Just don't you tell anybody else about it," Richard warned.

Rob gathered his co-conspirators, going after the best image processing minds at JPL, and I loaned him a couple of camera experts from my science team. They worked through every problem they could think of. Was there a place to mount the camera? What about cabling? What if the lander was swinging on the parachute, and the pictures were blurred? Were two pictures enough, or did we need more? Suppose there was dust on the optics? One by one they raised questions, and one by one they found what might be answers. The thing might actually work.

After two weeks, Pete found out about it. He was *not* happy. But trust had become a big thing on the *MER* team, and if somebody as smart as Rob thought he could solve the Gusev wind problem, then Pete—like Richard—felt he had to trust Rob to go forward with it. Rob named the idea DIMES, for Descent Image Motion Estimation System. Pete let him keep going.

Now that it was out in the open, DIMES became a much tougher sell to JPL management and NASA Headquarters than it had been to Pete. Those guys didn't know the details of how a descent imager might solve our wind problems. They just knew we had a busted schedule, a bloated budget, an overweight lander, and way too much work to do. Adding something this big to our list of things that might not work didn't seem like a good idea to anybody outside the project. The whole thing had the smell of desperation about it.

Pete, bless him, stood his ground. Even though Rob had kept him in the dark about DIMES until he'd had a good story to tell, once Pete heard the story, and once he cooled down, he responded in typical Pete style, backing Rob solidly. DIMES wasn't officially part of EDL yet, but Pete let the work go forward. With Pete behind it, the DIMES camera would go on the spacecraft, and the DIMES software would be written, no matter what anybody else thought about it.

Whether DIMES would actually be used at Mars, though, depended on whether Rob and his guys could prove it would work.

At the same time that the scientists were arguing over landing sites and the DIMES team was getting their story together, Adam Steltzner was taking a crash course in parachute design.

Adam looks like Elvis. Not the old, fat Elvis, but the young, dangerous one, with hair swept back in a DA and long rakish sideburns. Gold earrings shine in each ear. Adam was the lead mechanical engineer for the EDL team, and even though he'd never designed a parachute in his life, the job of getting the MER chutes to work suddenly fell to him.

Adam barely graduated from high school. In fact, he's still not sure that he did. Adam was born and raised in the laid-back town of Sausalito, California, and as a high-schooler he'd been even more laid-back than most. Through his sophomore and junior years he'd earned a steady stream of F's at school, getting stoned, hanging out, and working hard only at his theater courses. He'd pulled his act together, barely, his senior year, and he may have done well enough to graduate. But he never went to pick up his diploma, for fear that it wouldn't be there.

With academics behind him in the early eighties he found a job in a health food store in Mill Valley, and began playing bass in new wave bands, gigging around the Bay Area. But then, somehow, an unfamiliar kind of curiosity began to come over him. On his way to a show he'd see the stars in the sky, noting the pattern of the constellations. On his way back home, the stars had moved. Why? Intrigued in a way he'd never been in high school, he went to his local community college, the College of Marin, and signed up for a course in astronomy.

Astronomy is basically applied physics, though Adam didn't know that at the time, and the astro course required simultaneous registration in a course on "physics for poets." So Adam enrolled in that, too. And then, on his first day in physics class, something clicked. Before long he was signing up for courses in algebra, trigonometry, essentially finishing his high school education. With strong recommendations from his teachers at Marin, he applied successfully to UC Davis. Four years later he had a degree in mechanical engineering from Davis, through-the-roof grades, and free-ride grad school offers from MIT, Stanford and Caltech. Time at Caltech and the University of Wisconsin netted him a Ph.D. and a job at JPL, and the next thing he knew he was working on *MER*.

The original plan for the parachutes had been the same as for the airbags: use the *Pathfinder* chute. But by the time Adam showed up on the project, that idea was already dead. The lander had grown so much that there was no point in even testing the *Pathfinder* design to see if it might work. Simple calculations were enough to prove that the *Pathfinder* chute was too small to keep us from crashing. Something had to change.

The obvious fix, Adam knew, was to make the chute bigger. But there were limits to what he could do. The parachute had to be packed into a cylindrical canister at the top of the lander, and that canister was surrounded on all sides by hardware that couldn't be redesigned. There simply wasn't enough time to make big changes. Adam managed to enlarge the canister diameter from eight inches to ten, but that was as far as he could go. Whatever parachute we were going to fly would have to fit into a ten-inch can.

Our parachute, like *Pathfinder*'s and *Polar Lander*'s before it, was based on the chute that flew on *Viking* thirty years ago. It's called a "disk-gap-band" design, and the *Viking* testing had showed that it works well for the thin air and supersonic velocities you face when landing on Mars. Everybody since *Viking* has used the same basic design.

A disk-gap-band parachute looks like an upside-down bowl stacked on top of a tuna can that's had both its lids removed. The upper part is called the disk, an inverted bowl of fabric with a small vent hole in the center. The lower part, the band, is suspended below the rim of the bowl, a cylindrical bracelet of fabric with the same diameter as the bowl above it. The disk provides the atmospheric drag necessary to slow the lander down, while the band provides side-to-side stability, keeping the chute from wobbling and swaying as it plunges downward through a turbulent, unstable atmosphere. The disk and the band are separated by the gap.

Adam had plenty of knobs to turn in this parachute design. He could make the disk bigger, giving the chute more drag. He could make the band taller, making the chute more stable. He could fiddle with the size of the vent hole, the width of the gap, the kind of fabric used or the length and strength of the lines that connected it all together. But there was one great constant: whatever he did, it would all have to fit into that little ten-inch canister when the time came to pack it. Adding something to the design always meant taking something else out.

With changes made to enlarge and strengthen the *Pathfinder* design, it was time to do some tests. The test site was a National Guard gunnery range south of Boise, Idaho: wide-open country, with sagebrush plains against a backdrop of snow-flecked mountains. It's the kind of place you can drop big things from the sky and not have them land on anybody's head.

Unfortunately, it wasn't going to be enough for Adam to hook the new chute to a fake *MER* lander and drop the whole works from a helicopter. A helicopter drop provides nothing like the supersonic speeds our chute would experience at Mars. The point of the tests would be to stress the chute as much as it would be stressed in flight. In order to compensate for the low speed of a helicopter drop, the chutes in Adam's tests would have to land a huge vehicle, one that weighed eight thousand pounds. It was shaped like a giant lawn dart.

There are only a few helicopters in the world that can lift eight thousand pounds. One of them is the twin-rotor CH-47 Chinook, a fat monster of a machine used locally in Idaho for fighting forest fires. The thing costs a fortune to rent.

Dropping parachutes from helicopters requires patience. You need calm winds and good visibility. You need to avoid the afternoon thundershowers that can roil the skies over southern Idaho in the summertime. Your hardware can be ready and your helicopter on call, but if the weather isn't right you can't test anything. Days can slip by as you wait for conditions to change.

In the summer of 2002, Adam and his team were living in the Boise Radisson, rising each day at 3:30 A.M. in hopes that the postdawn winds would be low enough for a drop. A week went by, then a week and a half, with no chance to do a test. Three thirty was too early for Adam to go for his usual morning run, so he learned to go to the test site in his running gear, working out his frustrations on trails near the gunnery range after each cancellation. Other problems less likely than bad weather came up: Concerned airport officials, alert to security threats after the 9/11 attacks, had to be soothed over the sight of what looked like an eight-thousand-pound bomb being trailered daily past their facility.

At last, a calm and clear morning came. Adam and his team arrived at the range as the sky was growing light. They radioed the helicopter pilots, and soon a huge red and white Chinook came over the horizon, dust billowing as it touched down. One end of a two-hundred-foot cable was slung from the belly of the Chinook, the other end attached to the top of the giant dart. Adam and his team backed away. Only a two-man ground crew remained, their helmets and goggles protecting them from the debris kicked up by the helicopter.

The two big rotors spooled up and the Chinook lumbered skyward, pulling the line taut. After a quick radio exchange the pilot applied full power, and the helicopter slowly lifted the dart-shaped test lander free from its trailer.

From three hundred yards away, Adam watched the liftoff and climb of the Chinook through his binoculars. The pilot flew a slow oval pattern, laboriously gaining altitude on each leg. There was no swinging of the dart, no motion that could compromise a clean release as they approached the drop altitude. This pilot's good, Adam thought. He set the binoculars aside, preparing to track the high-speed descent by eye once the drop began.

Four thousand feet above the ground, the Chinook came to a dead hover against a clear blue sky.

"Here we go, baby," Adam said out loud, as the test lander fell away. "Heeeere we go."

The chute's deployment mortar fired, fabric streaming out behind the dart. The parachute billowed, then snapped open to make a perfect orange-and-white bowl.

And then, with a *pop* that was audible thousands of feet below, it exploded. In an instant the disk was riddled with holes, the band shredded, in ribbons. Small pieces of orange and white fabric streamed off and fluttered toward the ground as the dart fell, the wind screaming and whining in the vibrating parachute lines.

Adam stood motionless, watching, as the dart hit the ground. He felt as if the wind had been knocked out of him. The backwash from the chute slammed the fabric to the ground a second later, twisting it into a spiral and tangling the lines around the protruding fins of the dart.

Adam is an optimistic man by nature. He didn't panic. He wasn't even all that surprised. The first thing in his mind was that it had been a botched test, that they'd overloaded the parachute. To get the stresses right, you have to open the chute at just the right moment. Open it too late, when the dart is falling too fast, and you can overstress the chute. That must have been what had happened. The dart had instrumentation in it, and a little electronic data logger, to measure and record the stresses. Adam grabbed the logger from the dart, tossed it into his rental car, and sped back to Boise with his team.

They hooked up the logger in the business center of the Radisson, and quickly ran the numbers. The chute, they had figured, should be able to take a stress of thirty thousand to thirty-two thousand pounds. The test, said the data, had stressed it to twenty-nine thousand pounds, less than it had been designed for. And that stress had ripped the thing to ribbons. Something wasn't right.

They flew an identical chute the next day, under perfect conditions again. Again the chute exploded.

They had a problem. Adam called Richard Cook with the news, and it echoed quickly through JPL and up the line to NASA Headquarters. We were a year from launch, and we didn't have a parachute that worked. Whether DIMES would work or not was moot if Adam and his team couldn't solve this problem.

At the end of June, Adam hurriedly pulled together a parachute workshop, inviting every parachute expert he knew of in the world to help figure out what had gone wrong. They scrutinized the video of the tests, and they pored over the drawings. A likely explanation quickly emerged.

The design of the parachute had assumed that the band was shaped like a perfect cylinder, and reinforcements had been sewn into the band to guard against the enormous stresses that would hit it when the chute first snapped open. In reality, though, the band had been bowed out slightly in flight, like a barrel. When they ran the numbers, they realized that this barrel shape meant that the opening stresses had pulled *outward* from the sewn-in reinforcements rather than pushing into them as intended, ripping away fragile stitching and unzipping the parachute.

Simply fixing the reinforcements wouldn't be enough, the experts decided. The failure had been so catastrophic that the whole parachute would have to be strengthened. This could be done, of course, but the rule of finite canister size applied: Stronger fabric meant fewer square yards of fabric, and there was no way around it. Adam would have to shrink his chute, paying the price of adequate strength in stability, drag, or both.

There was no way to know what exact combination of changes was most likely to work. The right way to do it would have been to build a chute, test it and then make changes, iterating until the answer emerged. But like so many other things on the project, there wasn't time to do it the right way. So Adam did it the fast, expensive way instead, building three different chute designs at the same time and hoping that one would work well enough.

The strongest new chute, Adam said, was "a brick," reinforced heavily everywhere. He knew it wouldn't perform well, but he also knew it wouldn't shred. Another was reinforced only where absolutely necessary, and would perform very well, with good drag and good stability. The third was between the other two, a compromise design.

The new chutes wouldn't be ready until September. Adam now had three times as many parachutes to test as he'd had in Boise, and a lot less time to do it in. Trail-running for weeks as he waited for perfect weather was a luxury he could no longer afford.

But how do you test a parachute indoors? It would take an enormous wind tunnel, something truly gargantuan.

NASA has one. Ames Research Center, up the coast, has an aerodynamic research wing that boasts the world's largest wind tunnel. The thing is enormous, with laminated-wood fans powered by engines that total 135,000 horsepower, sucking air through a test section 40 feet high, 80 feet wide, and 120 feet long.

Adam and Wayne had talked months before about using Ames's big tunnel, but they had rejected the idea out of hand. Nobody, they concluded, would let them do something as crazy as firing a live deployment mortar inside a wind tunnel. But desperate situations require desperate measures, and they gave Ames a call.

The reply, to Adam's astonishment, was an enthusiastic yes. The Forty-by-Eighty, as it was known at Ames, was becoming an anachronism, rapidly being made obsolete by the improving ability of computers to predict aerodynamic flow around aircraft. Hardly anybody needed giant wind tunnels anymore, and on the verge of being shut down, they were desperate for paying customers. When Wayne asked about the mortar, the Ames guys just laughed. Their tunnel was made of quarter-inch steel, they said—they'd had entire airplanes inside the thing, with the engines running. Nothing our puny mortar could do would hurt it.

In late September, Adam went up to Ames to test the first of his three new parachutes. This was the "middle" design, which he thought represented the best compromise between strength and performance. If any parachute was going to get us to Mars, Adam figured, this was the one.

The setup was simple. Squarely in the middle of the wind tunnel's test section stood a tall, streamlined pylon. The chute canister and mortar were mounted atop the pylon, pointed downwind and poised to fire when the wind was at the right strength. The mortar would blast the parachute straight toward the huge fans that would be sucking air at high speed through tall, protective slats at the far end of the tunnel. A bank of lights shone overhead, providing the bright illumination needed for high-speed video recording.

Adam donned his headset, and after a short briefing in the control room he climbed up the catwalk to his observation post for the test. His perch was an old camera mount, a 4×4×3-foot Plexiglas bubble high on the tunnel wall just downwind from the pylon.

Voice chatter from the control room subsided as the fans slowly came to speed. This was so much better than a helicopter drop, Adam thought as he waited; it was so much easier to get the conditions right. As the wind reached the perfect velocity, the countdown began.

"Five, four, three, two, one, fire!" Adam's head swiveled as the mortar blew past him, the chute streaming downwind from the pylon. The fabric slowly unfurled, pulsating and billowing.

"What . . . the fuck . . . is that?!?" Adam said aloud after a full ten seconds had elapsed. What was going on?

The chute wouldn't open. Instead, it was oscillating wildly in the wind, opening halfway and then closing again, over and over, like a giant sea creature swimming. Slowly Adam realized what he was seeing. He'd never seen it before, but he'd read about it, and he'd heard stories from other parachute designers.

His chute was squidding.

Squidding is a personality disorder of parachutes. Squidding is a monster that waits for parachute designers in the dark. It happens when something's not quite right in the design, when the forces that want to pop the chute open are repeatedly overcome by other forces that flap it closed again. Squidding had never been seen in a parachute like ours before, not in thirty years of testing. But now this chute, the one that Adam was betting would take us to Mars, was squidding.

"Uh . . . well . . . ," he said, stammering into his headset, not quite sure what to do, "might as well shut it down." The big fans spun down and the chute started to sag . . . still squidding and then, too late, finally struggling open as the wind subsided.

For five minutes, Adam was ill. His wasn't the only crisis on the project, he knew. The airbags were weak. DIMES might not work. Everything was too heavy. But nobody wanted to be the one whose crisis brought the project down, and Adam was no exception. There had to be a fix.

His mind started working again. This had been the middle chute, not the one that he knew would perform best. What about the best-performing one? If it didn't shred, maybe that one would work. The next day the tunnel technicians mounted a new canister, with a new mortar and the best-performing chute, atop the pylon.

If any of the three chutes would open without squidding, Adam felt sure that this was the one. The band on the middle chute had been made of a strong fabric, but one that was impermeable. Air couldn't flow through it, and since air had to flow into a chute to force it open, maybe the impermeable band had been what was holding it closed. The band on this chute was made of permeable fabric. Perhaps the flow of air through the band would make the difference, would be enough to force the chute open.

Adam climbed back to his vantage point, and the five-second countdown began. On "fire" the mortar blew perfectly, the chute streaming again past Adam's bubble. "Oh, God . . . come on baby, come on baby, come on baby," he chanted as the chute oscillated. Then slowly, so slowly that it made him ache to watch it, the chute filled and finally inflated.

It didn't shred. That was the good news, it didn't shred. But the thing had taken *way* too long to open, squidding for seconds before the final pop. His weakest chute, Adam now knew, was strong enough. That problem was solved. But it still didn't open fast enough.

The technicians drove the tunnel down as he descended the catwalk. With the chute still flying he walked into the tunnel, leaning hard into the wind. He squinted up at the chute. Something was wrong.

"That vent's too big," he said to himself. He turned to the engineer with him, a man from the company that had built the chutes.

"That vent's too big," he yelled over the wind. "It's too big. I know what these parachutes should look like, I've seen them before. It's too big!"

The chute sagged to the floor as the wind died. Technicians gathered it up and separated it from the pylon, carrying it to an inspection room just off the tunnel. Adam found a tape measure, and sure enough, the vent hole in the middle of the parachute was too big. Something hadn't been communicated quite right between JPL and the parachute vendor. This wasn't the chute design he'd wanted to test.

It was time to improvise. Finding a piece of red nylon the right size, a chute technician cut it to an annular shape and sewed it by hand into the vent hole, shrinking the hole's diameter. It was impossible to repack the chute without sending it back to the factory, so the technicians set up a "dis-reefing" test instead, with the chute streaming downwind from the pylon but held shut at the base of the band by a simple lanyard. With the tunnel at speed and Adam holding his breath, the lanyard was yanked and the chute released. *Zaboom* . . . the chute opened.

It was encouraging, but as he flew back toward Pasadena that evening, Adam realized his problem wasn't solved. If he shrank the vent hole in the best-performing chute, he was convinced he had a design that would work. It was strong enough to take the stress, and the smaller hole and permeable band had solved the squidding problem. But the only test that had worked had been a dis-reefing test, not a mortar firing, and that was a clear violation of "test as you fly, fly as you test." He had to build one more chute, then go back to Ames again and do it right.

The redesigned chute was built and delivered, and on the tenth of October, 2002, Adam's fortieth birthday, he went back up to Ames for one last test. It was a disappointment. The chute still opened too slowly. Even more frustrating, measurements afterward showed that the vent hole was still bigger than he'd wanted it to be. Miscommunication and aerodynamics were conspiring against him; his time was now almost gone.

At three o'clock that afternoon, dejected after a long day of testing, he emerged into bright sunshine from the interior of the wind tunnel. Abruptly his cell phone went off, the record of a call he'd missed during the hours inside the heavy steel structure. He pulled out the phone and speed-dialed his voice mail.

It was his wife, back in Pasadena, eight months' pregnant with their first child.

"I'm going in for emergency surgery," her short message said, her voice shaking. "Bye."

What?!? Adam sprinted to his car, tunnel and chutes forgotten, and sped to the San Jose airport. But by the time his plane touched down at

Burbank, two hours later, he was a father, his daughter delivered by emergency C-section. It wasn't the first or last time that *MER* took a toll on a family.

On October 30, leaving his new family briefly for one more test, Adam arrived at Ames with the "flight build" of our chute, an exact duplicate of the parachute we'd have to fly. This really was the final test, there was no time left to make any more changes. But on that day there was no drama. The mortar fired, the chute blossomed, and in less than a second it popped open to form a perfect, rigid bowl. Adam smiled with quiet satisfaction. We had our parachute.

While Adam had faced his squidding demons, the DIMES team had been out in the Mojave Desert, flying a prototype camera on their own rented helicopter, over the most Mars-like terrain they could find. Matching the pictures up back in the laboratory, on a computer just like the ones in our rovers, the DIMES software quickly calculated the helicopter's velocity, perfectly, time after time. Miguel's instincts had been correct. DIMES would do what we needed.

With a chute that worked, DIMES performing and airbags that were good enough, Gusev Crater seemed within reach. We told the story back at NASA Headquarters, and Ed Weiler made it official: One of our rovers would go to Meridiani, and the other would go to Gusev. The dreaded Elysium site was gone.

The EDL team, I learned, could drink a lot of beer.

9 ATLO

―――――――

THE REAL WORK OF sending a spacecraft to Mars happens in
ATLO: Assembly, Test and Launch Operations. ATLO is the business of building a spacecraft, making sure it works and getting it
off the ground.

ATLO is like a river, the sequential merging of thousands of small
tributaries. Tiny electronic parts arrive, and each of them is tested. The
same happens for the circuit boards the parts are destined to go on. The
parts are soldered onto the boards, and everything is tested again. Then
the boards go into a larger assembly with other boards, there's another
round of testing, and so on up the line. It's the same with every other
piece of the vehicle: motors, instruments, telecom, propulsion, everything.
ATLO is a nightmare of logistics, where each crucial tributary juncture
can be delayed when any one of the key components isn't ready on time.
Whenever that happens—and it happens daily—the ATLO flow has to be
readjusted, the tributaries shifting to some new configuration that'll still
get the hardware on top of the rocket before the launch window arrives.
If you run out of time, you lose.

Before we could start ATLO we had to be reviewed again, of
course. . . . It was impossible to do anything without somebody review-

ing it first. Our ATLO Readiness Review was held in January 2002, and it was a disaster. Our presentations were hurriedly prepared and ragged, and they gave the impression that we really weren't ready to start building flight hardware. Our costs were growing again, and when that problem was added to our schedule woes, the situation looked impossible. Even the JPL review board members didn't seem convinced we could make it. The Independent Review Team members were ominously silent.

As I walked to my car after the second day of the review, one of the IRT members caught up to me in the JPL parking lot. The team looked very tired, he said, and that was no condition for us to be in at the start of something as difficult as ATLO. I resented his unspoken implication that we were too tired to do our jobs right, but it was hard to disagree with his observation.

A few weeks later I was back in Ithaca when an e-mail arrived from Pete late in the evening. The IRT's report was in, and Pete was sending it around for comment. I scanned it quickly, and my fears were confirmed. Page after page was filled with strident warnings:

"Project risk has increased substantially . . ."

"IRT expects cost to increase . . ."

"ATLO planning and readiness is not complete . . ."

"Substantial increase in risk to mission success . . ."

And then, down at the end, the final blow:

"Reducing mission objectives mitigation option:

- *"Focus ATLO on single flight vehicle to assure a fully flight qualified and characterized system."*

They were going to recommend that NASA take one of our rovers away.

Again, the issue was money. Back when we'd had our budget "cap" of $688 million, the IRT had predicted that we would overrun it by something between $84 and $154 million. We had howled in protest, arguing that we couldn't possibly spend that much. But our mounting problems with the lander redesign, the airbags and the parachutes had continued to take their toll, and now when we added it all up, it looked like we were going to come in at something very close to $800 million. If that's how it worked out, it would be an overrun of $112 million—right smack in the middle of the range that the IRT had predicted months ago. The bastards had been right.

You could quibble a little bit one way or another, but the overall answer was clear. We needed something like another fifty million bucks to bring this thing home.

With a number that big, getting it wasn't going to be a simple matter of generosity on Ed Weiler's part. The man had clout, but he couldn't simply conjure up money. There really were only two ways that Ed could come up with fifty million. One would be to take it away from other important projects, from somebody who needed it badly. The other would be to sell off one of our rockets.

The rocket thing had me terrified. A Delta II costs $65 million, give or take. And Deltas are great launch vehicles, reliable and in high demand. So a very easy way to solve our financial problems would be to sell one of our rockets to another project that needed one, and plow the proceeds back into our budget. We could send one rover to Mars, and use the other one for spare parts.

On March 26, 2002, Pete Theisinger, Firouz Naderi and I were summoned to NASA Headquarters to explain why we thought we should keep both rovers, and how much it would cost.

Weiler's opening comments set the tone. Back in the fall of '01 his instinct had been to kill off one rover, he said. He'd gone against his instinct and backed off then, but his instinct was telling him the same thing again now. "I want you to convince me why my gut feeling is wrong," he said. "I want you to convince me I should believe you."

Pete and Firouz were masterful, arguing dispassionately and clearly why ATLO would be more likely to succeed if we went forward with two vehicles, and confessing honestly to how much it would cost. I got up and did my best to describe the science benefits of keeping two, how it would protect us if one of our landing sites turned out to be a dud.

Next up after we'd finished was Glenn Cunningham, the head of the IRT, who presented the same scalding report that Pete had sent me.

"Okay, bottom line," Ed said when Glenn was done. "What's the best way to assure a successful landing?"

"In my opinion," replied Glenn, looking straight back at him, "it's to concentrate on building one rover well."

Ed leaned back in his chair and nodded. "That's what I thought."

My heart sank.

The JPLers and the high-ranking Headquarters guys went into executive session, kicking the rest of us out of the room. I walked across the hall, followed by Glenn, and slouched against the far wall, glaring at him.

"I hope you're pleased with yourself," I said abruptly. I'd been feeling bitter for weeks about the IRT report, and suddenly I was very angry.

He looked at me blankly, startled.

"You've just killed off one of our rovers," I blurted. "You've just doubled our risk and cut our science return in half."

It was a childish outburst, and I regretted the words as soon as I said them. Glenn was a good man, with far more flight project experience than I had, and he was simply doing his job of providing an honest opinion to NASA. He'd been right before, about the size of our overrun, and there was a damn good chance that he was right again now. But I couldn't take back what I'd said. I walked down the hall to be by myself.

Half an hour later the meeting broke up, the JPLers rushing out of the room talking on their cell phones. Ed's conclusion, to my amazement, had been that he'd find us another $50 million from somewhere. The final decision on whether we'd launch one or two rovers wouldn't be made until May 15, when the payment for the second rocket came due. But until then, we were allowed to keep going with two rovers. Weiler had gone against his instincts again, and against the IRT's recommendation. I silently blessed him.

I headed for National Airport, feeling weak with relief. There, luckily, I ran into Glenn and apologized for my outburst. He accepted graciously, and he wished us good luck. He obviously thought we'd need it.

So we were still in business, at least for a while, with a muted approval to move forward with ATLO.

Our ATLO leader was Matt Wallace. Born in the great state of New Jersey, Matt is a graduate of the United States Naval Academy. When the Navy told him that his eyes weren't good enough to fly jets, he'd gone after the next best thing: fast-attack submarines. Nuclear power school and sub training landed him a position on the USS *Albuquerque,* out of Groton, Connecticut, where he quickly qualified as engineering officer of the watch, running the sub's nuclear reactor and reactor control system. It was a time when the Soviet navy still roamed the seas, and the *Albuquerque*'s job was tracking and trailing Soviet submarines. Matt cherished his time in the sub force, loved being officer of the deck, with responsibility for a $400 million piece of machinery. But it was intense duty, with most of his time spent at sea, and after three and a half years it was time for a change.

Matt left the Navy at the age of twenty-eight and enrolled in graduate school at Caltech, earning a master's degree. Along the way he met a good-looking nurse from Glasgow, married her and found a job at JPL. He worked spacecraft power systems at first because of his nuclear background, and led the development of the power system for the *Sojourner* mini-rover on *Mars Pathfinder*. It was *Pathfinder* where Matt got his first taste of the high-stakes world of ATLO, and that was it for him. I asked Matt once what he liked about ATLO, and he replied: "I guess I need a certain degree of danger to get motivated."

Matt's original plan had been to start ATLO with the first REM, the Rover Electronics Module, on the fourth of February. The REM is the heart of the rover, the big stack of electronics boards that makes everything work. It was obvious why ATLO couldn't begin without a REM: the other parts of the rover can't be tested if there isn't a REM there to run them. But by the beginning of 2002 it was obvious that the first "flight" REM, one of the ones going to Mars, wasn't going to make it by February.

So Matt decided to start ATLO with the engineering model of the REM instead. This thing wasn't going to Mars; in fact, it wasn't going anywhere. The engineering model was a prototype, a piece of hardware that was supposed to be "just like" the flight REM, but that you could experiment with, make changes to. Starting ATLO with an engineering model was cheating, really, since we wouldn't be "building flight hardware" at all. But Matt had no choice. The schedule was going to hell, and we'd already delayed ATLO twice, setting off cries of consternation both times. It was time to get going.

ATLO began with delivery of the engineering model REM to JPL's Spacecraft Assembly Facility.

SAF is one of the landmarks of space exploration, the place where some of the great spacecraft in history have been put together. The heart of SAF is the high bay, a sprawling, high-ceilinged room kept constantly at the exceptional levels of cleanliness required for spacecraft assembly. Immediately adjacent to it is a control room, a warren of workstations and conference tables. Flight hardware enters and leaves the high bay through a giant airlock at its eastern end. Workers enter via an "air shower," a fixture like a shower stall where high-pressure air jets blast loose particles that might contaminate the spacecraft off your clothing, sucking them loudly through a grate in the floor. Everyone in the high bay dresses in the JPL ATLO uniform: a robin's-egg blue smock, matching booties and a floppy white hairnet.

On Friday evening, March 8, the rover electronics team brought the engineering model REM, enveloped in silvery plastic inside its transfer case, to the big SAF airlock doors, and wheeled it inside under Matt's watchful eye. Suited-up ATLO technicians opened the case and began taking pictures while the engineers turned to head home to their families.

"Where do you guys think you're going?" asked Matt, stopping them short. "We need to do the receiving inspection."

"It's eight o'clock at night on a Friday," one of them said. "Why don't we just wait 'til Monday? Are we going to be working the weekend?"

"You'd better believe we're going to work the weekend!" Matt replied sharply. He had the whole schedule in his head, and he knew we were in trouble. With hardware finally on the ATLO floor, it was time to set the pace that he knew we'd have to maintain all the way to the Cape.

In the weeks that followed, work inside the high bay accelerated as new bits and pieces of the rovers began to arrive, each cabled to the REM and run through its initial testing. Each new piece was handled with care bordering on reverence by the ATLO techs, like relics in some strange religious ritual. There is a "JPL way" of doing ATLO, developed through forty years of making mistakes and learning from them, and those of us who didn't know it were schooled quickly.

Everything is prescribed: how to hold a tool, who may hold it and who must watch. For each procedure, even the tightening of a single screw, there is a written set of instructions, a person to do it who has achieved proficiency in that task via training and certification, and someone else to watch everything and verify that it's done properly. Nothing is left to chance. It felt tedious at first, even oppressive, until I realized that the procedures had been born from past failures, and that they were our best defense against failures of our own.

As pieces came together, the tests grew more complex. During any major test, there's an intricate interplay between the shirtsleeve engineering team in the control room and the blue-suited technician team out on the ATLO floor. Nervous engineers who've just delivered their flight hardware sit in the control room at consoles under the direction of the test conductor, anxiously watching the action in the high bay through the long window that lines one wall of the room. Veteran ATLO techs, some working on their ninth or tenth mission, carry out the actual test, always keeping a close eye on their precious, embryonic spacecraft. Voice communication crackles over headsets, on a net known as VOCA, with the same kind of spare, precise language and protocols used by air traffic controllers and pilots. It's a strange, heady mix, with NASA-style cool underlaid by get-it-done passion and, sometimes, a whiff of desperation.

Schedule dominated everything. It was a sword that hung over us, it saturated the air we breathed. We lived in constant fear that some catastrophic event in the high bay would be the one that shattered our fragile schedule, that would keep us from getting to Mars. As Matt put it to me one day, "You never know when we're gonna hit the wall, which test, which failure is going to push us over the edge."

Tension bound the team together. Matt had chosen his people well. He didn't just go after the right technical skills. More than anything else, he had picked his people for their attitudes, for their willingness to work long hours and to subjugate their own interests in favor of those of the team. Morale was everything in ATLO, it was our schedule multiplier: if people love their work, they'll do it harder, longer and better. Matt knew this instinctively, and the attitude he projected outwardly flowed down through his team.

Inside, though, Matt was in turmoil. He did all the ATLO scheduling himself, and he knew better than anyone just how desperate our situation was becoming with each slipped delivery and each delayed test. He never shared the worst of it with the team, never let them know how little margin was left. I don't think anybody other than Matt, and maybe his wife Sandra, ever really knew just how bad it was. I was desperate myself to understand the depth of our schedule problems, but the only way I had of estimating it was by reading Matt's emotions, by judging how stressed and uncomfortable he seemed. With each passing month I grew more nervous.

Much of Matt's job, I came to realize, was psychoanalyzing everyone who was delivering hardware to ATLO, me and my science team included. Who can you trust to deliver when they say they will, and who should you mistrust? Who do you pressure when things get tight, and who do you give some breathing room? What's been their history on past projects? How good does their hardware tend to be once it arrives? It's not much help if somebody hits their delivery dates with hardware that's a mess, in need of weeks of testing and rework. Matt held the ATLO schedule in his head, but he also held a psychological profile of every member of the team.

Of all the miracles that Matt had to work, the one that seemed hardest to me was how to build two spacecraft in less time than it would take to build one. It sounded nonsensical, but that was what we had to do if we were going to prove the IRT wrong. How could it not be harder to build two than to build one?

The answer, it turns out, depends on what's slowing you down. If you're limited by how much money or how many people you have, then yes indeed, building two is harder than building one. But Weiler had given us the money we needed, and Charles Elachi, who was now the JPL director, had used that money to get us the people. With those worries gone, the next big limitation was how much hardware we had. There's great strength in being hardware-rich: With lots of hardware, you've got lots of pieces on your chessboard. Matt was the master of this game, and he played the pieces he had with consummate skill.

Think about it: Hundreds of tests have to be run on spacecraft like these, some of them lasting for days. The spacecraft are identical, so some of the tests need to be run on only one vehicle. With just one spacecraft, the tests would have to take place sequentially, each one starting only after the one that preceded it ends.

With two spacecraft, though, you can run two different tests at the same time, one on each vehicle, picking up time in your schedule. Or if something breaks before some critical test, you can pull a part off the other spacecraft and swap it in, rather than waiting for the broken part to

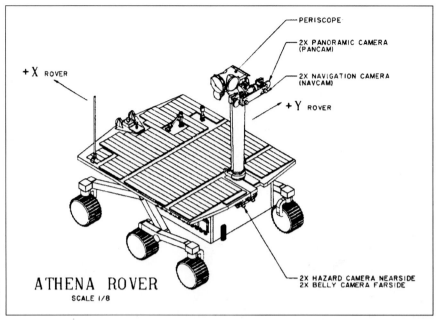

The design of the original *Athena* rover.

An artist's concept of the 2001 Mars lander, with Pancam and Mini-TES bolted to the lander deck and the little *Marie Curie* rover running around in the foreground.

The *Mars Sample Return* mission, as it looked before we came to our senses. That's our rover in the foreground, hiding behind a rock so it won't get toasted when the Mars Ascent Vehicle takes off.

A very early artist's concept of what a *MER* rover might look like, still parked on the lander that delivered it. Notice the lack of solar array winglets...not to mention the fact that there's no arm!

A somewhat later artist's concept of a *MER* rover. Still no solar array winglets, but it's starting to look like a real piece of hardware.

A *MER* rover, more or less as it actually ended up looking once all the design work was done.

Some of the *MER* project team. Left to right: Richard Cook, John Callas, Barry Goldstein, Charles Elachi, Richard Brace, Pete Theisinger, me, Mark Adler, Jennifer Trosper.

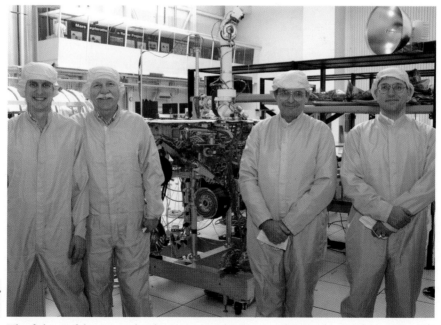

The fathers of the original *Athena* rover, in the Spacecraft Assembly Facility at JPL. Left to right: Me, Sam Dallas, Jake Matijevic, and Barry Goldstein. That's *Spirit* behind us.

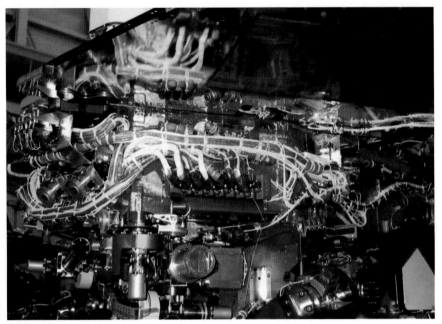

The front end of a *MER* rover, exposed wiring and all. It's a fiendishly complicated machine.

One of the test rovers, outdoors for the first time in the parking lot next to ISIL. The JPL parking pass probably wasn't necessary, but you can never be too careful.

JPL airbag engineer Kevin Burke being consumed by his work.

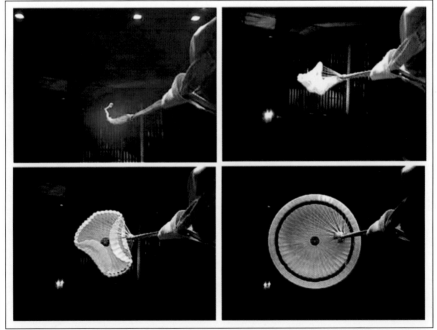

Our first successful parachute test, after many failures. This one took place just eight months before launch, in the big wind tunnel at Ames.

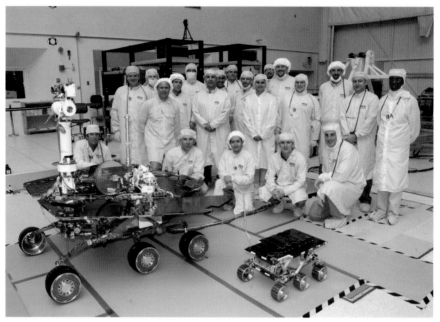

A rover family portrait, with some of the *MER* ATLO crew in the background. That's *Spirit* on the left and *Marie Curie*, the engineering model of the *Sojourner* rover, on the right.

Spirit's first drive, with a bunch of very nervous engineers looking on.

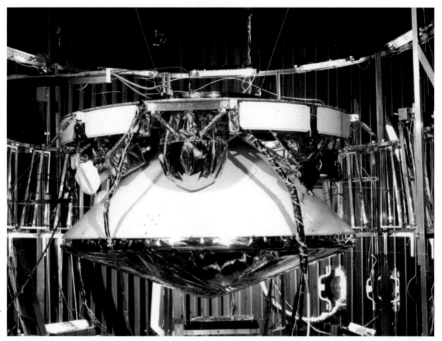

NASA/JPL

A fully assembled *MER* spacecraft, hanging in the big thermal vacuum chamber at JPL.

BEN THOMA

ATLO techs handling solid rocket motors in the PHSF high bay at the Cape. I thought I had an exciting job until I met these guys.

ATLO at the Cape. When you put it together for the last time, everything's got to work.

The mummies worship the rover god.

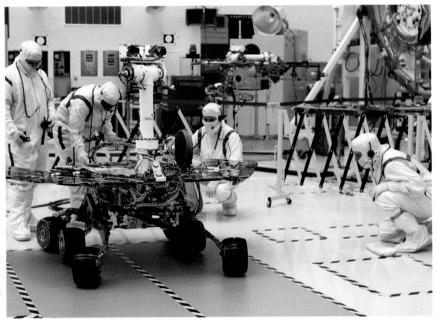

Spirit in the foreground, all tricked out and ready to go to Mars. You can just see *Opportunity* in the background, without her wheels on yet.

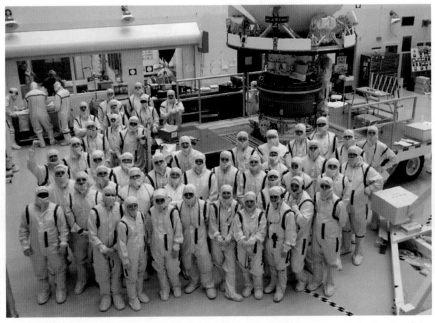

The *MER* ATLO team in the PHSF high bay. That's *Opportunity* at the back of the room, stacked onto the third stage of the Delta II and ready to go to the pad.

Pad 17-A at dawn, with *Spirit's* Delta II almost ready to fly.

The Delta II looks delicate from a distance, but up close it's a brute.

At long last, launch day for *Spirit*.

June 10, 2003: *Spirit* takes flight.

Launch day for *Opportunity*: Barry and me at Pad 17-B.

July 7, 2003: *Opportunity* heads for Mars.

Hitting the top of the martian atmosphere at Mach 25.

Seconds before impact, airbags inflate explosively around the vehicle.

RAD motors fire, cutting the downward velocity to zero just an instant before impact.

Anxious faces in the CMSA as we wait for news of *Spirit*'s landing. From left to right: Nagin Cox, Miguel San Martín, Pete Theisinger, me.

We're down, baby! Wayne Lee in the CMSA celebrating the successful landing of *Spirit*.

be fixed. These tricks work only if you have enough people and enough money to run all those tests and do all those hardware swaps, but we did. Even with two rovers, our schedule was on the verge of falling apart. With just one, it would have been impossible.

The scariest part of all was the flight REMs. Using the engineering model REM worked for a while, but sooner or later we had to get flight REMs into the rovers. REMs are the most complicated parts of the whole vehicle, and their delivery was delayed months beyond what Matt had anticipated.

If he took every bit of margin out of his schedule, Matt calculated that the latest date that the first flight REM could arrive was the fifteenth of July. Beyond that date, all of JPL's past experience said we wouldn't be able to make it to the pad in time to launch. But Matt was convinced that his team was the best that JPL had fielded in forty years, that these guys simply could do things that no one before them had ever done. So he gave the REM two more weeks. He set August 1 as the drop-dead date by which we had to have the first flight REM. Any later than that, and Matt was convinced we weren't going to make it.

The first flight REM arrived in SAF on August 9.

As summer slipped away into fall, the rovers began to look like rovers. We called them *MER-1* and *MER-2*. Each had a dedicated team to take it through the latter parts of the ATLO test program.

My own life settled into a kind of routine. Every Wednesday evening at 5:45 I'd catch a US Airways flight to Pittsburgh, and then another to LAX, arriving in Pasadena a little before midnight if I made my connection. Thursdays and Fridays I'd spend in ATLO, testing the science instruments and working with Matt and his team. A 7:00 A.M. flight out of LAX on Saturday would get me back to Ithaca for dinner, and an abbreviated weekend with Mary and the girls. The next week I'd do it again. It was exhausting and exhilarating. After all the years of hope and planning, seeing our rovers start to become real was deeply moving. I do my best to play the steely-eyed space explorer dude, but the first time I saw *MER-2* move, crawling slowly forward over blue plastic mats on the high bay floor, it brought tears to my eyes.

The best part of the long hours that we spent in ATLO was watching the slow breakdown of the natural barriers between the scientists and the engineers. With each shift in SAF I learned more about what Matt and his team were up against, how hard their job really was. And we'd talk long into the night about the science we'd do at Mars once we got there, the engineers hungrier than I ever knew to learn what the mission they'd worked so hard for was all about. Slowly a kernel of trust began to grow between the people who wanted the science to be perfect and the people who had to make it all work.

In the ever-shifting flow of Matt's ATLO plan, the rover we called *MER-2* came together before *MER-1*. Each vehicle was beginning to take on its own personality, and *MER-2* became our temperamental firstborn.

The nomenclature was complex. Whichever rover launched first would be designated *MER-A,* bound for Gusev Crater. The second would be designated *MER-B,* and would go to Meridiani. By fall we started to suspect that *MER-2* would become *MER-A,* with the better-behaved *MER-1* becoming *MER-B.* But a last-minute crisis, even at the Cape, could reverse the order. It felt strange to stand in SAF next to a rover and not be sure which side of the planet it was destined for.

MER-2 was definitely the problem child. Many of the really complicated tests had to be done on both rovers, and *MER-2* tended to hit each one first. Every serious test failure, it seemed, happened on *MER-2*. *MER-1* became the lucky rover, the one that benefited from all the mistakes we made on its sibling.

December brought the most crucial test of all: thermal vacuum. It was the most complicated thing anybody at JPL had ever seen somebody try to do with a spacecraft that was still on the ground.

Thermal vac was our best attempt to simulate Mars on Earth. It took place "up the hill" in the parlance of JPL, in an enormous test facility at the top of the hillside on which JPL is built, hard against the base of the San Gabriel Mountains.

The centerpiece of the facility is the thermal vacuum chamber, a hollow steel cylinder standing on its end, ten feet in diameter and twice as tall. Its inner walls are painted a deep black and cooled with liquid nitrogen that flows through embedded pipes. Lights in the upper end of the cylinder simulate the Sun, modulated for our tests to the reduced solar intensity at the distance of Mars. Giant vacuum pumps suck the air from the chamber, and tanks of gas backfill it to simulate the thin martian atmosphere. There's no red dust inside, and no wind, but otherwise the conditions are as much like Mars as we can make them.

Thermal vac was designed to take each rover through almost everything we'd ask it to do on Mars. The only thing we left out was driving, since the cramped quarters inside the chamber made that impossible. We'd unfold the solar arrays. We'd deploy the mast and stand the rover up. We'd use all the cameras to take pictures inside the chamber, and use Mini-TES to take spectra of heated rock slabs mounted against the chamber walls. We'd unstow the arm and take pictures of a target with the Microscopic Imager. The test would run around the clock, and inside the chamber we'd simulate the twenty-four-hour, thirty-nine-minute martian day. The test would take as long as it took.

The chamber door closed for *MER-2* thermal vac on Wednesday, December 11, 2002. My notes from one day early the next week are typical:

> Monday, 0400 PST: It's the beginning of day five, on what should
> be the biggest day yet for the payload. It's dark outside now, and
> cold. The control room is empty except for a few hardy souls,
> and those who are here look battle-weary. Outside the lab a big
> liquid nitrogen truck is discharging its load, the boil-off billow-
> ing around the delivery man, who's slumped in his chair and
> asleep behind his protective face mask. Yellow/white lights rotate
> outside building 248, lighting up the vapor and providing the

only sign that there's a powered-up spacecraft inside. Simulated Mars time happens to be synched up fairly well with Earth time right now, so inside the chamber it's dark too, and very cold. We'll be waking the rover up in another hour or so for a simulated UHF pass.

1130 PST: Seven and a half hours later, and nothing useful has happened. The rover's awake, the targets are at temperature, but we're dead in the water. One problem is that we don't know where the targets are, because we didn't have time to survey them in before the start of the test. Another is that we can't get any data off the spacecraft. There's a bunch of old data still in the flash memory, and the rover keeps trying to send it to us, over and over . . . it won't let us see the pictures we've just tried to take of our maybe-they're-where-we-think-they-are targets. There's no apparent solution, so the conclusion was to wipe out all the data products on board and start over. Hours of work gone.

1300 PST: Still no progress. The delete command won't work, so now we're going to reformat flash completely. Flight Software says "the major data manipulation commands appear to be totally broken." Uh huh. The clock keeps running.

1400 PST: We're starting to talk about who to send home to get some sleep. It's bound to be another hour or two at least before we're taking Mini-TES data, and even if everything works perfectly we've got about 17 hours worth of stuff to do once we start. Realistically, I don't see us finishing this first set of measurements until noon tomorrow. We've been here for ten hours already. Outside the rain is coming down so hard that you get soaked to the skin walking between the test lab and the building next door.

1430 PST: We just heard that *MER-1* failed a standup test over the weekend. This is scary stuff. If standup fails on Mars, we're a lander mission, and a poor one at that. It is not a good day on the *MER* project.

1530 PST: Lights are on in the chamber, heaters are on, and we're about to initialize the PMA, the big mast that points Pancam and Mini-TES. Here we go . . .

1605 PST: Images are down, or at least most of them. Our aim was off. Shooting again.

2055 PST: We've slowed to a crawl. The issue, again, is getting data off the spacecraft. We've been waiting for a simple Mini-TES scan for over half an hour. Nobody understands what's wrong, and we're all glowering at Flight Software.

0025 PST: We've taken Mini-TES data across the rover deck, somewhere in the vicinity of the calibration target, but we're damned if we can figure it out. We're still trying to work out which way is up and down and which way is left and right on the image. Somewhere in there should be the target, but we can't see it. What's going on?

0115 PST: We missed the target. Not by a lot, but by enough that we've got to re-do everything.

0200 PST: We think we've hit the target, but we can't see it . . . it just blends in with the background. This shift started 22 hours ago; I think it's time to get some sleep.

And that's how it went, day after simulated martian day. We simply couldn't get the rover to do what we wanted it to. Worst of all was the flight software that handled data on board the rover, writing to and reading from the "flash" memory that was our equivalent of a computer's hard drive. The write and read commands too often just didn't work. The command we seemed to use most was the one the Flight Software team called SHUTDOWN_DAMMIT, which was like hitting CTRL-ALT-DEL on a personal computer, killing everything off and starting over. As the test stretched out and the hoped-for Christmas break crept closer, morale sank.

The test conductor through much of *MER-2* thermal vac—the guy who ran the show whenever he was on console—was Leo Bister.

Leo, fortunately, is a calm man. When everything around him was in turmoil, when the rest of us were irritable from sleeplessness and swearing from the stress, Leo was the one imperturbable person in the room. In a test like this one, with the clock running and exhaustion clouding judgment, leadership with a steady hand is essential. Leo provided it. As I watched him at the test conductor console, I wondered if his composure flowed from his many years as a distance runner, the long, solitary miles leading him to a steady temperament. Maybe it was just a matter of stamina.

Leo has graying hair and a lean runner's build, and over the course of thermal vac he went from clean-shaven to fully bearded. When I asked him why, he said, "My wife hates that 'prickly' phase, so I figured that as long as I wasn't going to be home much anyway, now was the time."

On the seventeenth of December, the seventh day of thermal vac, Leo and I were both off shift and sleeping when we each got a disturbing phone message from Justin Maki, a JPL scientist who was working on all the *MER* cameras. At low temperatures—meaning at the kinds of temperatures we'd routinely see on Mars—pictures from the right Pancam looked "speckled," Justin said. What did this mean? I had to see the images, so I gave up on sleep and drove back in to the test lab. Leo was there when I arrived, and Heather and Miles, two Cornell students who had done much of the Pancam calibration work, were at the imaging console.

"So let's see some of this 'speckle,'" I said.

"You're not gonna like it," Miles replied. He hit a few keystrokes, and an image appeared on the screen.

It was hash, complete garbage. I could just barely make out the texture of the chamber wall, and one of the targets hanging on it. But the scene was overwhelmed by a bizarre pattern of static. If I hadn't already known what I was looking at, it would have been completely uninterpretable.

"Are they all like this?" I was shocked—this was something we'd never seen before.

"No, some of them just have a few speckles. But some are worse than this."

"What was the temperature when you took this one?" I couldn't take my eyes off the screen.

"The chamber was down around minus fifty-five," Miles replied, "but the camera was a lot warmer than that." Normal operating conditions on Mars. We had taken hundreds of perfect pictures with this exact camera and at these kinds of temperatures before we'd put the thing on the rover. What had gone wrong?

"Any of the other cameras doing it?"

"Nope. Not so far, anyway." Miles looked as baffled as I was.

I quickly ran through the implications in my head. Matt's schedule said we were a week away from the end of thermal vac, and six weeks or so from shipping *MER-2* to the Cape. Here was a sudden, unexpected problem that incapacitated Pancam, one of our most important instruments. Worse, the problem had to be somewhere in the electronics, and the electronics in every camera on both vehicles—the Navcams, the Hazcams, the MI, even DIMES—used the same design as Pancam. Only the right Pancam was speckling so far, and only at cold temperatures. But what if it was some degenerative problem, something that would get worse with time, and this just happened to be the camera that had succumbed to it first? We'd be blind on Mars. This was the kind of thing that could take the whole mission down.

The schedule implications were equally ominous. This was a problem we could fix only at JPL. Cape Canaveral is equipped with everything you need to get your spacecraft on top of its rocket, but there is no thermal vacuum chamber there. If the speckling appeared only at martian temperatures, there'd be no way for us to troubleshoot it in Florida.

I walked to the front of the cramped thermal vac control room, threading my way through the tight space between the consoles and the wall. Everybody looked very tired. I waited patiently for a moment while Leo finished typing a daily report.

"You got any ideas on this speckle thing?" I asked him. I knew that Leo had designed the camera interface, the scheme that we used to transfer electrical signals from the cameras to the REM.

"Could be a lot of things," he answered with typical caution. "I think this one's going to take some work." If even Leo didn't have any ideas, we were in bad shape.

We struggled through thermal vac for the rest of that week wrapping it up just two days before Christmas. Pictures from the right Pancam were speckled to the end, especially when the chamber was cold. We broke chamber and straggled home for the holidays with the knowledge that something was very wrong with our rover and that we didn't have much time left to fix it.

We chased the problem through early January, but there wasn't even a consensus among the engineers as to where we should look. Some suspected the electronics in the camera, others the electronics in the REM. Some suspected the cabling between them, or even just a loose electrical connector. The most optimistic view was that it was a problem with the test setup, something we'd never see on Mars. The most pessimistic was that the camera design itself was flawed.

With *MER-2* disassembled again for rework after thermal vac, there was no way to test the whole rover together as we searched for a cause. We put the REM by itself into a thermal chamber and took it down to cold temperatures. No speckle. We put the whole *MER-2* rover deck into a different thermal chamber and took *it* to cold temperatures, Pancam, mast and all. No speckle there either. Clearly the whole rover had to be fully assembled and cold for the problem to appear. Unfortunately, that wasn't scheduled to happen again until January 3, 2004, on the surface of Mars.

With the camera itself among the suspects, I made the painful decision to build up a brand-new right Pancam from scratch, and to put the Pancam team through the weeks of work required to calibrate and test it. Nobody knew that the camera was the culprit, but if we discovered a month down the line that it was, we'd have no alternative but to replace it. Building its possible replacement now was an expensive but necessary insurance policy.

Leo worked the speckle testing with us, concentrating on the signals that ran between the camera atop the mast and the electronics deep inside the rover.

"It feels to me like an interface problem," he told me one afternoon as we puzzled over it. "It's an LVDS interface, and those tend to be twitchy, especially at low temperatures."

I gave him a blank look.

"Low voltage differential signal," Leo explained, hoping that would help. "It's a standard way of transmitting lots of data very fast, but it tends to be finicky. It works fine over short distances with simple cables, but you don't typically use it with long lines and lots of interconnects. That could be where the problem's coming from."

I thought immediately of the incredibly complex cable that ran from the cameras down to the rover body. It was built in multiple segments: one running from the cameras to the joint at the top of the mast, another down the mast and through the swiveling joint at its base that helped the mast turn, another around the joint that we'd use to deploy the mast to its vertical position, and yet another from the front bulkhead of the WEB into the heart of the rover. Each cable had a different design, and each fastened to the next with a mechanical connector.

Back during the test when we had put the whole rover deck into a chamber, Leo had brought along an esoteric piece of equipment, something called a time domain reflectometer. They're simple things, really—linemen with the phone company use them to locate breaks in telephone cables. They send a high-speed pulse down an electrical cable, and then they look for pulses that bounce back, like an echo. If there's a break in a cable, or just a change in its properties, you can see a reflection from that discontinuity. If you know how fast the pulse is traveling, it's easy to figure out where the discontinuity is. Leo had used a time domain reflectometer once, long ago, in his days as an electrical engineering student, and had not touched one since. But this smelled like a cabling problem to him, and TDR was the tool of choice for cabling problems whether you were a phone company lineman or a spacecraft engineer. He collected some TDR data on our cable, and it looked okay. He filed the data away in his office.

A couple more weeks went by, still with no progress. Time was almost up, and we were getting desperate. Hoping one day for new inspiration, Leo went back to the TDR printouts that were still sitting on his shelf. Everything looked fine, as it had before, with no breaks or short circuits in the lines.

Then, as he scanned the data more closely, something seemed subtly wrong. For one short segment of the cable, the part that wrapped around the base of the mast, the impedance didn't look quite right. Impedance is a simple measure of how much a cable resists the flow of electrical current. Leo knew what the impedance was supposed to be in that part of the cable. The TDR data said it was about 25 percent higher than it should have been.

Could this be the problem? It wasn't much of a difference, but Leo's instincts were telling him that something was wrong, and he decided to follow them. He grabbed his TDR rig, took it down to SAF, where *MER-2* lay in pieces, suited up, and checked into the high bay. After an hour or two of testing, he had confirmed it: the impedance of that part of the cable was too high. Something was wrong.

It was late on a Friday afternoon. He made a quick phone call to Ball Aerospace, in Boulder, where the cable had been built. Monday morning they faxed him the drawings. Sure enough, they showed an error. The wires in the cable came in pairs, twisted around one another. Somehow, when it was built, the wrong pairs had been twisted together. This gave the cable the wrong impedance, the kind of thing that might make data transmission through it go flaky at low temperatures. Could this be what was causing the speckling?

Leo sent out the news, in an e-mail that I received in Ithaca during the last week in January. I was amazed at the kind of intuition that could allow him to sleuth out such a subtle problem from such esoteric data. But there was no way of being certain that this was the cause, Leo said, without another test. If we wanted to be sure, we'd have to replace the cable, put the rover back together, send it back up the hill, and do another thermal vac to see if the problem was gone.

The schedule impact was terrifying, but Matt decided there was no choice but to delay shipment to Florida so we could do it. It wasn't just the camera speckle; there had also been problems with the mechanical deployments during thermal vac, and we had to make sure that those had been fixed, too. So back up the hill we went.

The second MER-2 thermal vac began on Friday, February 14— Valentine's Day—with rover standup and deployments. I arrived at the test lab at 6:00 A.M. on Saturday, in cold, wet, raw weather, to be greeted by Rob Manning, with a look on his face that said I'd better sit down. Rover standup, Rob explained, had gone fine. But when the time came to deploy the mast for the speckle test, it couldn't be done. The reason, I was alarmed to hear, was that a short circuit had developed in the shoulder joint of the arm, and the short also affected the motor that deploys the mast. If we tried to drive the mast up in those circumstances, there was a chance we could blow a fuse, or even burn something up inside the REM. The mast had to stay down.

This was a colossal screwup. It wasn't like nobody had known about the short in the arm. It had been found a week before, down in SAF. Anybody who knew about the short and who understood the system design would have realized that we couldn't drive the mast up under these conditions. Yet we had sealed up the chamber with the mast still down. We were rushing so badly now that nobody who knew about the short had talked to anybody who really understood the system. It was a major communication breakdown.

So now we faced a choice. One was to break chamber, deploy the mast by hand, go cold again, and do the speckle test. That would take two days, at a time when Matt's schedule showed just ten days of margin left until launch.

The other was to go ahead and do the test now as we were, taking the pictures with the mast stowed. Nobody could think of a convincing reason why leaving the mast down should make a difference. But the cable where Leo had found the problem was the one that wrapped around the base of the mast, and it'd be twisted into a very different configuration with the mast stowed than it would with the mast deployed. As Rob put it, even if we saw no speckle, the test would "leave a lingering stench."

We caucused in a cold wind around a metal picnic table for half an hour—me, Matt, Barry, Rob and Richard Cook—and we decided to move forward with the test. If we saw speckle we'd open up the chamber and start searching again, our chances of getting to Mars probably gone. If we saw none, we'd head back to the picnic table.

The Pancam guys got to work, and within an hour the result was clear: the speckle was gone. Leo had most likely been right. We retreated again to the picnic table, this time with Pete Theisinger dialed in by cell phone. Nobody was comfortable, and we all wanted to open the chamber, deploy the mast, and do the test again. But when we balanced what was left of the speckle risk against all our other problems, it was time to move on. We'd know for sure that we had fixed the problem when *MER-2* got to Mars.

On February 1, the space shuttle *Columbia* broke up over Texas, killing the seven crew members aboard. It was a stark reminder that nothing is routine in the business of space exploration. Compared to the *Columbia* tragedy, our little tribulations felt suddenly insignificant.

10 THE CAPE

A T 5 A.M. ON February 22, 2003, a truck convoy left JPL, carrying the *MER-2* rover to Florida. In a nice dramatic touch, a magnitude 5.4 earthquake hit southern California just forty minutes
before the trucks rolled. Trucks and spacecraft survived the quake unscathed, and Rob Manning blew a stirring bugle call as the convoy pulled
away.

When you think of Cape Canaveral you usually think of Kennedy
Space Center, the home of the Space Shuttle. But most of the rockets that
launch from the Cape, including the Deltas that were slated to send us to
Mars, actually lift off from Cape Canaveral Air Force Station, just down
the coast from KSC. Like launch sites the world over, both KSC and
CCAFS are huge. Whenever you're dealing with extremely large explosive objects, you want them to be very far away from one another whenever the fuse on one is lit.

Our new home at the Cape was the PHSF, the Payload Hazardous
Servicing Facility, on the Air Force side. *Hazardous* is part of the name because the PHSF is set up for dealing with the many kinds of unpleasant
chemicals and devices that spacecraft tend to have in and around them.
Our RADs are solid rocket motors, and solid rocket motors can do very

bad things if they go off by accident with a bunch of people standing around. Each rover carries several radioactive substances in it, including the crucial radiation sources in both the Mössbauer and APXS instruments. You've got to be pretty careful when you're dealing with radioactive materials. And the hydrazine fuel that gets loaded into our propellant tanks before the whole stack goes out to the pad isn't pleasant stuff to work with either. The PHSF is built for dealing as safely as possible with these kinds of things.

The PHSF is a tall, windowless monolith of a building, surrounded on all sides by a lot of Florida swampland. Its main distinguishing feature is an enormous door on the east side, facing toward the Atlantic Ocean a few miles away and tall enough to admit the upper stage of a rocket with a big spacecraft on top of it. On foot, you enter the PHSF via a small door on the south side, stepping first into an unexceptional-looking anteroom, where a stern-faced security guard inspects your badge carefully and then, if you pass scrutiny, takes it from you and mounts it on a rack on the wall. This makes sense; once you're inside the high bay you're inside the high bay, and nobody needs a badge in there. And if anything bad ever happens, they'll want to be able to look at that rack to know who was inside when it happened.

You step through a door to your left, picking up your radiation monitor as you go, and into a changing room that's ringed by ordinary metal lockers. Suiting up for the PHSF high bay is not like the simple smock-and-booties procedure back at JPL. Here you wear a head-to-toe "bunny suit" made of Nomex, the flame-resistant fabric that's used in the suits that Indy car drivers wear. The first thing to go on is your head covering, a balaclava-like deal that leaves only your eyes and nose visible. (After a few days in the PHSF high bay you get very good at identifying people by the color of their eyes and the shape and size of their nose.) Next comes the coverall, a one-piece white jumpsuit with heavy olive-green straps sewn across the shoulders and down the thighs. I never found anybody who knew for sure what the straps are for; the best theory we could come up with was that they're handles for dragging a flaming body out of the rubble if something went boom. Once the coverall is on and zipped, next come the Nomex boots, zipped up the back and fastened below the knee with an elastic bungee cord. You quickly learn that the bungees go slack after hours of walking around, so from your second day onward a strip of shiny orange kapton tape goes around the top of each boot, snugging everything down tight. A metal grounding strap goes onto one wrist, a long telephone-like cord with an alligator clip hanging from the end, so you can ground yourself once you're inside and not create tiny electrical sparks when you're near a sensitive piece of equipment. Surgical gloves go on your hands, with more kapton tape around the wrists. And with that, you're ready to head inside.

You step onto a metallic grate. A sensor, detecting your presence, activates a set of stiff brushes that slide noisily through the steel slats of the grate, cleaning the soles of your feet. You step next across a shiny, sticky blue mat, like a giant piece of flypaper, that captures any particles on your boots that the brushes didn't get. Then comes the air shower, bigger than the one back at JPL; your ears pop slightly as the room pressurizes and the air jets whoosh when the door closes behind you. There's never much talking in the air shower. It's your last moment to reflect before what comes next, and you can't really hear much over the air jets anyway. The jets go quiet, you exchange a last look with your partners and you step out into the high bay.

The PHSF high bay is where it all happens at the Cape. High overhead a bank of lights illuminates the room with a vivid, yellowish glow. A flashing blue light, like the one on a police cruiser, tells you that a spacecraft is powered on. Flight hardware is scattered around the room, each with a knot of white-suited ATLO techs around it. Power and data cables snake across the floor, marked with bright yellow and black safety tape. Overhead there's an enormous yellow crane that can be used to move spacecraft parts, or even to move an entire flight system once it's all stacked and ready to go out to the pad. Wireless headsets provide communications to the control room in an adjacent building. Hardware flows through the high bay, hour by hour and day by day, the many tributaries of the ATLO river merging together there in Florida for the last time. In all my life, there's never been a place I felt more privileged to be.

The whole feel of ATLO was different at the Cape. There was a tension in the air beyond anything that any of us had felt back at JPL. Security was tighter. Cleanroom procedures in the high bay were tougher. Tempers ran closer to the edge. Everything the ATLO guys did in Florida was for real, and most of what they did was for the last time. When somebody mated an electrical connector or torqued down a fastener at the Cape, that was it . . . either it was going to work at Mars or it wasn't. And the schedule pressure, which had been bad enough back in California, grew worse with each new problem and each slipped delivery. It made for a tense existence.

There wasn't much time for relaxation. What passed for civilization, and the only place we had to blow off a little steam, was Cocoa Beach, a strip of beachfront hotels and surf shops just south of the Air Force station. The ATLO guys had moved to Cocoa Beach in force when the first spacecraft shipped, most of them renting semi-seedy beachfront condos for the six months or so that they'd be stuck in Florida. When I was there I holed up in a place that was grandiosely named the Royal Mansions, a modest quasi-resort at the north end of the strip. Each day I'd rise at 6:30 A.M., an unnaturally early hour for me, grab a big cup of coffee at the 7-Eleven down the street and start the long drive up the coast to the PHSF. You spend a lot of time in your car when you're at the Cape, and I had filled the glove compartment of my rented Chevy with Springsteen CDs that I'd brought down from Ithaca with me. Every day I'd pull into the PHSF parking lot at 7:30 A.M. with music blaring and coffee cup drained, ready for twelve hours of getting ready to go to Mars.

The main job for the science team at the Cape was wrapping up the final integra-tion of the instruments with both rovers. Some of the ATLO tributaries weren't due to join the river until both of the rovers had already gotten to Florida. And two of these, in an agreement that I had carefully negotiated with Matt, were the sensor heads for our APXS instruments, the instruments we were counting on to study chemical elements in rocks on Mars.

One part of each APXS is an electronics board that lives inside the rover, and those we had delivered months before, back when the rovers were still at JPL. But the other part of each APXS is the sensor head, which lives out at the end of the arm. It's the part you actually touch against a rock on Mars, and it takes a really long time to get it ready to fly.

The reason it takes so long is calibration. Calibration is essential for any instrument you send into space. You're going into an unknown environment, measuring things that no one has ever encountered before. So how do you know you can trust what your instrument's telling you? The best way is to make some measurements with that very same instrument, before you launch it, of things that you know and understand well. Doing that is called calibration, and without it we'd never be able to figure out what our readings on Mars meant.

Of all our instruments, it was the APXS that took the most time to calibrate. We needed to make measurements with each APXS sensor head of every chemical element that we'd ever hope to detect on Mars. We also needed to make measurements on a whole bunch of different kinds of rocks, to make sure that we could get their compositions right. If we couldn't do it on Earth, we'd never be able to convince anybody, including ourselves, that we could do it on Mars.

The problem with all of this is that APXS measurements take a really long time. A decent measurement takes at least an hour, and a really good one can take most of the night on Mars. For calibration, where everything is for keeps and even "really good" isn't enough, we decided that each measurement should be forty-eight hours long—two whole days. With all those Earth-bound samples that we had to measure, at two days a pop, calibrating the APXS sensor heads took months. There was no way we could get the calibration done and deliver the sensor heads to JPL before the rovers left for the Cape. So we had to ship them straight from Germany to Florida.

On February 18, just as calibration was supposed to be wrapping up, I got a terse and disturbing e-mail from Rudi Rieder in Mainz:

> On Friday one of the alpha channels of FM1 (the instrument to be delivered in March) showed excessive noise. . . . Disassembly of the detectors revealed that one of the detectors has a crack. . . .
>
> We believe that flying these detectors bears a very great risk: if for any reason one of the remaining detectors becomes noisy we may lose a very large fraction of data, both alpha and X-ray, thus rendering the measurements useless.

This was serious trouble. The APXS instrument makes two kinds of measurements, one of alpha particles and one of X rays. We use X rays to detect most chemical elements, and we use alpha particles to go after carbon. The failure Rudi was describing threatened both the alpha and X-ray parts of the instrument. Obviously we wouldn't get decent alpha measurements if our alpha detectors were cracked. That was bad enough. The really bad news was that the noise that came from the cracked alpha detector was so severe that it had wiped out the X-ray measurements as well. We could replace the one cracked detector if we wanted to, but the real problem was that we didn't understand why it had cracked in the first place. If the same thing ever happened on Mars, it would take out the whole instrument. Realizing that we had a fatal flaw in our design, we had to go searching for a new kind of detector.

Within days, Rudi announced that he had found some new detectors. In fact, he had quietly and cleverly pursued some alternative alpha detectors months before, just in case we ran into a problem like this. Typical Rudi. But just finding them wasn't enough. The job of testing them and getting them into the instrument, in one hell of a hurry, fell to Rudi's young apprentice, Ralf Gellert.

Ralf has always described working with Rudi as "interesting." Rudi is the Mad Genius of Mainz, a white-haired wizard who can conjure up brilliant instrument designs from almost nothing. But while Rudi is brilliant he's not always practical, and somebody has to be around who can actually build the things that Rudi conjures. That somebody is Ralf. Night after night, for a couple of years, Ralf could be found hunched over his workbench in Mainz late into the evening, soldering iron in his hand and cigarette hanging from his lower lip, working out the bugs in Rudi's APXS design.

Ralf was born and raised in northern Hessen, and he's tall, dark-haired and ruggedly handsome. He's also quiet and modest by nature, nothing flashy, and a little bit shy when talking about most things. He smiles easily. But if you tangle with Ralf on a technical issue, it's another story. The guy is a Ph.D. physicist who knows how to build things that really work, and he does not suffer fools gladly.

Ralf is also not big on writing things down. At his core he's a hands-on lab rat, the kind of person who's never read a hardware or software manual in his life. He just goes into the lab and does what he needs to do, keeping all the details of whatever he's working on in his head.

We procured some of the new alpha detectors. Working for three weeks, to the point of physical exhaustion, Ralf got them into the instruments and began the final testing. He did almost everything himself in Mainz, as was his habit, but he couldn't do vibration and thermal vac; he simply didn't have the right facilities. The quickest way to get vibe and thermal vac done, we decided, was for somebody to carry the sensor heads to JPL, and to run the tests there.

"I'll just deliver it to JPL myself," Ralf said of the first sensor head when we talked about it on the phone, "and then go lie in the sun."

"You sure?" I asked. It seemed there ought to be some way of getting the hardware to JPL without making Ralf haul himself all the way across the ocean.

"I'm sure. You know, for Germans, it's something else to lie under a palm tree, even if it's just at the Saga Motel." I could hear the smile on his face as he said it. Ralf's single, and being away from home for a while would be no hardship. After weeks of nearly round-the-clock work to get Rudi's new alpha detectors in, I could understand why he wanted a break, even if it was just at a cheap motel in Pasadena.

Ralf brought the first instrument to JPL on a Friday, March 7, and by noon on Saturday the vibe testing that simulated the violent shaking of a Delta launch was done. Everything survived. Thermal vac came next. I was at the Cape with Göstar Klingelhöfer at the time, trying to help him deal with some problems we were having with the Mössbauer sensor heads, and I followed Ralf's progress as he worked through the thermal tests back at JPL via phone calls that I made every day from my office in the building next to the PHSF.

After four long days the thermal vac testing was done, too, and Ralf was ready to hand-carry the first APXS sensor head to the Cape (where, he knew, more palm trees were waiting). To make things a little more interesting, the second sensor head had just arrived at JPL, also hand carried from Germany, this time by Oleg Bogdanovski. Oleg was a Russian research assistant working in Mainz whose travel burden had been eased not by palm trees, but by the fact that he had a girlfriend to visit at Caltech.

On Thursday, March 13, I was suited up and in the PHSF high bay with Göstar when I heard the tolling of the telephone. It's impossible to hear a normal phone in there over the roar of the air-conditioning, so somebody had rigged the high-bay phone with a big metal ringer that clangs like the bell on a train when somebody calls. John Wirth, one of the ace ATLO dudes, picked it up.

"Hey, PI man, it's for you!" John shouted across the high-bay floor.

Göstar shot me an uneasy glance across the lab bench. If somebody goes to the trouble to track you down on the high-bay phone, it probably isn't to deliver good news. I walked across the floor with my usual hypercaution, being especially careful not to trip on the yellow power cords that wound across the floor next to the *MER-2* lander.

"Hello?"

"Hey Steve, it's Joel." Joel Rademacher was the lead payload engineer back at JPL, and he'd been working through the testing with Ralf. "We've got a problem."

"With the APXS?" I asked.

"Yeah."

"I thought all the tests had gone okay?"

"That was the first one," Joel explained. "This was the second one, the one that Oleg just brought over. We finished vibe and started up the thermal cycling this morning, and then something really bad happened."

"Is Ralf still there, or has he left for the airport already?"

"He's still here."

"Put him on."

I waited nervously, suddenly warm and uncomfortable in my bunny suit. Over the phone I could hear the vacuum pumps of the test lab back at JPL clacking away. Finally Ralf came on the line.

"Hey Ralf, what's up?"

"Well, I was about to go to the airport with the first instrument this morning." His voice sounded a little shaky. "But then I thought I should check on the thermal test. So with my luggage in my hand I went up and had a last look at the vacuum chamber, where the second instrument was chilling down. Just as I got there, someone came to me crying and screaming, 'Ah, we smell something, and there's a peak in the temperature!' "

"Wait a minute, slow down," I said, trying to follow. "The techs were standing outside the chamber and they *smelled* something?"

"Yes. And I also smelled it."

"The smell was coming from inside the chamber?" I was confused; the chamber's supposed to be sealed.

"No, no," Ralf explained. "The smell was coming from the GSE."

The ground support equipment: the electronics outside the thermal chamber that control the sensor head inside the chamber. At least that made sense, but it was hardly reassuring, since the GSE and the sensor head were connected by a cable that ran through a bulkhead in the chamber wall. If there was a smell coming from the GSE, then it probably meant bad things for the instrument, too. And that "peak in the temperature" that Ralf had mentioned had to be in the sensor head itself. You don't put temperature sensors on GSE.

"So it smelled like what, something burning?"

"Yes."

"What was it?"

"I don't know."

Ralf sounded awful. I tried to picture him, and to imagine how he must feel, slumped against the wall with the phone in his hand, lab techs staring at him and the vacuum pumps rattling away in the background. His work of the last two years had just literally gone up in smoke. I wanted to go easy on him, but I had to find out what was going on.

"Have you looked at the instrument yet?"

"No. The damn thing is still inside the chamber. It'll take five hours to warm it up."

"Is there anything you can do from the outside to figure out what happened?"

"Not much," said Ralf. "The only thing I know right now is that there's a short from minus twelve volts to ground somewhere in the sensor head."

Ouch. "So what do you think?"

"I think the thing is dead," he replied flatly. "We'll have to fly the flight spare."

The flight spare. We had one, built for an occasion just like this, but I knew that Ralf didn't want to fly it. I didn't either. The spare wasn't nearly as good an instrument as the one Ralf had been testing. It hadn't been calibrated. And, I knew, it had been put together in a real hurry. If the flight sensor head, which Ralf had assembled with all the care he could muster, had some fatal flaw in it, what kind of shape was the spare in?

There was nothing Ralf could do while the thermal chamber was warming up, and Matt had made it clear to me that we had to get the first sensor head to the Cape immediately if his guys were going to have any chance of getting it onto *MER-2* in time. So we hung up, agreeing to meet the next morning when Ralf got to Florida. He headed for the airport to catch his red-eye to Orlando.

That night, while Ralf worked his way east with our one "good" sensor head, I lay sleepless in my room back at the Royal Mansions, mulling it over. Was the good one really good? If the one in the chamber had suddenly died for some unknown reason, why should we trust the one that Ralf was bringing to Florida? I knew that Ralf was ready to drop off the one sensor head that he thought was good and then head straight back to Germany where he could do an autopsy on the dead one after he'd gotten some rest. But I also knew that I'd never be able to get the first one past Matt and Richard and onto *MER-2* unless we knew what had killed the bad one. If the thing could somehow develop a problem that resulted in a burning smell coming from the ground support equipment, could we even be sure that it wasn't capable of hurting something inside the rover?

An e-mail from Rudi that arrived during the middle of the night left no question where he stood: "I do not want Ralf to rush back to JPL now and try to repair anything," Rudi said. "I rather want him to stay in Florida and see to it that the remaining sensors make it safely onto the rovers. An investigation into the cause is necessary and will be done, but at home in Mainz, where we have proper facilities and tools." But I knew this wouldn't fly with Matt and Richard. And the more I thought about it, the more it didn't fly with me either. We needed to find out what had gone wrong right away.

Göstar picked Ralf up at the Orlando airport early the next morning, and brought him to the Cocoa Beach Holiday Inn where all the Germans were staying. He'd barely had time to shower when I arrived. We got together in the lobby, huddled on a brightly colored tropical print couch. Ralf looked as bad as I had ever seen him, haggard and worn. It was hard to have to tell him that I wanted him to head straight back to JPL and take apart his dead instrument, instead of getting the rest he needed and then flying home to Germany. But I had to, and I did.

"This makes sense," he said after a long silence. "But it is cruel for me."

"I know, Ralf," I replied. "But the instrument you just handed us may have a fatal disease, and we can't be sure until you do the autopsy on the other one."

Ralf grimaced. I could see that he needed some time to get used to the idea. He spent the morning lying on the beach, catching an hour or two of restless sleep. We got him the last seat on a plane back west that same afternoon, and he made his flight. He arrived at LAX at 9:00 P.M. Pacific time, in a torrential downpour, and drove back to the Saga. He'd had no real sleep in almost forty-eight hours.

At 9:00 on Sunday morning, Ralf met up at JPL with Joel Rademacher and Ron Burt. Ron had recently retired from JPL, after years as one of their top quality assurance engineers. QA engineers are not exactly beloved by the people who build spacecraft at JPL. It's QA's job to be the workmanship watchdogs. They're the ones with the sharp eyes and the clipboards who monitor every procedure that happens around flight hardware, from a crane lift of the biggest spacecraft to the torquing of the smallest screw. Everybody knows that QA is important, but it's only human to feel a little resentful toward somebody whose main job is to watch you and make sure you don't screw up. The Germans were particularly queasy about the whole JPL QA process, and Ron was the one QA guy whom they felt comfortable with. So we called him out of retirement and brought him in on a Sunday to help with the autopsy.

Ralf settled himself at a lab bench with Ron by his side, slipped a grounding strap onto his wrist and took the ruined sensor head out of its static bag. Hooking the instrument leads to an ohmmeter, he began the slow process of disassembly. Piece by piece it came apart with no change on the ohmmeter. The short was still there, somewhere inside the instrument.

Then, as Ralf removed the last of four screws that held together the tiny stack of circuit boards that lay at the heart of the sensor head, the short vanished.

He froze for a moment, unsure of what had just happened, and then squinted carefully at the instrument. He seemed to see a tiny glint, like a reflection from a whisker of metal, just at the threshold of visibility between the cooling plate for the X-ray detector and a pin that he knew was supposed to be at minus twelve volts. Could this be what had caused the short? Ron brought in a microscope, and they put the instrument under it. Sure enough, there was a minuscule sliver of metal there, strung like a tiny tightrope between the plate and the pin. They shook the cooling plate assembly lightly and the sliver didn't budge; it had been welded in place by the intense heat of the short circuit. They pulled it off with tweezers, dropping it onto a piece of kapton tape to preserve it. Ralf worked his way carefully through the rest of the instrument, checking for shorts and opens. There was nothing; everything else looked normal.

Could the instrument have survived this? Ralf carried the whole circuit diagram for both the sensor head and the GSE in his head; in fact, Ralf's head was the only place such a diagram existed. There was only one electrical component in the path of the short inside the instrument: a tiny component called a resistor that was there to impede the flow of electrical current. That resistor, Ralf knew, could handle up to one watt of power before anything bad would happen to it. And it was paired, Ralf also knew, with another resistor in the GSE that would burn up if you ever hit it with anything more than about a quarter of a watt. If there was a short circuit inside the instrument, it could create a sudden, potentially damaging burst of current that would flow through both resistors. But if the quarter-watt resistor in the GSE burned up first—creating the odor the lab techs had smelled—it might have broken the circuit and cut off the flow of current before anything really bad happened to the resistor that was actually inside the instrument.

A glimmer of hope that he might be able to save his creation began to form in Ralf's mind. He and Ron took the rest of the instrument apart, piece by piece, searching with the microscope for more metal slivers. There were none. They put the whole thing back together and tested it. It worked perfectly.

Ralf called me in Florida, with relief in his voice as he gave me our first piece of good news in weeks. Ron thought that the one sliver of metal was a fluke, Ralf said, probably a tiny shaving that had been stripped from a screw when he had done the final assembly back in Germany. There's no way you can inspect for something like that, since the thing you'd be inspecting for is deep inside the buttoned-up instrument. What you do instead is vibe and thermal testing, which is exactly what had uncovered the problem. The instrument we already had in Florida had passed vibe and thermal with no issues, so Ron couldn't see any reason why we shouldn't fly it. And if it was good enough for Ron, then it was going to be good enough for Richard, and for Matt, and for me, too. It really was okay to send the sensor head we had in Florida to Mars.

Knowing that it would take days to test our back-from-the-dead sensor head, and knowing that we were out of time at the Cape, Ralf was ready to get on a plane and take the thing back to Germany. We still could send the flight spare to Mars on *MER-1,* and our "Lazarus" instrument would be a good thing to have in the laboratory. At least it was well calibrated.

But I desperately wanted to send that well-calibrated instrument to Mars, and after I hung up the phone with Ralf I went to find Matt in his office to talk to him about it. And in that one conversation, all the trust that Matt and I had built up over the months of ATLO paid off. I *really* wanted to get the Lazarus instrument onto *MER-1,* and I had to have some time to test it before we did. I told Matt so, in very plain language. Matt knew he probably had the time I needed in his schedule, though just barely. And he also knew me well enough now to realize that I wouldn't be asking for something like this, so late in the game, unless it was really important. Matt assented. I called Ralf back, and I told him to get ready to bring the instrument to Florida.

Even with Matt's approval, it still wasn't a slam dunk. As soon as word got out that we wanted to fly an instrument that had been through a massive test failure, the JPL QA guys descended in force. They wanted circuit diagrams to prove that no other components had been damaged, and of course none existed . . . because Ralf never wrote anything down. Rudi quickly sketched out some drawings in Germany and faxed them to JPL, and the QA guys went for it. Joel vibed the instrument again, and began the thermal test. Ralf stayed with the instrument at JPL until he was convinced that the danger was past, and then flew back to Florida to help Matt's guys put the first sensor head onto *MER-2*.

Three days later, the thermal testing of the Lazarus instrument at JPL was done. With Ralf now at the Cape, it fell to an unfortunate test engineer named Jose Guzman to hand-carry it to Orlando, on a red-eye through Philly. I met Jose at the airport, and I took the instrument from him, wrapped up inside an ungainly looking cardboard box. I thanked him and wished him well, and I got into my car, placing the box very carefully on the seat beside me. I thought for a moment about trying to put the seat belt around it, but that didn't seem likely to work. So I drove very carefully out Route 528, keeping an eye on that precious cardboard box the whole way. And an hour and a half later, back at the PHSF, I had the pleasure of delivering the very last piece of our payload to the Cape.

About three weeks later, Matt's team discovered a potentially fatal flaw in something called the Telecom Services Board, deep inside each rover. This didn't pose a huge problem on *MER-1,* which NASA was about to name *Opportunity.* But on *MER-2,* to be named *Spirit,* the solar arrays had already been stowed for launch when they found the problem. This meant that fixing the TSB would require firing the pyrotechnic devices that were used to release the arrays. The ATLO guys fired the pyros, opened up the rover and went in and fixed the board. They had no idea what havoc this was about to create.

11 THE FUSE

O N MAY 13 I was back in my office in Ithaca when the phone rang.

"Hi Steve, it's Pete. We've got a problem."

Oh shit, what now? "What's up?"

"Down at the Cape on Saturday they were doing some chassis isolation impedance measurements on the cruise stage. They expected to see three ohms, and instead they saw 3.8 kiloohms."

English, Pete. "What's that mean?"

"It means we probably blew the SPG fuse in *Spirit*."

The single point ground fuse was something I knew about. It was actually pretty useless. It had been put into the design to protect touchy hardware against short circuits during ATLO. Bad things can happen when you have a lot of technicians working around a complicated spacecraft. Somebody can drop a screwdriver into powered-up electronics, and *bzzzt*, you've shorted it out and toasted something. The fuse had been put into the design to protect against that sort of thing. But what the designers who put the fuse in hadn't realized was that the guys who were responsible for the most easily toasted components in the rover were a step ahead of them—they had built in their own protection against shorts. So the fuse wasn't even necessary. But it was in the design now anyway, like it or not.

"What made the fuse blow?" I asked.

"We don't know yet."

"When did it blow?"

"We don't know that either, but probably last month. We have data that says the fuse was good up to the twenty-fourth of March, and we have other data that shows it was blown by April eighteenth. So somewhere in there, something bad happened."

The dates sounded familiar. "When was the TSB rework done on that rover?"

"They blew the solar array pyros on April fifteenth. That makes it a possible culprit." What Pete was saying was that pyro firings they'd had to do to get at the *Spirit* Telecom Services Board might have been what had blown the fuse.

"So that's where we are," said Pete. "Not for the faint of heart. I'll keep in touch as things develop."

This was bad. You don't accidentally blow a fuse on a $400 million spacecraft two months before launch and expect nobody to notice. In fact, the news was already reverberating at both JPL and NASA Headquarters. Nervous over all the IRT's warnings, and hyperaverse to risk after the *Columbia* tragedy, the decision makers on both coasts were haunted by the thought of another very public Mars failure. This was the kind of thing that, if we couldn't clear it up, could make them decide not to launch us at all. To me, a decision never to launch our rovers would be as bad as failure in flight—either way, we wouldn't get to Mars. But NASA and JPL management didn't necessarily see it that way. Failure in flight, they had learned, could bring public humiliation. A decision not to launch, on the other hand, would just be seen as prudent management. It wasn't really a money issue; by this time we had spent most of the money anyway. But there was no question: We'd be kept on the ground forever if we couldn't prove that we hadn't screwed up the hardware.

There were two questions we would have to answer before we'd have a chance of flying.

The first question was why the fuse had blown. Maybe it had blown when they had fired the solar array pyros, though nobody had a good explanation for that yet. That thought was scary enough, since we would have to fire those pyros again on Mars. Even worse, maybe it had blown because we had some kind of spooky now-you-see-it-now-you-don't short circuit somewhere inside the rover. Intermittent shorts are the kind of thing you can chase forever and never find. But if you have an intermittent short lurking in your spacecraft and it reappears in flight, it can kill you. We'd never be allowed to launch if we couldn't figure out why the fuse had blown, and prove why it had blown to a skeptical audience.

The second question was whether it was safe to fly now that the fuse had been blown. We were just two weeks away from the start of our first launch window. More to the point, we were just five weeks away from the *end* of our first launch window. Five weeks was not enough time to open the lander up, open the rover up, replace the fuse, put things back together and get out to the launch pad. So we needed to be sure that *Spirit* would work right with her fuse blown. If we couldn't prove that, then she wouldn't fly at all.

There was one sure way to tell if it was safe to fly with a blown fuse, and that was to test the hell out of a spacecraft that had a blown fuse. Fortunately, we had already done that, and we had the records to prove it.

The fuse had been blown by accident several times during ATLO, and each time that had happened it had taken a while to replace it. Nobody had thought much about it at the time, but when the ATLO guys went back through their logs they realized that a lot of their testing over the past year actually had been done with the fuse blown for one reason or another. It turned out they had simulated every phase of the mission, including launch, cruise, landing and all the key deployments that had to happen on the martian surface, with the fuse blown. They had run all of the telecommunications systems, Earth to Mars and Mars to Earth, with the fuse blown. In short, at one time or another they had made one space-craft or the other do with a blown fuse almost everything it knew how to do.

There were a few odds and ends that hadn't been tested with the fuse blown, but in a matter of days we ran *Spirit* through practically all of them. I made sure that we checked out all the instruments, and all the data looked good. We even did crazy stuff like put a camera into a test setup back at JPL where we could flip it back and forth between "fuse good" and "fuse blown" configurations, once per second, just to see if we could get it to do something weird. Nothing happened. Before long we were convinced that *Spirit* was as good a spacecraft with her fuse blown as she had been with it intact.

Why the fuse had blown, though, was a bigger worry. After much in-vestigation a theory began to emerge, and it was just what Pete had sus-pected: that firing the solar arrays' pyrotechnic cable cutters had caused a short circuit.

Almost every spacecraft has moving parts, and ours have a lot more than most. Moving parts have to be locked tightly in place during launch. Rockets shake their payloads violently as they roar into space, and you don't want an arm or a mast or a solar array flapping around and breaking itself as that happens. But once you're on the martian surface and it's time to get to work, you have to release all the "launch locks" that held your hardware down, so that everything can move the way it was designed to.

Any launch lock that's supposed to release something critical like a solar array has to be 100 percent reliable. If it doesn't let go of the hard-ware when it's supposed to, your mission is over. So what our spacecraft use, in many places, are things called pyrotechnic cable cutters. Cable cut-ters have been on spacecraft for decades, and they're simplicity itself. Think of a cable cutter as a guillotine powered by a shotgun shell. An ex-plosive charge is contained at one end of a short metal tube, like the bar-rel of a sawed-off shotgun. Jammed into the tube, right next to the explosive, is a hard metal plug. The plug is blunt on the end that faces the explosive, but it's honed to a sharp blade on the other end. The blade is aimed at a thin metal cable that holds the hardware in place. When it's time to release the hardware, an electrical pulse is sent to the explosive charge, detonating it. The explosion pushes the plug down the tube, the blade cuts the cable and the hardware is set free. Very simple.

It was starting to look like the fuse may have blown back in April when the ATLO guys had fired the cable cutters that released the solar arrays. They had never planned to fire those cable cutters on Earth. But by the time it had become clear that they needed to fix the Telecom Services Board inside each rover, the solar arrays on *Spirit* were already folded up and held in place with their launch locks. So the only way to get at *Spirit*'s TSB was to fire the cutters and open the arrays.

Only *Spirit* had a blown fuse. And, tellingly, only *Spirit*'s solar array cable cutters had been fired. So a lot of circumstantial evidence was pointing to those cutters as the things that may have done it.

But why would firing cable cutters blow a fuse?

A fuse is designed to conduct electricity. But if the electrical current that flows through it is too powerful for too long, the fuse is designed to blow. Blowing is very simple: so much current flows through the fuse that it melts, breaking the connection and stopping the flow. You put a fuse into a circuit to protect sensitive components that might be damaged by too much current. And if something happens that lets too much current flow, it's the fuse that blows first, breaking the circuit and protecting everything else from harm.

When the ATLO guys looked hard at how the solar array cable cutters fired, a theory for how the fuse might have blown emerged. Here's how it went:

When it's time to fire a cable cutter, a switch inside the rover, called a relay, flips. For thirty-two milliseconds—thirty-two thousandths of a second—current flows through the relay to the cable cutter. And then the relay flips back again, cutting off the current.

Thirty-two milliseconds is an eternity for our fuse. If you put a current that high through it for that long, it'll blow, no question. But nobody expected the current to flow for nearly that long. As soon as the current hits the explosive charge in the cable cutter, the charge detonates, and the explosion should break the circuit. It takes less than a millisecond for the charge to go off, so the current should flow for less than a millisecond. And we knew that the fuse should survive a pulse that short without blowing.

But here's the problem: What nobody had figured on was that the explosive used in the cable cutter has a metallic element called zirconium in it, and zirconium conducts electricity. When the charge exploded, the circuit was indeed broken, but only momentarily. An instant after detonation the explosive gases cooled, letting a zirconium-rich residue—capable of conducting electricity—solidify inside the cable cutter. So instead of conducting electricity for just the time it took to fire the explosive charge, the cutter would let current flow for almost the full thirty-two milliseconds that the relay was closed. And thirty-two milliseconds was long enough to blow the fuse.

The theory was easy enough to test. The cable cutters that had released the solar arrays back in April had been tossed away after they were fired. But if the ATLO guys could find them, all they had to do was see if they still conducted electricity or not. If the zirconium-rich gunk had condensed inside them, it should still be there, and they should still conduct. It took a while to track the cutters down, but they were found. And sure enough, they still were conductive. This was the smoking gun, and it was good news. The fact that they still conducted electricity convinced us that it was the firing of the cutters that had blown the fuse, not something much more sinister inside the rover.

So that was it. We had blown the fuse when we had opened the solar arrays, and the spacecraft worked fine with the fuse blown. We had to convince JPL management and NASA Headquarters that we had our story straight, but it looked bulletproof. We were still on our way to Mars.

All the action on the fuse was happening in Pasadena and in Florida, and things were moving so fast that when I was home in Ithaca it was tough to follow everything that was going on. The guy I really had to talk it all through with was Barry. Barry is an early riser. He gets up at about 5 A.M., and he's in the office by seven. I had learned a long time ago that if you want to have a nice quiet talk with Barry, the best thing to do is to call him on his cell phone at 6 A.M. Pacific, while he's driving to work. So the morning after the cable cutters were found, that's what I did, planning to pick his brain on how the story was going to be told up the line.

Instead, what I got was a high-speed half-hour soliloquy. Barry is wound pretty tight anyway, but I had never heard him this worked up over a technical issue. The words tumbled out of him all the way down the 210 freeway, his voice betraying both his fear of something that could kill our mission and his irrepressible engineer's joy at being confronted with a tough problem in desperate need of a solution. Yes, we knew now how the fuse had blown, and yes, we knew that the spacecraft worked fine with a blown fuse. But as they'd done the investigation, a far more sinister problem had emerged. There was a fundamental flaw in the design of our pyrotechnic system, and it was so bad that it could end the mission.

The problem was simple: the fuse wasn't the only thing that would see thirty-two milliseconds of high current when a pyro was fired. Most of the pyro firing circuits on the vehicle also had resistors in them called ballast resistors. As current flows through the ballast resistors they heat up, like the filament in a lightbulb. If you heat them a little, they're fine. But if you heat them too much, you can burn them up, just as had happened to the resistor that got toasted in Ralf's APXS ground support equipment. In essence, each resistor can act like a little fuse if you hit it with too much current. And like the fuse, the ballast resistors in the rover were designed to take only a brief millisecond burst of current. What Barry didn't know was whether they could survive a high current for a full thirty-two milliseconds.

So here was the nightmare scenario: Over the past year of testing, the ATLO guys had fired a whole bunch of pyros on both vehicles. Most of the pyro circuits had ballast resistors in them. And any pyro that developed a conductive residue inside it when it fired would subject its ballast resistor to a thirty-two-millisecond blast of current, which was way more than the resistors were designed to handle. A pyro might have worked fine during a test. The first time it was fired, all its ballast resistor had to do was survive long enough before burning up that the pyro would go off, and that took less than a millisecond. But with its resistor destroyed by that first firing, the circuit wouldn't work the next time we tried to fire it—in space.

And there was no way to measure which resistors might have been destroyed, because they were buried deep inside the rovers. We were out of time.

This was a very ugly new development, and it didn't take long for it to make it up the management chain. Unless we could prove somehow that every single pyro firing circuit on both rovers was going to work when we asked it to, there was no way we would ever fly.

The day I talked to Barry about the pyros was the day I started wearing my reindeer antler full time. I'm embarrassed to admit it, but I'm terribly superstitious about space missions. Yes, I know, it's utterly illogical. There's no way that what you wear can have any effect on a piece of hardware hundreds of millions of miles out in space. Still, I've had some bad experiences.

People in the space business love T-shirts. There's been at least one T-shirt for every space mission that's ever flown. And every time I have gotten a mission T-shirt—every single time—something bad has happened to that mission. Hardware has failed in flight. Launch vehicles have exploded. Spacecraft have vanished. Instruments have returned garbage data. For every space mission T-shirt I have ever owned, there has been some tale of woe. So at the very start of the MER mission, I made the decision never, ever, to wear a MER T-shirt.

The reindeer antler took superstition to another level. Years ago, Mary and I had taken a wonderful, long bicycle trip up the coast of Norway into Lapland. Somewhere in the very far north, at a little roadside stand, I had bought a simple necklace, the tip of a reindeer antler strung on a thin leather cord, from a Lapp craftsman. That same night, asleep with the necklace around my neck, a rockfall had hit our tent, a granite block the size of a brick crashing through the tent fabric inches from my ankle. My ankle intact, I got it into my head that my reindeer necklace was a good-luck charm, and I had worn it subsequently during several risky ventures in far-flung parts of the world. It wasn't clear whether good fortune in the mountains translated into good fortune in ATLO, but you can't be too careful about this kind of thing. I had worn my good-luck charm all through thermal vac, and we had eventually come out of that more or less unscathed. So if we were going to make it to the pad now, I figured I needed it full-time. I strung it on the strongest leather cord I could find, and I tied it tight around my neck.

On May 21, nine days before the start of the *Spirit* launch window, we met with JPL director Charles Elachi to update him on where we stood, since he'd have to make the recommendation to NASA Headquarters on whether or not to launch. JPL had been hearing bad rumblings from back East, and Gene Tattini, the former Air Force three-star who was Charles's deputy, summed them up: "What Weiler has been saying, to everybody who will listen, is 'You guys aren't going to fly for two years.' And I think he means it."

The reality, though, was that it was worse than that. The option to fly in two years that Gene was talking about wasn't much of an option. The fortuitous alignment of Mars and Earth in 2003 made it the best launch opportunity for the next eighteen years. In fact, that great geometry in 2003 had been a big part of why NASA had picked us to fly in the first place. But in 2005, the next chance to go to Mars, the geometry was terrible. If we launched in 2005, we'd arrive at Mars when the planet was far from the Sun, and when it was almost as far away from Earth as it ever gets. Solar power would be bad, and communication to Earth would be awful. The mission was so bad in 2005 that it wasn't clear that it made sense to fly it at all.

The meeting with Charles lasted for seven hours. The engineers convinced him that the spacecraft would work with the fuse blown. They convinced him that firing the cable cutters was what had blown the fuse. The ballast resistor thing was still a mess, but they convinced him that we might still be able to survive it if a bunch of things broke our way.

At the end, Charles laid it out: "I came into this meeting today with the attitude that I'll tell you guys to terminate and launch in two years." I caught my breath. Even Charles had been thinking about killing us off? This hit hard; through everything up to this moment, Charles had been the one manager on either coast who had never said aloud anything to suggest that his faith in us had been shaken. Was he saying this now just for effect, or did he mean it? I couldn't tell. He went on, "Now, I feel that we ought to continue. That doesn't mean that we'll launch. But I didn't hear anything today that says that I should stick to my original intent and terminate this thing."

We were still alive. But we were in very deep trouble.

The next day we took it to Weiler, on the phone. The conversation went to hell from the start.

Charles began, "Ed, I want you to know that I'm following a very methodical process of getting to the launch decision."

Ed broke in immediately. "Charles, we haven't even started and you're already assuming we're having a launch! We're not having a launch unless—"

"Ed, Ed, wait a minute. I've already decided that I'm not going to recommend a launch. I'm not going to come to you with a launch recommendation unless we can resolve three things. First, what was the root cause of the blown fuse? Second, did it create a stress that could have damaged the vehicle? And third, can we operate safely in either configuration? If we can't resolve all of those questions, we're not going to launch.

"I went into yesterday's meeting convinced that we should terminate right now, and launch in two years." Charles was ignoring for the moment how bad 2005 was. "But after I listened to everybody, I came to the conclusion that we ought to proceed with integration with the launch vehicle.

"That doesn't mean we're going to launch! But I'm convinced now we can operate with or without the fuse. That's resolved. On root cause, there's a high probability that we know the cause, but a little more work needs to be done over the next couple of days. And on the resistor stress issue, additional work is definitely needed. So I recommend that we go ahead with the processing at the Cape. The baseline is that we won't launch. But if appropriate, we can change our minds and decide to launch at the time of the Mission Readiness Review next week."

We had just a few days to get our story straight and convince both Elachi and Weiler that we were ready to fly. If we couldn't do it, we were done. And the launch was delayed until the eighth of June at the earliest. A third of our launch window was gone before we'd even headed for the pad.

It was time to get back to Florida. On Tuesday the twenty-seventh, three days before the Mission Readiness Review, I pulled up at the PHSF in my rental car at one o'clock in the morning. It was dark and muggy, with mosquitoes swarming in the lights. At 2:15 A.M. the big doors of the high bay slid open and six of the ATLO guys, still dressed in their white flame-retardant bunny suits, wheeled the spacecraft out into the night. They had been busy in Florida while all the talking was going on in Pasadena and Washington. *Spirit* was mounted atop the third stage of the launch vehicle now, the whole thing swathed in silvery protective plastic. Blue lights flashed, and a black-clad SWAT team member with a bomb-sniffing dog worked his way around the vehicle. Our machine was worth hundreds of millions of dollars, it was explosive and it was radioactive; nobody was taking any chances. The ATLO guys hooked it up to a big light-green semi, and off it crept into the darkness, at three miles an hour.

It's quiet at night in the Florida swamps, and it was a peaceful trip out to the launch pad. With my windows rolled down I could hear the waves lapping along the causeway, and as I drove I could see the lights of the other pads, strung out northward along the coast. We pulled up at Space Launch Complex 17 at 5 A.M., with a tiny crescent moon hanging over pad 17A. Searchlights blazed around the pad, lighting up the huge scaffoldlike gantry that surrounded our sleeping Delta II.

The truck inched up to the base of the gantry and parked beneath it. The time had come for what the launch team, without a trace of irony, calls an "erection and mate procedure." As the sun rose over the Atlantic, a crane atop the gantry lowered a hook, and slowly our spacecraft was lifted to the top of its rocket. Almost sixteen years since the start, we finally had hardware on the launch pad. I tried to savor the moment, and briefly to forget all the doubts that still hung over us.

The last three days before the review were terrible. There were more than a hundred pyro firing circuits on each vehicle, and the ATLO guys had to exonerate every one of them. There were only three ways to do it: A circuit was safe if it didn't have any ballast resistors in it. There weren't many circuits like this, but fortunately the circuit for the solar array cable cutters was one. So they hadn't burned up any resistors back when they blew the fuse, because there weren't any resistors in that circuit to burn up.

A circuit was safe if it had never fired a live pyro. The ATLO guys hadn't fired all the pyros on both vehicles, so this covered a bunch of them.

A circuit was safe if they could find the actual pyro that it had fired, test it and prove that it didn't conduct electricity.

Matt and a bunch of his team spent those three days and nights working almost around the clock, running through circuit diagrams and test logs, tracking down all the circuits that were safe because they had no resistors or because they had never been fired. The rest of Matt's team went on a desperate bicoastal Easter egg hunt, searching through drawers, boxes and bags at JPL and the PHSF to find every fired pyro they could get their hands on.

By the day of the review, Matt and his team had cleared every pyro circuit for flight but four. On *Spirit*, nobody could find pyros that had been used during ATLO to release the arm and the high-gain antenna. So those circuits were in question, and that was a big problem. If we couldn't release the arm on Mars, three instruments and the RAT would be useless. And if we couldn't release the high-gain antenna, we could communicate directly with Earth only at excruciatingly low rates over the low-gain. On *Opportunity*, the story was even worse. The two missing pyros there were the ones that released the rear wheels. If those failed on Mars, *Opportunity* would become a lander mission.

We told the story to the review board, and their verdict was predictable. If we couldn't find the rest of the missing pyros, our only hope was to try to test our way out of this mess. That meant taking ten resistors for each suspect pyro circuit, wiring them up just like they're wired in the rovers, and hitting each of them with a thirty-two-millisecond current pulse that was even more powerful than what they'd see in flight. Only if they survived would everybody be convinced that the ones on the rovers were okay, too.

The problem was, nobody was very confident they'd survive.

The weekend was a frenzied one at JPL. The first news came on Saturday, when some West Coast ATLO guys found the two pyros that had been fired months before to release the rear wheels on *Opportunity*. They had been in a bag in the Spacecraft Assembly Facility that somehow hadn't been searched before. A quick test showed that the pyros were open—that they didn't conduct electricity. *Opportunity* was in the clear and ready to fly.

But of course it was *Spirit* that was out on pad 17A, and on *Spirit* we still had the problems with the arm and the antenna. There was no choice but to suck it up and actually test the resistors. Everybody was scared to death, but by late Monday it was done. Against all expectations, they passed. The resistors were—just marginally—tough enough.

It was time to go to NASA Headquarters and tell our story. The West Coasters flew east and the Florida crew flew north. We converged in MIC-5, just two floors below where the original headquarters shootout had been three years before, at 7:30 A.M. on June 3, to tell Weiler we were ready to fly.

The mood was not congenial. Ed's job at a meeting like that is not to be nice, and it is not to be a cheerleader. The reason for this is pretty simple: It would be his neck on the line if we failed, not ours. "I want you all to realize that I am completely comfortable not launching this thing," Ed said when we gathered in his conference room. "Your job is to convince me otherwise."

The resistor story was clean, and we knew it. We laid it out, and he bought it. He had to. The fuse thing finally seemed to be behind us.

But then Ed spotted another problem.

Over the course of ATLO, we had powered up the transponders that we use in each rover to transmit data to Earth about six hundred times. All but two of those times they had worked. But on each rover, just once, the transponder had refused to send any data when it came up. The two times this had happened, the ATLO guys had done what anybody does when their expensive high-tech hardware doesn't work: They turned it off and turned it back on again. And both times that fixed the problem. Nobody understood why.

Understandably, Ed didn't like this. His job before approving a launch is to probe for potentially fatal flaws, and the transponder looked to him like a candidate. None of us had anticipated this, mostly because we'd been so focused on the fuse and the pyros, and we didn't have very good answers for him. After almost three hours of asking probing questions and hearing responses that he didn't like, Ed couldn't take any more. "Charles, you know that this could take out the whole mission, and yet you're coming to me with a recommendation to launch. How can you do that?"

Charles couldn't take any more either. He snapped back, "Because when I make recommendations like this, Ed, I don't listen to paper-pushing Washington bureaucrats. Instead, I listen to the advice of experienced JPL engineers."

"Dammit, Charles," Ed responded, "if some of us 'idiot paper-pushers' here in Washington had asked a few more dumb questions back in '98, maybe we wouldn't have lost two spacecraft!"

Things were getting out of hand. It was sounding like Ed was ready to slip us to 2005, or worse. I jumped in and reminded him just how bad 2005 was compared to 2003. It backfired.

"Careful, Squyres," said Ed, looking me in the eye. "You may not be doing what you want to do. You're convincing me that I should never launch these things at all. In fact, I think they'd look pretty damn good over in the Air and Space Museum."

And with that, they went into a closed-door executive session, just the brass from JPL and Headquarters.

Richard, Matt and I paced restlessly in a nearby conference room. There was nothing to do; we tried to occupy our minds by inspecting the multicolored "fever charts" that were thumbtacked to the walls, depicting NASA's latest take on the health of each of their Space Science missions. Green meant life was good. Yellow meant that a project had problems that might be solvable and red meant problems so bad nobody knew how to solve them. There wasn't much green on that wall, and where it appeared was for mostly projects that hadn't really gotten started yet. There was an awful lot of yellow and red up there, though, and it was easy to see where the nervous mood that seemed to permeate NASA Headquarters must come from. I wondered morbidly what came after red—maybe black?— as the executive session dragged on to an agonizing twenty minutes, then thirty. Finally, after forty-five minutes, Ed emerged, marching by me without a word. I crept back into the conference room, ready for the worst.

But we were okay. Chris Scolese, Ed's deputy, was still in the room, and he explained what had happened. Chris is an engineer, and he has managed space flight projects. What Chris knew is that practically every spacecraft that's ever flown has had some kind of weird problem that popped up once or twice during testing, never to be seen again. You have to take some risks in this business, and the risk we were taking with the transponder was lower in Chris's judgment than the risks we'd already decided we were willing to take on launch day and landing day. Chris had told Ed that he thought we should fly, and Ed had accepted Chris's advice. But it had been a tough call by both of them.

The last step was for Ed to take it to Sean O'Keefe, the new NASA administrator. I headed back to Florida, and when I touched down in Orlando there was a crisp, joyful voice mail waiting for me from Pete: "You're goin' to Mars, buddy."

I slept well that night. It was the first good night of sleep I had had in weeks.

The next morning, as I pulled into the parking lot at the PHSF, I ran into Chris Voorhees, one of the youngest of the ATLO guys, carrying a big cardboard box full of bowling shirts. A few weeks back, the whole ATLO team had been inspired to have flashy black-and-yellow bowling shirts made up. The shirts were very slick, with the owner's high-bay nickname embroidered on the front and a drawing of the rover on the back, instrument arm extended and clasping a bowling ball. Everybody was going to Shore Lanes down in Merritt Island that night to try out their new shirts, and they invited me to come along as the token scientist. It was time to unwind a little.

I pulled up at the bowling alley in a downpour. After a couple of games, it was pretty obvious from the numbers up on the scoreboard that we were a group that didn't get out much. But beer flowed, chicken wings were consumed and a few bowling pins fell. It was the loosest I had seen everybody in months.

Near the end of the third game. Mark Boyles's cell phone went off. Mark was our mission assurance manager, which meant he was the guy who was stuck with keeping track of everything that was wrong with each spacecraft. I watched as he took the call, pacing up and down by the pinball machines, his face growing dark. He clicked off.

"Shit."

"What is it?" I asked him.

"We screwed up the resistor tests. The current was too low."

"We have to do the tests all over again?"

"Yep."

Oh God. "How much were they low by?"

"About an amp."

That didn't sound like much. "What'll happen if we test them at an amp higher?" I asked.

"They'll blow," replied Mark. He sounded very sure.

Two terrifying days passed. We were back on death watch again. It was awful; we had been so close. Nobody wanted to do the resistor tests again. If we failed, it wouldn't prove that the resistors in the rover were bad. But we all knew that a failure would be enough to ground us. Nobody would say it, but we all felt the same way: Somehow we had to find a way to quit testing, and to get our birds out into space where nobody could stop them.

Thursday morning, news came from JPL. They had found the pyro that had released the instrument arm on *Spirit*, and it was open. All that left was the high-gain antenna. Meanwhile, out on pad 17A, the launch crew was fueling the second stage of our rocket.

Pete and I huddled in his makeshift Florida office, trying to figure out how to convince Elachi and Weiler that we could make the mission work even if the high-gain antenna never released. Could we get by without the antenna at all, and just use the orbiters to relay data? Could we point the antenna at the Earth by turning the rover? It sounded desperate, but convincing everyone that it could work might be what it would take to get us off the ground.

And then Matt burst into Pete's office. One of his guys had found the last pyro—the one that had been fired to release the antenna. It had been inside a bag, inside another bag, in a locker in the PHSF. They had tested it, and it was open.

It was all over. We really were going to Mars.

12 LAUNCH

THE LAUNCH OF *SPIRIT* finally came in hot, hazy sunshine on June 10, 2003, after two days of scrubs forced on us by the thunderstorms that rumble across central Florida almost every summer afternoon.

A rocket launch is a frightening, beautiful, violent thing to watch. Earth's gravity cannot be overcome without great exertion, and when you actually see something ascend into space with your own eyes, the thing that surprises you about the process is that it *ever* works. Rockets are an outlandish way to reach into the heavens, really, and it's astonishing to me that humans have mastered them to the point where they're almost always successful.

Of course, it's that "almost" that's the issue. There's not much for scientists to do at a rocket launch, or even for most of the engineers who built the spacecraft. Once you hand your precious creation over to the launch team, about all you can do is wait for things to take their course and hope for the best. It's like a strange game of roulette, and if you compare the real record of launch vehicles to a game of roulette in Vegas, the odds go something like this: Put down your four hundred million bucks, spin the wheel and as long as it doesn't come up double zeroes, you get to keep your money and keep on playing. Hit that "00," though, and everything's gone. Around JPL, it's what's called "a bad day at the launch pad."

Our vehicle was the Delta II, built by Boeing. From a distance, the Delta's a sleek, elegant-looking machine. Up close, it's a bit of a brute. The first stage is a slender blue-green shaft. Its one main engine is fed by highly refined RP-1 kerosene and super-cold liquid oxygen, or LOX, delivering more than two hundred thousand pounds of thrust. Put differently, the thrust from just that one engine is enough to hurl a hundred tons of hardware upward off the surface of the Earth. The first stage is ringed at the base by nine slim, white, solid rocket boosters, each capable of putting out more than a hundred thousand pounds of additional thrust. The second stage sits atop the first, and when called upon it burns a different, volatile mix: hydrazine and nitrogen tetroxide, two very nasty chemicals. One final stage, firmly affixed to the spacecraft, uses a single solid rocket motor to kick the spacecraft out of Earth orbit and on its way to Mars.

Despite appearances, the Delta is really a flimsy thing. Rockets are made that way on purpose. Weight is the coin of the realm in the rocket business, and every ounce of structure beyond what's necessary to hold the thing together reliably means that much less payload you can put into orbit, or on the way to Mars. The stuff on top of a rocket is called "payload" for a reason: The more load the rocket can deliver, the more the customer will pay. So rocket engineers place enormous importance on stripping every unnecessary bit of hardware off a vehicle. Poised for launch, more than 90 percent of a Delta II's weight is propellant.

The familiar backward-from-ten countdown of a rocket launch is really the culmination of an intricately choreographed ballet of humans, software and hardware that begins days before. A modern launch vehicle has tens of thousands of parts, all of which must function in just the right way—and in just the right sequence—to make the rocket fly. It's a problem of complex synchronization, and the technique that's been perfected through decades of experience is for the launch crew to follow a rigid, painstaking, by-the-book and by-the-clock procedure. All the team members—some from Boeing, some from NASA and some from the Air Force—have precisely defined roles to play. Procedures are honed, and crews kept sharp, by launch simulations that replicate every kind of emergency and contingency that the sim supervisors can devise.

On launch day, with the countdown clock at two and a half hours before liftoff, klaxons sound and the crew at the pad is evacuated. Twenty minutes later, with everyone at a safe distance that's measured in miles, ten thousand gallons of RP-1 begin to flow into the first-stage fuel tank.

Next, with fueling completed, loading of LOX begins at T minus 75 minutes. At T minus 33 minutes the hydraulic system that's used to steer the first stage's huge main engine nozzle is powered up, and the nozzle is put through a comprehensive series of slews to check out its motion.

Then, at T minus 20 minutes, the countdown stops. This is the first of two "holds" built into the count, and it gives everybody on the launch team a chance to catch their breath, to get synched up again if they're not and to assess, in a calm, measured way, their bird's readiness to fly.

The count resumes. Final checkouts of onboard systems, topping off of helium gas used to pressurize the fuel system and pressurization of the fuel tank itself follow in quick succession. There's another built-in hold at T minus 4 minutes, and the spacecraft atop the rocket is cut off from the launch complex's power supply, transitioning over to its own internal batteries. In the final few minutes, all of the pyrotechnic devices on the vehicle are switched from a safe state to armed and ready to fire. LOX is topped off, valves close and the final seconds of the countdown begin.

A split second before liftoff, RP-1 and LOX pour into the main engine's combustion chamber at tens of gallons a second, reacting explosively to generate the engine's thrust. Then, once the main engine thrust is steady, six of the nine solids ignite. When that happens, it's showtime. Solid rocket motors have no off switch and no throttle. Once they light, that's it—they just rock and roll until the propellant's gone, and there's nothing anyone can do to make them stop. Solids work, most of the time, and their beauty is in their simplicity. They have no pumps, no valves and no moving parts. But there's no control, either, and that makes them scary things. It was solids that took down *Challenger* in 1986, and solids that caused the worst Delta accident anyone's ever seen, a decade later, during the launch that followed *Mars Pathfinder*'s. That one took out a Global Positioning System satellite, along with a couple of dozen cars in the parking lot at LC-17. I've never been to a launch where I didn't breathe easier once the solids were gone.

The six ground-lit solids come off the vehicle a minute into flight and ten miles high, dropping into the ocean. There must be hundreds of them there now, littering the seafloor just off the Florida coastline. The three solids that remain light off in midair, burning for another minute before they too fall away, from an altitude of thirty miles. The first stage keeps firing for two minutes more, accelerating at a crushing eight g's until the moment called MECO—main engine cutoff—when both RP-1 and LOX are nearly gone. Propellants spent, the first stage separates from the rest of the vehicle and begins a long fall to the ocean from a height of almost seventy miles.

Next the second stage kicks in. By second-stage engine cutoff—SECO 1—less than five minutes later, Florida is receding in the rearview mirror at more than seventeen thousand miles an hour. A fifteen-minute coast takes the vehicle clear across the Atlantic, through sunset and into the night. Reigniting over the African coast, the second stage fires again for a short two-minute burst, adding another couple of thousand miles an hour to the velocity. Final shutdown of the second stage—SECO 2—and ignition of the brutal third-stage solid engine take place in darkness above the Kalahari Desert. Ninety seconds of violent acceleration later, pyros fire to separate the spacecraft from the burned-out third stage. And if everything has gone exactly the way it's supposed to, you're on your way to Mars.

I watched *Spirit* launch, with my family and a bunch of the Cornell faithful, from KARS Park, a recreation center near KSC that NASA maintains for its employees. There were hundreds of NASA guests there, crowded together along the marshy shoreline. Pad 17A was just visible through the haze, the tower rolled back and our rocket standing alone. An alligator cruised silently offshore. NASA had piped the audio from the launch control room through loudspeakers mounted on the telephone poles, so we could follow the count. As each milestone ticked by I distracted myself by chatting with launch guests: Sofi Collis, the little girl from Arizona who'd won the NASA contest to name the rovers, and one of the sewing machine ladies from Dover who'd made our airbags. Everything went by the timeline.

As we came out of the T-minus-four-minute hold, the tension level in the crowd began to rise, subtly but perceptibly. And somewhere around T minus 90 seconds, a sudden, intense realization came over me that it was really about to happen. Sixteen years of work and hope were about to come down to one brief moment.

And then, they did. There was a sun-bright explosion of white-yellow light at the base of the vehicle, and a billowing of steam. The rocket was just a thin, white-tipped, blue sliver from this distance, and it rose soundlessly and with what seemed like impossible speed from the pad. The crowd cheered; I just held Mary against me, quietly and tight, and watched it fly. I think it was a long time before I breathed.

A Delta II liftoff is nothing like the slow rumbling of the old *Saturn V* moon rocket, or even like the stately rise of a Space Shuttle. The Delta is a nimble hotrod of a rocket, and it leaps off the pad. Its sound, by the time it finally gets to you, is more of a high-frequency crackle than a throaty roar. You're miles away from the thing at ignition, so the sound can take fifteen, twenty, thirty seconds to get to you. By the time it does, the rocket is *away*—supersonic and moving out, head-tilted-back high in the sky.

Watching our first Delta fly was strange. This was the moment I had anticipated for so many years, for longer than my children had been alive. But as *Spirit* climbed away from us, my feelings were nothing like what I had expected them to be. Through all the years of hope and frustration, through all the proposals and all the crises, I had always drawn sustenance from the dream of actually making it to that moment in Florida. The launch of our rover was something I had played over and over in my mind for years, and in this endless mental film loop my feelings were always ones of pure, unalloyed joy as the rocket rose into a perfect blue sky.

The reality, on that hazy, muggy day in June, was very different. There was joy, to be sure—joy and a profound relief that nothing and no one could stop us from getting off the ground, ever again. But there was fear, too, and a strange sadness. *Spirit* was leaving forever, and the best we could wish for her was a one-way trip with certain death at the end. We had done our best to prepare her for the dangers she would face, but had we done enough? We would hear from her in the months ahead, reading her temperatures and her voltages, and viewing her pictures and spectra if she made it safely to Mars. But none of us would ever see her again with our own eyes, and that made me surprisingly sad. It was hard to let go.

The ground-lit solids fell away after a minute of flight, tiny white needles tumbling end over end in clean arcs toward the Atlantic. The Delta was gone from sight within another few minutes, and we quickly crowded under a blue-and-white plastic canopy to squint against the sunlight at the barely visible screen of a cheap video monitor tuned to NASA TV. The rest of the events in the launch sequence ticked off with precise perfection—separation of the air-lit solids, MECO, first stage sep, second stage ignition, SECO 1, SECO 2—each event putting behind us a risk our spacecraft would never face again. The Deep Space Network station in Tidbinbilla, Australia, on a cold winter night half a world away, picked up a signal from the spacecraft within seconds of when we expected it. In just minutes we knew from the telemetry that we had a happy, healthy rover on its way to Mars.

We had one evening to celebrate, the whole Cornell crew swilling down hot wings and beer at a sawdust-and-jukebox place down on the waterfront called Frankie's. And then it was time to get back to work.

The launch window for *Opportunity* opened on June 25, and it ran until the four-teenth of July. I wanted my whole science team to see one of our babies go, and the best way to make that happen was to hold a team meeting right before the launch so they could all hang around and watch the rocket fly. We gathered, more than a hundred of us, on the twenty-second and twenty-third of June, at a hotel in Cocoa Beach.

First up on the agenda, on the morning of the twenty-second, was Pete. He caught me in the hallway outside the meeting room moments before we were scheduled to start, wearing his "you're not going to like this" look. "I'm afraid I've got bad news for your troops," he said. "We're gonna be delayed, by at least a couple of days and maybe more. The Boeing guys need to scrape a bunch of cork off the outside of the rocket."

Cork?

As a rocket ascends through the Earth's atmosphere, it's subjected to intense aerodynamic heating, brought about by the friction of the air against the fragile skin of the speeding vehicle. The heat concentrates in strange ways on the rocket, with an intricate pattern that's determined by how the complex and evolving shock waves created by supersonic flight impinge on the vehicle structure. Even with today's supercomputers it's not an easy business to calculate aerodynamic heating on a launch vehicle, and back when the Delta was first designed it had been pretty much impossible. Rather than take a chance that the heating could burn through the rocket's thin skin, the original Delta team had taken the classic engineer's "kill the problem" approach, putting a coating of thermal insulation on the outside of the vehicle.

The insulator they chose was cork. It seemed to me like a bizarre material to use on a launch vehicle. Cork on a spaceship? It was so nineteenth-century. Where's the gutta-percha, it made you think, where's the beeswax?

But cork made a lot of sense when you thought about it a little bit. It's a good insulator, it's lightweight and it's inexpensive. It had worked fine back in the seventies, and when something really works in the rocket business, you tend not to mess with it unless you have to.

But something had gone awry with the cork on our rocket.

Boeing glues the cork to the outside of their rockets at a plant in Pueblo, Colorado. The stuff goes on in big sheets, stuck onto the outside of the rocket with a polyurethane glue. The glue is supposed to be smeared on with a notched trowel, just like what you'd use to lay tiles in your bathroom, and it's supposed to be smeared with the grooves that the trowel leaves behind running vertically down the side of the rocket. The idea is that if the rocket gets rained on, the grooves behind the cork will make little channels that'll let the rainwater seep out. Sometimes even rocket science isn't rocket science.

Well, it seems that somebody had been having a bad day in Pueblo, and the glue hadn't been smeared on right. To be more precise, it had been smeared on with a trowel that didn't have notches, and it had been smeared on horizontally instead of vertically, making little ridges that trapped water instead of letting it flow out.

You wouldn't think it would be all that big a deal for a rocket to get wet. A little moisture doesn't seem like much compared to everything else it has to survive, until you remember that the stuff that fills the rocket's huge oxidizer tank, just millimeters beneath its skin, is liquid oxygen, at a temperature of 183 degrees below zero centigrade. That's cold, and when you load LOX into the tanks any water that's nearby freezes in an instant, expanding into ice as it does. If that happens beneath your cork, the cork will start cracking off the side of the rocket. And that was exactly what had happened on our rocket when they had run a pre-launch LOX loading test on it.

So the bad news that Pete had to deliver was that a bunch of unfortunate Boeing guys were going to have to spend the next several days cleaning up the mistake that had been made in Pueblo, scraping all the cork off the insulating "bellyband" of our rocket and replacing it. It would be unpleasant work. The bellyband is right where the nine solid rocket motors attach to the first stage, and a lot of the scraping was going to have to happen in very tight quarters back behind the solids.

So there would be no launch for several days at least, and there was nothing to do about it but wait. Most of the team decided to hang around, determined to remain in Florida long enough to see *Opportunity* fly. It wasn't like there was anything we could do to help get the rocket off the ground, though. So everybody pretty much just went tropical, hanging out by the pool, getting sunburned at the beach and generally trying to make the best of an unexpected situation.

Out at pad 17B, the less fortunate Boeing guys were working hard in the sweltering heat, scraping away old cracked cork and gluing fresh new cork in place. The polyurethane that they liked to use back in Pueblo couldn't be applied in the exposed outdoor conditions of a launch pad, so what they used instead was something called "white RTV." Boeing had used this stuff for years to do minor cork repairs out on the pad. They had never had to do a wholesale replacement of big areas of cork with it, but nobody expected that to be a big deal.

We were finally ready to fly on the evening of June 28. The cork problem had taken four days off our launch window.

Launch night was a circus. The *Opportunity* launch would be a milestone for us. The moment that that burst of flame appeared at the bottom of our second Delta, everything we had worked so hard to build would be on its way. So I wanted to send *Opportunity* off in style. Just south of Cape Canaveral Air Force Station is a public recreation area called Jetty Park. It's just a few miles from the Delta pads, and it affords some of the best views you can get anywhere of a Delta launch. We threw a huge launch night party there, letting everyone even remotely connected with the project know they were invited. A couple of thousand people came.

Things looked bad from the start. Skies had been blue most of the day, but as the sun set and the midnight launch time began to approach, a gusty wind came up, low clouds settled in and a light drizzle began. Lightning flickered around the horizon. I had the audio from the launch net plugged into one ear as I circulated among the party guests, which let me track the progress they were making out at the pad. I tried to be as upbeat as I could, but when you've been to a lot of rocket launches you can almost smell a scrub coming. This one smelled like a scrub.

As the moment of liftoff approached I headed out to the jetty with Mary and the girls, scrambling up onto the rocks to get the best view we could. But no sooner had we gotten to where we could see the rocket than I heard the launch director's voice in my ear calling a scrub. The winds above the launch pad were blowing toward a populated area—toward us, as it turned out. When a rocket blows up it can release some pretty nasty chemicals, and with the wind doing what it was doing, if our Delta had exploded in the first few moments of flight, those nasty chemicals would have gassed the thousands of people at our party, among others.

Deltas can attempt to launch twice, and only twice, on any given launch day. The second attempt, if it happens, comes forty minutes after the first one. The launch crew recycled the count to try again forty minutes later, hoping for a shift in the wind.

We came out of the built-in hold at T minus four minutes once more, and again I climbed up onto the rocks with my family to watch it go. But seconds later a scrub was called again. This time two things had gone wrong at once. The wind direction was okay, but now there was too much wind shear right above the pad. Wind shear is an abrupt change in wind speed with height, and it's the kind of thing that can tip a rocket sideways while it's still moving slowly. The other problem was that some knucklehead in a fishing boat had wandered into the restricted zone downrange from the pad, and there wasn't enough time for the Coast Guard to chase him out. So we were done for the night. The party guests straggled out of the park, hot, wet and grumpy.

The next morning we learned to our shock and surprise that it had been a good thing that we had scrubbed. When the Boeing crew went out to the pad to inspect the rocket in the morning light, they found that much of the newly repaired cork had peeled away from the side of the vehicle. Nobody really knew whether or not the cork was necessary to keep the rocket safe, but we might have found out the hard way if we had launched.

The clock was ticking, but it was clearly going to be a little while yet before we would get off the ground.

At 3 P.M. on Sunday, June 29, the Boeing, NASA and JPL teams gathered in a big conference room in the Delta operations building at Cape Canaveral Air Force Station to try to figure out what had gone wrong with the cork. One by one the Boeing engineers reported in with their findings, and slowly a picture began to emerge.

The amount of cork that had peeled off the side of the rocket was something like 10 percent of the total insulation on the outside of the vehicle. Surprisingly, the newly applied white RTV adhesive had come completely unstuck from the rocket's skin, letting the cork peel away in big sheets. Boeing's best guess was that there had been some kind of oil or grease on the aluminum skin of the rocket when they had tried to glue the cork on, so the stuff simply hadn't stuck properly.

So what to do? If grease contamination really was the problem, a little lab analysis should prove it. Meanwhile, they'd strip away any suspect cork, clean the aluminum beneath it really well, and put some fresh cork down in place. We might be ready to launch on Wednesday, the eighth night of our three-week launch window.

As the meeting broke up, Pete point-blank asked one of the KSC engineers the question that we'd all had on our minds: What would have happened if we had launched the night before? The answer he got was honest and direct: "There's no way to tell. There is some unquantified possibility that if we had launched last night we would have had shock wave impingement on uncorked regions, had a failure of the tank skin and either had a loss of performance or a catastrophic loss of the vehicle."

Yikes.

We got the lab results on Monday morning, and there was no evidence of contamination. So much for that theory. Boeing's next idea was simply that this particular kind of RTV takes longer to dry than anybody had realized, and that it had still been wet when they had done the LOX test. This sounded odd. . . . They'd been using this stuff for years and they didn't know how long it took to dry? But they decided to add another twenty-four hours of glue-drying time to the schedule, pushing launch out to Thursday at the earliest. And then, after some consultation with the company that made the glue, even that didn't seem like enough time. A forty-eight-hour slip would push us to the Fourth of July, which would be a bad day to try to pull a launch crew together. So launch moved to Saturday, July 5.

By Wednesday afternoon, the situation had descended into farce. In order to find out if the white RTV was any good at all, Boeing had decided to do some pull tests, gluing "coupons" of material together in the laboratory to see how well they stuck. For some reason, though, they had set up these tests by gluing aluminum to aluminum, instead of gluing aluminum to cork. The glue manufacturer pointed out what seemed obvious—that this wasn't the right way to do the test. And so they had to glue up some new cork-to-aluminum coupons, losing another day in the process. Now we were looking at July 6.

This was starting to get bad, and nerves had begun to fray at JPL and NASA Headquarters. It seemed ludicrous that something as mundane as gluing cork to aluminum was starting to endanger a $400-million launch, but that was where we were.

On the afternoon of July 4, we had a telecon with eighty people on the line to discuss the cork situation.

First up was the news from the pull tests. It wasn't good. "The cork was peeling at ridiculously low stresses," said the Boeing engineer who had done the tests. "I'd pick up a coupon and the cork would literally just fall off as I lifted it. The bottom line is that the white RTV just doesn't have any peel strength to speak of." After all the years that they'd been using the stuff to patch cork out on the pad, it turned out that it wasn't any good.

In the meantime, though, somebody had located another glue that might work: red RTV. This sounded like good stuff. It took only twenty-four hours to dry, and according to the Boeing test engineer it had "peel strength like crazy." Why we hadn't been using it in the first place wasn't entirely clear to me, but at least there was new cause for optimism.

The pad techs had been busy while all this testing was going on, and there was now a full coat of cork back on the rocket. It was bonded down with the same white RTV that had peeled before, though, which made it all suspect. With the launch window disappearing fast, nobody wanted to take the time to strip it all off and replace it all with red RTV; that would take days. Instead, the decision was made that we'd load LOX and hope that this time around none of the cork peeled. If it did, we'd replace just the stuff that peeled with the new red RTV. Anything that stayed in place we'd leave in place, bad RTV or no.

And there was one more hope. Out on the West Coast, Boeing had cranked up their computers and tried to figure out whether or not the cork was even necessary in the first place. But the results of this thermal analysis were inconclusive. Boeing concluded that it was safe to fly without any cork at all, but a NASA review board that had been convened to hear their story wasn't convinced. "It's not a silver bullet," the head of the review board said. "There simply hasn't been enough time to really do the analysis carefully. We can't approve a launch with at-risk cork."

So we were stuck. We had at least one more LOX test ahead of us, on the night of July 4, with glue on the rocket that we knew was bad. And if any cork peeled there'd have to be another round of scraping and gluing, and maybe more.

The frustration was getting intense. Even worse was an unfamiliar feeling of helplessness. We had faced all kinds of crises through ATLO, but until now they'd always been things we could deal with ourselves. In this situation there was nothing that any of us on the *MER* team could do. We'd drive every day through the heat and humidity up Route A1A and the Cape Road to the meetings at KSC, and drive back down again to our hotels and our condos. We'd listen, we'd probe, we'd offer opinions. But the only opinions that really mattered were the ones that belonged to the launch team, and that wasn't us. All we could do was wait.

On the evening of July 4, the few stragglers from the science team who were still around gathered on the beach behind the Cocoa Beach Hilton to try to forget about cork and rockets for a while. Fireworks are legal in Florida, and locals and tourists alike had been stocking up heavily on both pyrotechnics and beer for weeks. By nightfall the Cocoa Beach strip looked like a war zone, with drunken revelers crowding the beach and hand-fired missiles screaming crazily overhead. Up the coast we could still see, just barely, the lights of LC-17 filtered through acrid white clouds of gunpowder smoke. It was hard to feel very festive.

On Saturday, July 5, NASA held another Launch Readiness Review, in the E&O Building at KSC. The latest LOX test had been a failure—the cork that had peeled away before had peeled away again, and even more had gone with it. But Boeing and NASA stuck with the decision they had made at the telecon on the Fourth: Leave the cork that seemed to be stuck in place alone, replace the stuff that had peeled with the new red RTV, and declare victory. Launch was rescheduled for Sunday night.

I felt a little queasy about this, and after the review was over I went and had a few quiet words with Omar Baez, the NASA launch manager. The fact of the matter was that nobody knew for sure that we could survive without the cork, yet 85 percent of the cork on our rocket was stuck on using a glue that we now knew didn't have "any peel strength to speak of."

Omar's answer provided a penetrating insight into the kinds of decisions that have to get made in this business. "The cork story isn't perfect," he admitted. "But the cork that's still bonded on with the white RTV has survived two LOX loads and hasn't peeled yet. The thermal analysis story isn't perfect, but it's enough to make you feel pretty good about it. In an ideal world you'd like to wait until both those stories were perfect. But there's just a week left in the launch window, and it's time to go."

And so it was.

Then, agonizingly, something else came up.

It's not a topic that NASA likes to talk about, but rockets can become wild, dangerous things when something goes really wrong with one. If a rocket begins to go seriously off course, it can be necessary to "render the vehicle nonpropulsive," as they like to put it, before it heads for a populated area. So every rocket—even the Space Shuttle—lifts off from its launch pad laced with high explosives. And every launch crew includes someone called a range safety officer. The range safety officer monitors the rocket's trajectory as it ascends. And if the rocket strays too far from its appointed path, it's the range safety officer's job to blow the thing out of the sky before it kills somebody. Range safety at Cape Canaveral is the job of the United States Air Force, and they take their job very seriously.

It takes a burst of electrical current to set off those explosives, so our rocket carried a set of batteries to provide that burst just in case it was needed. Two identical sets, actually, in case one failed. And, we learned, some tests out at the pad had just shown that one of the two batteries was registering a voltage that was just a wee bit lower than expected. The thing probably still would work if it had to, and even if it didn't work there was a backup. But rules were rules, and according to the rules the voltage was too low for us to launch. So Range Safety announced that they were no-go for launch. At a minimum, they wanted more time— something we were now running precariously short of—to open up the battery and find out why the voltage was low.

They opened it up, and the news was bad: One of the cells in one of the destruct system batteries had failed completely. There's no way the Air

Force would let us fly in this condition, and you really couldn't blame them. Launch was delayed to Monday night at the earliest.

So where could we get a new battery? If Boeing ordered one from the factory, the best case was that we'd try next to launch on July 11, just a few days from the end of the window. But there was one other possibility. Just up the coast, at another launch complex, there was a Delta IV on the pad in final preparations for flight. Rumor had it that the Delta IV used the same kind of battery in its destruct system as the Delta II did. Could we swipe that battery and maybe try to get off the ground before the eleventh?

We could, and we did. Boeing pulled the battery off the Delta IV, popped it into our Delta II, and tested it. Everything checked out. We converged on the E&O Building one more time, at 5 P.M. on July 6, for the final approval to launch.

We didn't get it. The Air Force guys were on the speakerphone, and they still weren't satisfied. They wanted to take the failed battery cell apart and understand why it had failed before they'd give us a green light to launch.

This was starting to seem awfully damn conservative to us *MER* folks as we sat there watching what was left of our launch window slip away from us. We knew that the new battery that had been put into the vehicle was good. We knew that if it was ever needed, it would have to work for only a fraction of a second. And we knew that it was part of a fully re-dundant system; there was a completely separate and functional backup battery if the first one failed. It was nice to see the Air Force taking their responsibility to protect the good citizens of Cocoa Beach and Titusville from fiery death so seriously, of course, but it was getting frustrating.

JPL, NASA and Boeing all pushed back hard on the Air Force guy, but he wouldn't budge. "I've got to follow the protocol; I can't give you a green light on this telecon." They called up the battery company, put an expert on a red-eye and flew him to Florida to try to figure out why the battery had gone bad. There would be no decision until the Range Launch Readiness Review at 1:00 the following afternoon.

There was one week left in our launch window.

The battery guy, fresh off his red-eye the next morning, took the battery apart and couldn't figure out what had gone wrong with it. The Air Force's conclusion was that they'd let us roll back the gantry and start the countdown anyway. But then, at T minus 13 minutes, they'd take a hard look at the battery voltage and see if it was where it was supposed to be. If it was, we were off to Mars. If it wasn't, we were grounded again.

We gathered again on the night of July 7 at Jetty Park. There was no crowd of thousands this time, it was down to just the hard core: a few members of the science team, some die-hard graduate students and the usual collection of space fans and just-happened-to-be-there tourists that show up at any rocket launch. The weather was good, and the countdown uneventful. This one was smelling like a launch, as long as the batteries cooperated.

I had the net plugged into my ear again, and I listened with particular attention as the destruct system battery check came at T minus 13 minutes.

"SSC report CRD battery voltages."

"Alpha is three-five-dot-nine-seven. Bravo is three-five-dot-three-six."

Thirty-five volts. Oh yeah. The batteries were right where they needed to be. We'd never know why the other battery had failed, but it wouldn't matter if we got off the ground tonight.

"RCR report range status."

"Range ready."

And range safety, finally, was green.

We went into the T-minus-four-minute hold, and the Cornell crew headed onto the beach for liftoff with bagpipes in hand. Two of the people on my team—Mary Mulvanerton and the curiously named Mr. Jon Beans Proton—play the bagpipes. Jon, in fact, plays them spectacularly. They had once startled the hell out of me by marching into a meeting on my birthday playing "Happy Birthday" on the pipes. Months before, Jon and Mary had gotten it into their heads that it would be a nice touch to pipe *Opportunity* off into space as she headed out over the Atlantic. They had drawn curious crowds nightly for weeks as they had practiced on the beach behind the Royal Mansions.

We stood on the sand in a tight cluster just above the surf. I had my arms around Nicky and Katy, calling out the countdown that I could hear in my headset to everyone around me. "Forty-five seconds."

"Thirty." I smiled as the bagpipes began to play, somewhere off in the darkness behind me.

"Twenty seconds."

And then, over the net: "Hold. We've had a hold. We've had a cutoff. Fill and drain closure valve failure."

What? "Hold," I shouted. "Hold!" The bagpipes played on hesitantly for a few more moments, and then deflated with a depressing groan.

What the hell had happened? Somebody had called a hold just seven seconds before liftoff! Voice traffic on the net was streaming by so fast I could barely follow it:

"The clock is stopped."

"ATC-3, main power disable reset."

"Reset."

"Prop 1, secure water flow."

"Closed."

"Prop 2, PSD purge press closed."

"Closed."

Holy shit. Something, it was clear, had gone seriously wrong with a LOX valve on our first stage, and they were shutting things down as fast as they could. In the final moments before a rocket launch, so many things are happening at once that computers play a major role in monitoring everything that's going on. For each event—like the closing of a valve—there are a set of "red limits" that describe how things are supposed to work. A valve is expected to take a certain amount of time to close. And if it takes too long, or if it doesn't close at all, then the computer concludes that the valve is outside its red limits, and a hold is called.

The traffic on the net kept coming, in the clipped, crisp phrases of people who know exactly what they're doing and have to get it done in a hurry:

"SSC, hydraulic external power off."

"Off."

"NSC, initiate spacecraft transfer to external power, and report complete."

Opportunity was perched atop a loaded bomb. I was the only one who could hear what was going on, so we sprinted back to the Jetty Park Bait Shop, where some of the students had jury-rigged a sound system and had patched the net audio into it. Things out at the pad were coming under control as we got there, with each explosive "safe and arm" system on the vehicle being switched over to safe.

"SSC, third-stage S&A safe."

"Safe."

"FSC, ground solid motor ignition S&A safe."

"Safe."

"Ground solid motor sep S&A safe."

"Safe."

The team ticked through the rest of the explosive systems on the vehicle. The bomb was defused. But my relief was tempered by the thought of another scrub, and another night gone from the launch window.

"SSC, perform T-minus-four-minute recycle."

Really??

I was impressed. They were recycling the count. There was only half an hour left until the second and last launch opportunity of the night, and a valve on the LOX tank was malfunctioning. But instead of packing it in, they were going to pull the count back to T minus four minutes and try to take one more run at it. These guys seemed to want to get us off the ground as badly as we did.

The LOX fill and drain valve, the computer had said, had been "slug-gish" in its behavior, and it hadn't closed completely when it was supposed to. The launch team caucused as the time slipped away. After almost ten minutes, they came up with a troubleshooting plan for the valve that was pretty much what anybody would do in a situation like that: fiddle with it.

"We'd like to go ahead and put in work cycling the LOX fill and drain valve closed to open three times . . . and we'd like to do that twice."

We clustered around the speakers under the bait shop awning, lis-tening.

"Cycling the fill and drain valve. Coming back open. Cycle number two. Coming back open."

It seemed like a hell of a way to fix a Mars rocket, but these guys sounded like they knew what they were doing.

"Okay, we just cycled another three times, and it's functioning cor-rectly. That's a total of six times."

The valve was working, or at least it was working well enough to sat-isfy engineers who knew more about it than just what the red limits were.

Omar Baez came on the net: "All stations, this is the NLM. At this time I'd like to poll the team with their readiness to proceed. The proper response is go/no-go." One by one, the stations called in "go." Omar was satisfied: "NASA is ready to proceed with the terminal count."

And then they cycled the valve six more times, just to be really sure.

It was a cool, professional bit of work by the launch team, payoff for all the training and all the sims. In less than half an hour they had safed the vehicle, diagnosed the problem and gotten the rocket ready to fly again.

At T minus four minutes they picked up the count again, and we headed back to the beach. This had to be it, but the whole business of getting *Opportunity* off the ground had become so crazy I didn't know what to expect anymore.

My family gathered together. At T minus thirty seconds the bagpipes began to play again, the beautiful strains of "Amazing Grace." I mouthed the final countdown silently, and then, at last, squinted in the darkness against the burst of flame at the base of the vehicle. *Opportunity* accelerated away from us with the same swiftness that *Spirit* had weeks before, arcing out over the water and lighting up the night sky.

And that was it. In moments the Delta was gone from sight, carrying *Opportunity* and everything that we had fought so long for with it.

Had we left a trail of peeled cork behind us as we soared out over the Atlantic Ocean? We'll never know.

PART THREE

FLIGHT

13 FINAL APPROACH

I WRITE THIS WHILE seated on US Airways flight 21, traveling to LAX on a one-way ticket. I kissed Mary good-bye this morning at the airport, and Nicky and Katy, awakened briefly from sleep, at home half an hour earlier. Unless disaster hits, we're facing a long separation.

The last month has not been a good one at Mars. Three weeks ago JAXA, the new Japanese space agency, officially declared the death of their crippled *Nozomi* spacecraft, shortly before it was due to arrive at Mars. The name means "Hope." Then, just four days ago, the British lander *Beagle 2* disappeared. It had been released six days before from Mars Express, its European-built mothership, bound for Isidis Planitia. It has not been heard from since.

The exact cost of *Beagle 2* is something of a mystery, at least to me, but it's reputed to be in the vicinity of $80 million. If true, that's less than our overrun was. With big ambitions and underdog status, *Beagle 2* had garnered international attention. But now it seems to be gone.

The *Beagle* story so far is eerily reminiscent of the *Polar Lander* experience: a low-budget spacecraft with no ability to communicate to Earth once its final approach and entry sequence has begun. Empty data packets in each relay orbiter pass after the expected moment of landing. Silence from the big radio telescope at Jodrell Bank, trying to pick a needle out of a radio haystack. A string of press conferences with an ever more discouraged-sounding team still trying to express optimism. Like the *Mars Polar Lander* team before them, the *Beagle* guys have no way right now to tell if they're dealing with a malfunctioning transmitter or a smoking hole in the ground. I sent their PI, Colin Pillinger, a note of encouragement a couple of nights ago. But he's been working full-time on his mission since 1997, and he could be forgiven now for beginning to feel the first pangs of despair.

Mars Express itself is now safely in orbit around Mars, poised for what should be a long and successful mission. But its success has been eclipsed, at least so far, by the disappearance of *Beagle 2.*

Spirit and *Opportunity* are both alive and mostly well, but their cruise has not been without difficulties. The biggest problems, to my mortification, have been with both of our Mössbauer spectrometers. The first in-flight health check of the instruments on *Spirit* took place in July, on a night when Mary and I had taken Nicky to see *Swan Lake* at the Saratoga Performing Arts Center. Pacing with my cell phone during intermission, Justin Maki back at JPL told me between bursts of static that most of the instruments looked fine, but that there was "something fishy" in the Mössbauer data. Justin's fishiness turned out to be a horrible distortion of the motion of the instrument's drive system, the critical device that's supposed to vibrate a little radioactive cobalt-57 nugget back and forth with great precision. If this happened on Mars it would scramble our Mössbauer spectra, making them difficult and maybe impossible to interpret. Göstar Klingelhöfer puzzled over the problem back in Germany for weeks, but to no avail. We cooked up various theories for why the drive system might behave in strange and unpredictable ways in zero gravity, and we tried to convince ourselves that they made sense. But it was hard to escape the notion that we probably had broken something during launch.

The health check on the *Opportunity* Mössbauer a few weeks later confirmed it. *Opportunity*'s Mössbauer drive was running just like it had back in Florida. It wasn't the zero gravity that had screwed up the drive system on *Spirit*; there had to be something wrong with it.

We spent months troubleshooting. Just like when we had a problem with something during ATLO, our approach was simply to try lots of different things and see if any of them worked. After about three sessions, spread over many weeks, in which we transmitted commands to the spacecraft and looked at how the instrument responded, a picture began to emerge. Something seemed to be blocking the motion of the drive system. The drive moves back and forth along its axis, and it seemed to be hitting something at one end of its travel. The breakthrough came when we adjusted some of the drive system parameters so that the total distance of travel would be shorter. The instrument began working again. Whatever that thing inside the instrument is, we can keep from hitting it if we use a shorter stroke. And even using that shorter stroke we can still get decent, though not perfect, Mössbauer spectra.

So, if the instrument works the same way on Mars that it's working now, we're mostly in the clear. But that's a very big if. What is the drive system hitting? A bent wire? A little glob of glue that somehow shook loose during launch? We've looked at pictures of the instrument that we took in Florida before we launched it, and we've taken apart a nearly identical instrument that we have back here on Earth. We don't have a clue what the damn thing could be. Whatever it is, though, it seems to have gotten in there as a result of launch vibration, and that's not a very comforting thought with the furious jolt of an airbag landing just six days away. The first post-landing health check of the *Spirit* Mössbauer will happen a week from today, if all goes well, and it will be a very scary moment.

There's also something wrong with the Mössbauer on *Opportunity*, and that one could be a lot worse. The drive system there seems to be working fine, but the spectra that we've gotten from the instrument's internal calibration channel are smeared out and useless. Mössbauer spectrometers are very sensitive to vibration, and the *Opportunity* calibration spectra look just like what you'd see if something in the instrument were shaking in an unexpected way. But we don't have any idea why this would happen. There's nothing on the spacecraft that should be moving, and we don't see the same problem on *Spirit*. Maybe something has shaken loose in the calibration channel itself, allowing it to rattle around when the drive system moves? It's a nice theory, but when we look at the calibration channel in our ground instrument, it hardly seems credible— the thing couldn't be built more solidly. So we're baffled, frustrated and worried.

If the problem lies only in the calibration channel, then we're okay with this one too once we get to Mars. We built a separate calibration target, mounted on the outside of the rover, just in case something bad happened to the internal calibration channel. But if the problem also somehow affects the instrument's main channels, which we can't check until after we've landed, then we're in very deep trouble. *Opportunity* is going to Meridiani Planum, the hematite capital of the solar system, and we're counting on the Mössbauer to be our main hematite detector.

Göstar, of course, feels sick about all of this. The main focus of his professional life for the past decade has been the design and construction of three flight Mössbauer spectrometers. Two of them are the sick instruments on *Spirit* and *Opportunity*. The third was on *Beagle 2*.

Spirit and *Opportunity* themselves are doing fine, though they've had a rough ride. In October, we weathered a solar storm of historic proportions. Every so often the sun erupts in a burst of X rays and protons. It's the kind of thing that you design a spacecraft to handle, within reason. Of course, it wouldn't be *MER* if we didn't get hit during our cruise by *the strongest solar flare ever recorded*. If plagues of locusts could happen in space we'd probably get hit by one of those, too. Protons can beat on space electronics pretty badly, and spacecraft all over the solar system started freaking out when that storm hit. The aurora in the night sky over Ithaca was beautiful.

We rode it out, helped substantially by the fact that our spacecraft are spinners. There are a couple of different ways you can keep a spacecraft stably oriented with sunlight on its solar panels. The elegant way is to use a set of star sensors, gyros, reaction wheels and thrusters to keep everything precisely pointed where you want it at all times, with the spacecraft electronics running the whole show. The brute-force way is to spin the whole thing about an axis that points roughly toward the sun, effectively turning the entire spacecraft into a gyroscope that never drifts off its axis. The brute force way works only for spacecraft that aren't expected to do much, but that describes ours pretty well until we hit the top of the martian atmosphere. So we spin 'em. And with such a simple attitude control system, the protons can knock you silly and you'll still probably come through it okay. We did.

And now, just to top it all off, there's a dust storm on Mars. It's always something. The storm first kicked up about two weeks ago, and Earth-based telescopes, *Mars Global Surveyor* and *Mars Odyssey* have all watched as it grew daily. So far the storm shows no signs of going global, and the skies over Gusev remained clear as of yesterday. It's cloudy at Meridiani now, but that won't matter for four weeks, which is a long time for martian weather. We'll worry about Meridiani when we get there.

Spirit is now six days out from touchdown in Gusev Crater. Last night was TCM-4, the fourth trajectory correction maneuver since launch. Each TCM imparts a tiny kick to the spacecraft's trajectory, nudging the projected spot of landing incrementally closer to the aimpoint on the planet. The navigation so far has been phenomenal, and last night's burn was truly minuscule. It consumed just twenty grams of fuel, in what one of the navigators called a "mouse fart." It'll be another day or so before the tracking data show where the burn actually put us. TCM-5 and

TCM-6, if we need them, are scheduled to happen just forty-eight hours and six hours, respectively, before we land.

DECEMBER 30, 2003

Well, it may have been a mouse fart, but it was a damn near perfect one. We've gotten enough tracking data now to figure out where TCM-4 put us, and it's just about dead-nuts on the Gusev aimpoint. After the burn the navigators looked carefully at the radio signal they were getting from the spacecraft, checking out things like its Doppler shift and its position in the sky relative to astronomical radio reference points like quasars. After a day and a half of tracking, their best guess now at the trajectory we're on would have us impact the surface just *eighty meters*—less than the length of a football field!—from the target point in Gusev Crater. This is unbelievably good. In fact, it's so good that there has to be a certain element of luck in it. Still, it's pretty amazing work. After a trip of something like five hundred million kilometers, to be within eighty meters of where you want to be at the end of the trip is like flinging something from New York to Los Angeles and hitting within a millimeter of where you aimed it.

Of course, the trajectory before we hit the top of the atmosphere is only part of the story. The eighty-meter number is how close to the target we'd hit if we're on the best-guess trajectory, and if the atmosphere does absolutely nothing bad to us once we hit it. That's not going to happen, and what the atmosphere really is going to do to us is now the big issue.

In fact, the atmosphere is starting to look pretty scary, and it's the biggest risk (at least, the biggest one that we know about) that we're going to face on Saturday night. The problem is the dust storm. The main storm appears to have begun blowing itself out over the past few days, and the core of the storm never got anywhere close to Gusev. However, all that activity has lofted some dust very high into the atmosphere, where winds have now swept it most of the way around the planet. It's not enough dust to obscure the sun or settle out on our solar arrays in any significant way. But it is enough dust to warm the atmosphere up, and that's very bad.

The physics is simple: The dust grains absorb sunlight, they warm up, and they radiate their heat into the gas around them, warming it up, too. As the gas heats it expands, and the whole atmosphere "puffs up" a bit. The end result is that the density of much of the atmosphere is lower than it normally is, by a pretty substantial amount. It's an interesting science problem if you're an atmospheric physicist, but it's a good deal more than that if your $400-million spacecraft has to ride through that atmosphere on a parachute four days from now.

The timing of events during our high-speed plunge toward the surface is critical. And the timing of some things, like when the chute opens, depends on the properties of the atmosphere. The vehicle is able to sense how much aerodynamic pressure it's experiencing, and it pulls the ripcord when the pressure that it feels is what it's supposed to be. But other events, like the firing of the RAD rockets right before impact, get triggered according to the actual distance above the surface, sensed by the radar. And therein lies the problem.

Everything should work fine if the atmosphere is the way we expect it to be. All of the events have been designed to happen at just the right altitude for a "normal" atmosphere. But when the atmosphere puffs up like this the deceleration that the vehicle feels is different, and the result is that the parachute will open closer to the ground than we want it to. The lower opening altitude gives us less time between when the chute opens and when the RAD motors fire. That's less time to slow down, less time for DIMES to work . . . less time for all the things that we're depending on to get us on the ground in one piece. The EDL guys are calling it "timeline erosion," and it's got everybody pretty thoroughly spooked at the moment.

There never was any way that we could know exactly what the atmosphere would be like on landing day, of course, so Rob Manning and

Wayne Lee and company put as much margin into the timeline as they could. That way we'll have some slack if things go bad. But things *have* gone bad now, and the latest calculations suggest that we haven't got much timeline margin left if the atmosphere over Gusev really is as warm as we think it is.

The whole thing came to a head this evening, in a meeting in room 875, the project conference room on the top floor of building 264, our new home at JPL. Pete, Richard and Matt were all there, along with all our atmospheric experts, all our parachute experts and a lot of nervous-looking JPL managers. Pete was very cool about it, and after listening to what everybody had to say he decided that we'd make a small change, opening the chute at a slightly higher dynamic pressure than we had originally planned to. Dynamic pressure drops as the vehicle falls toward the surface, so popping the chute at a higher dynamic pressure will make it open at a higher altitude, buying us back a little bit of timeline margin.

Sounds like a no-brainer, but there's a nasty twist. The higher the dynamic pressure you open your chute at, the more likely the chute is to squid instead of inflating properly. So this decision makes it more likely that we'll squid, and we don't know how much more likely.

The chute experts on the team are convinced that Pete's doing the right thing. Another way of saying that, I guess, is that they're more afraid of losing their timeline margin than they are of squidding. Their theory on squidding is that even if we do squid for a while up high, we'll inflate properly once we get lower down where the dynamic pressure is low enough for proper inflation. If this theory is right—and if we don't squid so long that we're too close to the ground when the chute finally does inflate—then we're okay.

The reason that this is all theory instead of established fact, of course, is that we never did a parachute test under real martian atmospheric conditions, at supersonic speeds in really thin air. We didn't, *Beagle 2* didn't, *Mars Polar Lander* didn't and *Pathfinder* didn't. The only project that ever did it was *Viking,* and their high-altitude chute test program cost them a hundred million bucks back in 1974. It'd be a quarter of a billion, easy, today. And that was too much money for us to handle.

So we'll goose the dynamic pressure a tiny bit, buy ourselves back a little bit of timeline margin, and hope it all works.

One of the many bad things about all of this is that everybody is so worried about the atmosphere that we're not going to do any more TCMs. The spot that we're headed for right now is a pretty dusty part of Gusev Crater, and I don't like it much. We can live with it if we have to, but I'd much rather be headed toward a place where there'll be less dust to obscure the rocks from our view. About a dozen kilometers east of where we're aimed is a place where the surface is much darker, swept clean by dust devils, which are little martian mini-tornadoes that occasionally pass through and vacuum dust off the surface. I'd love to do a TCM-5 to nudge our aimpoint closer to those dust devil tracks, so we can find some cleaner rocks. Problem is, we're so uncertain about what the atmosphere is doing that we're worried that an eastward nudge could smack us against the eastern wall of the crater if we land really long. So Pete has canceled TCM-5, and he's almost definitely going to cancel TCM-6, too, just letting it ride from here. So it goes.

The thing I keep wondering about in all of this, besides the obvious issue of whether or not *Spirit* will die on Saturday night because of the puffed-up atmosphere, is whether the atmosphere had anything to do with whatever became of *Beagle 2*. The Brits don't have any RAD motors in their design; they just ride their chute all the way down to the surface, and count on it to slow them down enough that their airbags won't burst when they hit the ground. I checked the data, and the atmosphere was very warm over Isidis Planitia on Christmas day. So their chute definitely would have opened lower than they wanted it to. Did they have enough timeline margin to handle it? Will we?

.   .   .

JANUARY 2, 2004

The new year is upon us, and the science team is beginning to arrive from all over the United States and Europe. Our home is in building 264 at JPL. It's a familiar place to us now.

Building 264 isn't much to look at from the outside, just a glass-and-concrete cube eight stories tall, set squarely in the middle of the JPL campus. But its ordinary appearance belies its true nature and its history. 264 has been mission control for some of the great planetary missions; indeed, it's where I worked *Voyager* flight operations as a young kid in my long-ago grad school days. Inside, we've taken over the fourth, fifth and sixth floors, and half of the eighth. Six and eight are primarily office space; most of the real action takes place on four and five.

Four and five are nearly identical, with four dedicated to *Spirit* and five to *Opportunity*. The two floors are so alike in their layouts that the "human factors" wizards that somebody at JPL decided to hire to help us with this kind of stuff have concluded we need to color-code them . . . the idea being that sleep-deprived scientists and engineers will see the colors of the walls and the chairs and the carpet and not get confused about which rover they're working on. I'm not convinced that it works; I've gotten so tired during some of our rehearsals that I haven't been able to remember what the color code is. But the human factors people are still trying. When you come off the elevator on the fourth floor now there's a wide red stripe on the wall emblazoned with the words MARS EXPLORATION ROVER 2003, and below it now a big sign reading MER A, SPIRIT, FLOOR 4. And just in case you still don't get it, right next to that are two huge *Mars Odyssey* images of Gusev Crater. Up on five it's blue, MER B, OPPORTUNITY, FLOOR 5, and Meridiani. I'm trying to drill "blue is for B" into my head.

The one real difference between four and five is that one corner of four is taken up by the Surface Mission Support Area—the SMSA—which will be mission control for both rovers once they're down on the surface. This is where the engineers analyze each downlink from the spacecraft, and where they transmit each uplink. Angled across the room are long rows of consoles, each topped with a prism-shaped blue sign declaring its function: TELECOM, POWER, MECHANISMS, MDOT and so forth. When things are really rocking in the SMSA there's an engineer at every console, poring over the data from their subsystem, reporting status and issuing go/no-go calls.

At the head of the room are the three stations of power: the flight director, who is the conductor of this peculiar orchestra; the mission manager, who approves all the key decisions and who is ultimately held responsible for the safety of the vehicle on each sol; and the ACE, who transmits the actual commands. All voice communication is on headsets over VOCA, and it's done with the same crispness and urgency that we used back in ATLO.

The SMSA may be where we will succeed or fail, and good-luck talismans have begun to appear of late. Perched beside the flight director's console now is Art Thompson's stuffed sheep, wearing its wireless headset and optimistically attired in the red/blue glasses it'll need to view 3-D images once we land. And in a place of honor atop the Mechanisms console is the same plastic SpongeBob SquarePants that was the mechanical team's high-bay mascot back in Florida.

For the scientists, most of the work gets done in the science assessment rooms. Those are color-coded, too, red on the *Spirit* floor and blue on the *Opportunity* floor. Each is divided into work areas that are packed with high-end graphic workstations, flat-panel monitors and huge digital projectors. There are so many power-gobbling, heat-dissipating electronic gizmos in each science assessment room that we have to use wireless microphones and a PA system just to be heard above the noise of all the cooling fans.

Five of the work areas in these rooms are for each of our five "science theme groups." The STGs are loosely organized groups of scientists that work together on some particular topic. For each rover we've set up one STG for geology, one for mineralogy and geochemistry, one for the physical properties of rocks and soils and one for atmospheric science. Then there's a fifth STG for each rover that looks at the bigger picture, thinking about what we should be doing as much as a week or two downstream. I'm letting most of the theme groups organize themselves, but for the long-term planning STGs I've picked natural leaders from the team that I know will keep us moving toward our goals—whatever we discover those to be once we've actually hit the ground.

The sixth work area in each room, at the front, is for the science operations working group chair. The SOWG chair is the leader of the science team on each sol. Everybody on the team will come into each sol with an idea of what they'd like the rover to do, and if you add up all the desires it'll always be much more than the rover can actually accomplish. So it'll be the SOWG chair's job to guide the team to a consensus on what the rover should actually do on that sol, finding a way to make everything fit and to squeeze as much science as possible out of the vehicle. The people I've picked to be SOWG chairs are all scientists who have spent decades working on Mars, and who have the experience and the ability to command the respect of the rest of the team.

Up on the sixth floor, above the fray, is the SOWG room: the place where we'll gather each day on this planet to decide what will happen tomorrow on that planet. The room is dominated by an enormous U-shaped oak table, ringed with big comfortable chairs for each of the key participants from the science team, with signs at each position to indicate their function. Behind that is another long oak table for the engineers and their many planning workstations. Microphones at every position feed a state-of-the-art sound system. The chairs and carpet are aqua, I guess so we won't confuse them with the color codes on the other floors. On two of the walls, six huge projection screens show the output of all the planning workstations. The other two walls are glass, to accommodate the overflow of onlookers that we expect to descend on us once we start using the place for real. It's a beautiful, showy facility, just oozing that coveted "mission control" feel. John Callas, the *MER* science manager, has been planning and designing the SOWG room for a year now, and he's been carefully hoarding the money he needed to build it for longer than that. It's known throughout the project, with a mixture of admiration, derision and envy, as the Callas Palace. To date the best use we've put it to has been watching game seven of the Yankees–Red Sox series back in October, but I hope that'll change soon.

Once a month, ever since launch, we've been putting our new facilities to use in operations readiness tests. The ORTs have been full dress rehearsals for flight ops. We actually built four rovers back during ATLO, and the two that aren't about to land on Mars have been our hangar queens, sheltered in a facility halfway across the laboratory from 264 that's called ISIL, short for In-Situ Instruments Laboratory. ISIL is a very cool place. Half of it is a clean room, with all the trappings you'd normally associate with flight hardware. The other half, separated from the clean room by a very important glass wall, is the dirtiest room at JPL. The floor is red soil and rock. The ceiling has wall-to-wall floodlights that can deliver the intensity of full sunlight on Mars. Sunglasses and coveralls are a necessity when you spend any time in there; otherwise you spend all your time squinting against the glare and come out covered in red dust. It's an eerie environment, and you could almost convince yourself that you were standing on the martian surface if it weren't for the camouflage netting that they had to hang around the room to keep the rovers from getting confused by the walls.

Each time we did an ORT, all through the summer and fall, we'd operate rovers in ISIL, driving them from 264 just like we'll do it on Mars. Our ORTs were training exercises, like the simulations that airlines use to train pilots or that NASA uses to train launch crews and astronauts. The ORTs gave us experience that was high on realism, but where fatal mistakes wouldn't lead to fatal results.

If the objective of the ORTs was to let us make mistakes, then they succeeded brilliantly. The first ones were comical in their ugliness and futility. Nobody knew where anything was. The first ORT was like the first day of middle school; you could almost hear people asking "Where's my homeroom? What's my locker combination?" If we wanted to have a meeting we had to rip plastic bags off the brand-new color-coded chairs just to have a place to sit. There was no food to eat as we worked through the night. Some people lost weight as the simulated sols wore on, and those who didn't survived mostly on pizza, eaten from boxes that piled up in the hallways.

The results that we got out of the rovers were just as ugly. Bad commands would trigger the fault protection software again and again, and the vehicles would just sit there motionless under the floodlights and do nothing for sol after sol. We were running these ORTs on hard Mars time, so a blown sol meant twenty-four real hours and thirty-nine real minutes wasted, over and over. One time we accidentally deleted three-quarters of what would have been a beautiful Pancam panorama from the rover's memory before we could get it to the ground. Other times we would have data products hit the ground, and then watch them simply disappear somewhere in all our computers before we ever got a chance to look at them. Driving was the worst; in all the ORTs put together I don't think the rovers moved more than a total of ten meters.

Even in the rare instances when ORTs went well for a few sols, the pace of operations was excruciatingly slow. I've known it would be like this for years now, but knowing it and experiencing it firsthand are two different things. The unfortunate truth is that most things our rovers can do in a perfect sol, a human explorer on the scene could do in less than a minute. Combine that fact with a bunch of blown sols and it doesn't make for a very pretty picture once we get to Mars.

If we'd had some bad ORTs early in the game, followed by good ORTs later, then maybe we at least could have convinced ourselves that our training was working the way it was supposed to. The reality, though, has been that we never did have a good ORT. Even during our very best one, back in November, only half the sols worked. Half. The other half were busted sols where nothing whatsoever got done. Batting .500 would be great in baseball, but at $4 million a sol it's not a very satisfying prospect.

It's hard to know how worried I should be about this. A lot of our busted sols can be blamed, I think, on the kludgy things we had to do to trick our earthbound rovers into believing they were on Mars. The rovers expect to be able to look up in the martian sky and find the Sun to figure out which way they're pointing. In an ORT, there's no Sun. The rover gyros expect to have to compensate for the rotation of Mars. In an ORT, they have to compensate for the rotation of Earth. We're pretending to communicate with the rovers through the Deep Space Network, with radio signals traveling hundreds of millions of kilometers at the speed of light. In an ORT, all the communication takes place through network cables strung under the sidewalks at JPL, and we fake the light-time delays. Every time we do something to fake a rover out and make it think it's on Mars, there's a chance to screw up, and we seem to do it often. "That was an ORT-ism, it wouldn't have happened on Mars" has become our soothing refrain. But the fact is that a lot of the blown sols haven't been ORT-isms, they've just been mistakes made by what is still a woefully inexperienced team. And there are also a whole bunch of other problems—like ones related to cold temperatures, which we simply can't fake in ISIL—that we haven't been able to practice on in the ORTs at all. Some of those are waiting to bite us on Mars.

So overall I'm nervous. The fundamental problem, as it has been since the start of this project, is that we haven't had enough time. If there were another six months of cruise ahead of us, and if there were half a dozen more ORTs to go, then I think we'd be okay. As it is, I'm expecting things to be pretty ragged—at best—when we first hit the ground.

But it's time to go now, so we'll go.

14 GUSEV

JANUARY 3, 2004

LANDING NIGHT. IT RAINED all day yesterday, and today the skies over the San Gabriel Mountains have been clear and blue. JPL looks like the circus has come to town, with tents strewn across the mall and TV news crews scurrying about with their cameras and their lights. I set off a smoke alarm in my apartment trying to iron a shirt, so I'm going to look rumpled on television tonight.

It's about three hours before landing now, and I'm alone in my office, enjoying what may be the last quiet moment I'll have for a while. Above me on the wall are two pictures. One is a collage of images that Katy gave me for Christmas: Nicky onstage as the Swan Queen in *Swan Lake,* Katy at a horse show, Mary and me arm in arm at my parents' fiftieth anniversary. Family stuff. I miss them already.

The other picture is a yellowed image of the Norwegian explorer Roald Amundsen and his crew on the deck of their ship *Fram,* taken in Tasmania in 1912, just after they had returned from the South Pole. Amundsen reached the pole first and returned to civilization safely, and

I've always thought of his expedition as the standard against which other voyages of exploration should be judged. It's not for any contributions he made to science; he made almost none. It's simply for how he tried something so audacious and pulled it off with such success. On the polar plateau, where others had died of scurvy and starvation, some of Amundsen's men actually *gained* weight. The worst medical crisis of their polar journey came when one of them developed a toothache on the way back from the pole. The keys to Amundsen's success were meticulous planning, obsessive attention to detail and giving himself lots of margin for error in all that he did. Our goals are very different, and goodness knows there aren't any lives at stake tonight. But our success or failure will hang on how well we've been able to do the same things.

I'm surprised by how relaxed I feel. I was expecting to be very tensed up by now, worried and unable to sleep in the nights leading up to landing. Instead, the last few days have been among the most relaxing since the start of the project. We've done all we can, and there's nothing left now but to throw the dice. I guess that how you feel on a night like this ultimately comes down to what you believe. And I believe in these rovers, because I believe in the people who built them.

We'll find out very soon whether I'm right.

I gather up my wireless headset and my laptop, and I stuff a couple of extra headset batteries into the pocket of my jeans. This isn't the night to have my batteries go dead. I step out of my office and head for the elevator, and the CMSA.

The Cruise Mission Support Area is where we've flown the spacecraft from ever since launch, and it's where we'll land it from tonight. I say "we'll land it," but the truth is that we're essentially spectators this evening. The radio signal from *Spirit* takes ten minutes to get to us from Mars now, traveling at the speed of light, and if we saw something in the telemetry that we didn't like, reacted to it instantly and fired a command up to fix the problem, it'd take another ten minutes for the command to get there. Compared to that twenty-minute round-trip time, the actual landing, from when we hit the top of the atmosphere to when we're bouncing on the surface, takes just six minutes. So spectators are all we can be for most of the evening, following events that unfolded at Mars at least ten minutes before.

The CMSA is in JPL's Space Flight Operations Facility, the building just up the hill from 264. It's like a smaller version of the SMSA, with the same consoles and the same blue signs at every seat. Mine says PRINCIPAL INVESTIGATOR, and it occurs to me as I settle into my chair and slip my headset on that it's probably a good thing that there's assigned seating tonight. I'm in the back row with the management types like Pete and Richard. None of us back here have consoles, because none of us have any subsystems to monitor. We're the spectators watching the spectators.

As I scan the faces around the room, everyone's all business. There's no obvious tension in the air yet, but there's no joking either. Cell phones go off a couple times a minute; I've spent so much time with these guys that I know who's getting a call by recognizing their ring tones. The mission manager tonight is Mark Adler, who looks pretty natty in a nice conservative blue shirt and a tie. Most of the rest of us are our usual selves, though, with me in the same tattered good-luck jeans that I wore for both launches. As mission manager, Mark gets the honor of picking tonight's musical selection, something that has become a tradition during critical events on Mars missions. He smiles as Bobby McFerrin's "Don't Worry, Be Happy" starts to pulse over the VOCA net.

It's 18:50 Pacific, and we've got two-way lock on the spacecraft through the seventy-meter Deep Space Network antenna at Goldstone. Jason Willis, who is the flight director for the landing, is conducting the final subsystem poll to see what kind of shape the spacecraft is in before the final turn to the entry attitude. Attitude Control, Thermal, Prop, Power and everybody else call in nominal. *Spirit* says she's ready to go. We should see the turn begin at about 19:05.

19:07: Telecom reports that they're seeing the turn. Slow and steady.

19:26: The turn is over, and Attitude Control says we're within one degree of the proper attitude for entry into the martian atmosphere. That's close enough. The propulsion system on *Spirit* has done its job. The Prop guys smile briefly, for the first time this evening.

The next step, and it's a pretty big one, is HRS venting. The electronics inside the rover generate a lot of heat, and when *Spirit* is all buttoned up inside the aeroshell during cruise she needs some way to get that heat out so that the electronics don't get too warm. There's a cooling system on the spacecraft, the Heat Rejection System, that uses a liquid coolant that runs through the rover. The coolant picks up heat there and then gets pumped out to radiators on the cruise stage where the heat is radiated away. Before the cruise stage separates from the aeroshell, all that liquid has to be vented out into space. It's not a complicated thing, but it's our first pyro firing, and that makes it a very big deal. Everybody in this room is a veteran of the pyro problems we had back in Florida, and we're more than a little anxious about it.

At 20:06, Rob Manning comes on the net. "HRS venting right . . . about . . . now." Rob's callout is based on the predicted time for the venting, plus the one-way light time, but it takes a few minutes before we start to see any consequences of it on the ground. There's silence around the room as the subsystem leads scan their data for anomalies. After a few minutes it's clear that there are none: Our first pyro firing of the night worked. We're eight minutes from cruise stage separation.

"This is where it gets serious," says Richard quietly, as the anxiety level begins to ratchet upward. The tension is evident in the faces around the room, and I can feel it growing now in myself. Everything this team has done to prepare for EDL is about to be put to the final test, with what feels like the entire world watching us. Even the usually unflappable Pete is pacing nervously. As he walks by me he mutters, to no one in particular, "I hope this works."

At 20:14 comes the first truly big event, cruise stage sep. Jason counts it down: "Cruise stage separation in five, four, three, two, one, mark."

There are several long minutes of silence, and then Polly Estabrook, the voice of EDL Telecom, comes on the net. "Flight, we see cal tone one and cal tone two. Cal tone one was thirty seconds long. We have a rover battery low temperature fault."

Low temperature fault? I hold my breath for a long moment, until Jason states matter-of-factly that the fault is expected. Polly continues.

"We got eight seconds worth of cruise stage separation tone."

"Copy that," says Jason in reply. "Mission, Flight."

"Go Flight," replies Mark.

"We can confirm cruise stage separation. We're back in lock and receiving tones." There's a brief flurry of applause, momentarily breaking the tension. Our cruise stage has done its job, and it's been cut free from the spacecraft. After years of design work, a year of ATLO and seven months of cruise, it's about to become a martian meteor.

We're ten minutes from entry now, and Adler's breaking out the good-luck peanuts. The story is that during one of the *Ranger* missions to the Moon, back in the early sixties, somebody at JPL brought some peanuts into the control room to snack on. That *Ranger* succeeded, the first to do so after many failures, and since that day peanuts have been a fixture during critical events for JPL missions. So Mark's handing around little blue Planters packages to everybody. I tear mine open and begin to munch immediately, the one and only thing I can do tonight to improve our chances of a safe landing.

Jason conducts the final preentry poll. Just like before the turn, everybody calls in nominal, Jason replying to each report with a crisp "Copy that."

"Mission, Flight."

"Go Flight."

"All subsystems are reporting ready for EDL. Both primary and secondary complexes are configured and green, and *MGS* is green." We have six minutes until we hit the top of the atmosphere.

The door to the CMSA swings open, and in come the heavy hitters: Weiler, Elachi and Sean O'Keefe, the NASA Administrator. All the other

NASA Headquarters and JPL managers are off in the next room, watching us through the glass, but these guys want to be where the action is, and if they want to be here, they get to be here. They take their places in the back row, where everyone is on their feet now. Weiler is chewing gum furiously.

20:26: Three minutes to entry. Out at Mars it's all over, one way or another.

One minute to entry. Altitude 121 miles. This is it. Projected on the screen at the far end of the room are two displays. One is the crucial Doppler curve: an evolving green line against a bright white background, beginning to bend perceptibly now as our spacecraft starts to fall down the martian gravity well. *Spirit* is traveling more than 12,000 miles an hour.

The other display shows the acceleration tones, the signals from the spacecraft that indicate that it's beginning to feel the friction with the martian atmosphere. We've been on accel tone one for a while, but now they begin to tick upward: accel tone two, then accel tone three. We're in the atmosphere.

Things start to happen quickly, and I press my headset to my ear, straining to hear every callout: accel tone four, then five, then six. Chatter on the net has stopped, and now it's just Wayne Lee, the voice of EDL, reciting the critical events as they're anticipated to happen. "Expecting parachute deploy in five seconds, four, three, two, one, mark."

We're down to thirty-five thousand feet, still falling very fast. The chute should be out by now. The room is hushed, everyone desperate to hear something from *Spirit*. But there's a lag as the computers on the ground struggle to pull the feeble tones out of the radio chaff, and nothing's changing on the tone display. Where's the chute? I'm gripping the edge of the tabletop in front of me.

Polly calls out: She sees the chute deployment tone! The room bursts into applause. "So far, so good," says Pete next to me. Come on, sweetheart.

Heat shield deploy tone. Another big one. The heat shield is away; the lander should be descending on the bridle. And now, radar solution tone, the radar sees the surface. We're almost there. . . .

And then, silence. We should be down, we know we're down, but there's no way to get a signal. If all has gone well, the airbags are bouncing and tumbling now, the antenna pattern wheeling wildly across the martian sky. And if all hasn't gone well, the result from where we sit looks the same. Either way there's no signal.

Pete says quietly, "Whatever's happened has happened." I exhale. Boy, that went by fast.

The room is silent, everyone waiting to hear from Rob or Polly. My eyes scan the room, looking at tension in the faces, listening on my headset for some sign that we've survived. Everything hangs on whatever happens next.

And then, Rob Manning's voice comes over the VOCA net, clear and steady: "We have a signal indicating bouncing on the surface."

"Son of a bitch!" yells Pete, and the room explodes. There are hugs, cheers, tears. I'm pounding my fists on the tabletop, eyes pressed tightly closed, my emotions abruptly overwhelming me. Are we really on Mars? After so many years? The room's in pandemonium, any trace of NASA cool completely gone. I gather myself together enough to push through the cheering bodies and find Barry, wrapping him in a bear hug, thumping him on the back. It's been a long, long road for us.

But seconds later, everything stops. Wayne is back on the net, reporting loss of signal, and the room goes slowly quiet again. I'm not surprised by silence now from Mars; if anything, I was surprised that we'd heard anything while we were bouncing. I simply cannot believe that we've lost our spacecraft. The first couple of bounces are the most dangerous ones, and if we survived those we should have survived the others, too, unless something very strange happened. And Manning sounded awfully sure that he had seen a signal.

Minutes go by. Adler turns to me and says, "It's amazing that the bouncing tone got through." He's clicking his ballpoint pen, over and over. Polly, Rob and Wayne are all clustered around the EDL Telecom console, searching. It's been more than ten minutes now since we heard from the spacecraft.

"We see it! We see it! We see it!" It's Polly this time, crying out with none of the cool that Rob had had earlier. The CMSA erupts again in cheers, hugs and relief. It's been so long since impact that if we're seeing a signal now we have to be alive and motionless on the surface. We did it. . . . We're really down, really on Mars.

If there's a signal from the spacecraft then there's work to be done, but the room is in chaos. Jason stands up, ignoring his headset microphone, yelling over the din. "Everybody, please, can we keep it down?" He's trying to regain control of his Mars landing.

Polly calls out, over the VOCA, "We're in lock, we're in lock." Goldstone's actually got signal lock on the rover. This is quite a telecom system.

Things seem to be good on Mars, but they're out of control in the room. At the moment of impact the CMSA doors swung open and everybody who was outside the room came in; the room has twice as many people in it now as were here before we touched down.

Pete takes charge. "Everybody not part of the ops team, please . . . out!" The behind-the-glass folks retreat quickly back to where they came from, and slowly a sense of order is restored.

As things begin to settle down people start to pull out their cell phones, quietly calling friends, loved ones. This is a big moment, and not just for the people who worked on the project directly. I look around the room at my companions, the engineers who made this thing work. I've never felt so close to this many people in my life.

Jason's back on the VOCA, talking through the limited data with his subsystem leads. We won't get any more information on what has happened until the first *Mars Odyssey* relay pass, in a couple of hours, so the work here is almost done. Satisfied, Jason calls over to the SMSA. "Surface, Flight. *Spirit* is yours."

And with that, EDL for *Spirit* is over. It's time for me to get across the street. As I'm packing up my stuff to go, Pete leans across the table, still with a big grin on his face. "I got your instruments to Mars, guy . . . now the rest of it is up to you."

Two hours later now and we're in the SMSA, waiting for the first data from the surface. This, if it works, will be a UHF data relay through the *Mars Odyssey* orbiter, the first time that the radio link from the rover to the orbiter has ever been used. It's another crucial moment, because it will tell us whether or not the critical deployments on the lander and the rover worked. The key ones are the lander petals, the solar arrays and the Pancam Mast Assembly, also known as the PMA. All of them involve pyros and drive motors. If the PMA deployment didn't work, Pancam and Mini-TES will be useless. And if either of the other two didn't work, we're done.

The room is full. All the subsystem leads are at their consoles, waiting to jump on their data as soon as the first bits hit the ground. At the front of the room, as flight director, is Chris Lewicki. Chris is just twenty-nine years old, very young to be "Flight." But the guy is a major hotshot . . . scary smart and poised beyond his years. He's the right man for the job tonight.

I'm all the way at the back of the room, at the imaging console with Justin Maki. Justin's console is where the very first pictures will come up, and Lewicki has to keep clearing people away from the area so that we'll have enough room to work. We've got two screens in front of us. The one on the left, which only people who are gathered around the console can see, is where the images will first be displayed. There we can look the images over, get a sense of what we're seeing and do simple tasks like enhancing the contrast. The right screen is piped to big projectors at the front of the room, and it also goes out live on NASA Television. When pictures first pop up, Justin will process them quickly on the left screen and then, once I'm satisfied with how they look, he'll drag the good ones over to the right screen. This is our first downlink from the surface of

Mars, and we don't even know whether our cameras are still working. Nobody's forgotten our speckle adventures during thermal vac. We want to make a good first impression on the world, so my job is to sort through the pictures in real time and pick out the ones that look good enough to display.

Time passes slowly, giving us all a chance to worry some more. Chris is talking to everyone on VOCA, taking us through the expected series of events, and what possible anomalies to look for in the data. The feeling is much like what it was before entry a couple of hours ago, and with good reason. We know that we made it to the surface, but that's about all we know.

"Flight, *Odyssey* ACE, we have APID 46."

"Here it comes!" shouts somebody across the room. *Odyssey* is sending us data from the rover. The link worked.

"Thank you very much," says Lewicki calmly over the VOCA.

Kevin Burke is at the mechanisms console right in front of me, and he'll be the first to know how the deployments went. There are no pictures yet.

"We're iron cross," says Kevin as the numbers begin to appear on his screen, "and we are fully retracted." The petals are open and the airbags are pulled all the way in. "Woo hoo! My man, we're here!" He's grinning.

His data continue to stream in. "Flat deck!" he calls out as he gets the first tilt readings. Next he scans his contact switches: "Closed closed closed closed . . . solar arrays are deployed. All of 'em!"

I spin around back to the imaging console. The first picture should be down soon. "Here we go, Navcam subframe," says Justin, barely able to suppress the emotion in his voice. And there it is on the screen, our first image from Mars. But it's instantly familiar: the Pancam calibration target and part of the solar array. There's nothing martian to be seen.

"Put it up," I tell Justin, and he drags it over to the right screen. There's a quick burst of applause. Not until I see it up on the big screen does it occur to me that the vantage point from which it was taken proves that the PMA is deployed. Our hardware is on Mars, and it's working.

"Right rear Hazcam," says Justin. Here we go.

"Oh my God," I gasp. Justin drags it to the right screen, both of us already realizing how pointless my job is if all the pictures are perfect. I stand, pump my fist in the air and point to the front of the room. "Look!"

Cheers erupt around the room. The foreground is the same familiar view of the lander that we saw in the Hazcam images from all the ORTs. But this time there's no wall in the background, no camouflage netting. Instead, a beautiful horizon curves across the top of the scene, bent into an arc by the fish-eye lens of the Hazcam. Small dark rocks are scattered across the terrain. The resolution's low, so it's hard to say yet just what we're looking at. But we're sure not on Earth anymore.

The images start to flow in. By now it's obvious that there's nothing wrong with any of the cameras, so I tell Justin to just throw everything up on the screen as soon as it hits the ground. Justin does a quick contrast enhancement on our first full Navcam frame, drags it to the right screen and the crowd goes nuts again.

This one was taken before the PMA had been deployed, with the camera looking along the axis of the mast toward the front of the rover. The clarity is remarkable. I can see the rover deck in the foreground, with screw heads and solar cell wiring clearly visible. Small dark rocks are scattered across the scene, casting shadows in the afternoon sun. Toward the right-hand side of the image is a bright patch with no rocks that looks like some kind of a depression . . . maybe a little impact crater filled in with dust? My eyes scan upward toward the horizon and I see another of these little dust-filled hollows, and maybe another. And then, on the distant horizon, a tiny hill. Part of my mind is studying the landforms, already analyzing what the images reveal. But most of my mind is simply marveling at the fact that this thing I see on the screen in front of me is the same piece of hardware that I stood next to less than a year ago at JPL and in the PHSF high bay in Florida. It's *MER-2,* it's *Spirit* and it's alive on Mars.

Soon the screen is filled with overlapping images, and the nature of the scene around the rover begins to emerge. The terrain is smooth in every direction, with no big boulders or clusters of rocks anywhere. In one direction—we don't know which way is north yet—there are hills on the horizon. It's impossible to tell how big they are or how far away they might be.

The most puzzling things are in the immediate foreground, right in front of the lander. There's something very big just to the left of the forward egress ramp, and something else just like it to the right. They look like rocks, but it'd be a very weird coincidence to have the only two big rocks around us be immediately next to the lander. Are they rocks? Or are they pieces of airbag with rocks hidden under them? Or are they maybe just lobes of the airbags that haven't fully deflated yet? The airbags deflated nicely on *Pathfinder,* but maybe ours haven't. . . . Goodness knows they're a different design. Whatever those things are, they're definitely blocking the forward egress path off the lander. Unless they deflate, we're going to have to find another way off this thing.

Once we do get off the lander, though, the terrain looks like it'll be pretty good. This stuff is going to be very easy to drive on. Suddenly it hits me: Is this what a martian lakebed looks like? Could we really have been so fortunate that we landed on bare lakebed deposits, with nothing burying them? Maybe the reason that the rocks are so small is that the lake sediments never consolidated into solid rock, so that when there's an impact they just bust into tiny pieces. It's impossible to tell what's really going on until we see things in color and at higher resolution, but this is looking awfully damn good so far.

Lewicki comes on the VOCA, his voice in my headset just audible above the commotion in the room: "PI, this is Flight."

Who, me? "Go Flight."

"Could you give us your professional opinion at this time?" Chris is grinning.

"It's beautiful!" is all I can find to say.

The night on Earth wears on, and the reality of living on Mars time starts to set in. With everything at Gusev working the way it's supposed to, we're transitioning abruptly to Mars surface operations. My work shift each sol, including this one, is supposed to start at 15:30 Mars Local Solar Time, or LST, late in the martian afternoon. That's after the rover has pretty much wrapped up its work for the day and has sent data to the ground for us to think about. My shift is supposed to go until something like midnight LST, working through the early part of the martian night to plan what's going to happen on Mars the next martian morning. Today 15:30 LST at Gusev works out to 9:31 P.M. Pacific, so I'm due to work until about 6 or 7 A.M. Earth time. Tomorrow my shift will start thirty-nine Earth minutes later, at 10:10 P.M. Pacific. It's confusing, and the best way to deal with it is simply to cut the cord with Earth and start thinking exclusively in Mars time. My Earth-time watch is in a drawer back in my apartment, but a watch that's been modified to run on Mars time is always with me. And with blackout shades on the windows of building 264 and Mars clocks on the walls, I can't really tell what time it is on Earth anyway, unless I go out of my way to figure it out.

Still, this is going to take some getting used to.

Through the martian evening the subsystem leads continue to dig through the engineering data that came down during the *Odyssey* pass, working out more details of how *Spirit* is doing. Not all the news is perfect. The current coming from the solar arrays is less than we had expected, which probably means that there's more dust in the atmosphere than we had anticipated, left over from the storm that hit a few weeks ago. And the rover is facing just about straight south, which means that for part of the day, until we can turn to a different heading, the high-gain antenna's view of Earth will be blocked by the head of the PMA. But these aren't big problems, and overall things are about as good as I could imagine them being at this stage of the game.

In a quiet moment back in my office I take a look at my e-mail, finding, to my surprise, dozens of congratulatory messages from complete strangers. It seems odd that they're mostly from Australia, Hong Kong and Korea, until I realize that that's the part of the world where people are actually awake now. While I'm scanning them my phone rings, and it's Colin Pillinger, the PI of *Beagle 2*, calling from the U.K., where it's very early in the morning. We have a brief and pleasant conversation in which he offers his congratulations and I wish him success in regaining contact with his spacecraft. It's a gracious and classy thing for him to do, especially under the circumstances.

A couple hours later, Mike Malin comes into the fourth-floor science assessment room with Tim Parker, another geologist on the science team. The DIMES images came down during the *Odyssey* pass, and Mike and Tim have been comparing what they see in the DIMES pictures to old images from Mike's camera on the *MGS* orbiter. There's a pattern of impact craters that they've matched up, identifying generally where we hit the ground. To my delight, we've landed long, flying over the dusty center of the ellipse and coming down by sheer good fortune right in the dust devil tracks where I wanted to be. An awful lot of things have gone right tonight.

Of course, we can't tell where we are from the DIMES pictures alone, since they were taken before impact, and before we did all our bouncing and rolling. But they tell us where to start hunting, and as Mike and Tim have gone through the Navcam images, they think they've spotted enough landmarks on the horizon to work out roughly where we came to rest. If they're right, those hills that we saw, which are to the southeast, are at least a couple of kilometers away. Everybody is talking already about going there, of course, but two kilometers is too far for us to drive. There's a good-sized crater only a few hundred meters to the northeast, so maybe that's a place we'll be able to get to.

The sol comes to a close, and it's time for dinner and sleep, as soon as I can get through a couple of news interviews. I'm on the mall at JPL, where it's suddenly morning outside, talking to a reporter when my assistant Diane comes up. There's some kind of deep emotion written on her face, but I can't tell what it is. She looks tired. Good news, bad news? She's saying something to me about *MGS*, but I can't catch it. I wrap up the interview and walk quickly over to her.

"What's wrong?"

"Nothing's wrong. The overnight *MGS* pass just came in. Some Pancam images are down." Oh my God. I run back to 264 and up the three flights of stairs, not waiting for the elevator. Trying to catch my breath, I walk down the hallway to the Pancam room, expecting it to be mobbed.

Instead, the room is dark. I step inside and there's Jim Bell, the lead scientist for Pancam, all alone, slumped in a chair and staring at a monitor. There are tears in his eyes. On the screen is Mars, with colors so perfect and details so sharp it's like being there yourself. He looks up at me.

"It works, man. . . . It works."

This is so good. I can't believe how good this feels. But I'm not going to get all misty-eyed. I'm cool. I go back outside, do a couple more interviews, and then head home and microwave up some frozen spaghetti. I'm sitting on the couch in a darkened apartment in my underwear, eating my spaghetti, when suddenly the events and emotions of the day come tumbling down on me. Pancam is really on Mars, after all these years. The whole damn thing is on Mars. I dissolve into tears.

SPIRIT SOL 2

Today was the Mössbauer health check. I came in early so I could be here for the downlink, hoping to learn that the landing hadn't done any more damage to the instrument's drive system. My confidence wasn't high. Mechanical damage to an instrument is a scary thing, and it's been impossible to really figure out what went wrong during launch. If the landing shock made the damage much worse than it already is, we could be in real trouble.

Göstar Klingelhöfer and the rest of the Mössbauer guys were in their room on the fourth floor, right around the corner from the SMSA, while we waited for the data to come down. I hung out in the SMSA at the Science Downlink Coordinator console during the downlink, looking over the SDC's shoulder at the "packet watch" display, the program that searches for data packets from the spacecraft and identifies which instrument they came from. After sitting through what seemed like an hour of Hazcam and Navcam packets, finally the Mössbauer packet counter began to increment. I walked quickly around the corner to the Mössbauer room.

"It's coming," I said.

Daniel Rodionov, a Russian postdoc from Göstar's lab, was at the Mössbauer workstation, impatiently reloading his database every few seconds. Finally the file that had the data we were looking for appeared. Daniel's next step was to transfer the file from the workstation to the laptop that the Germans had brought with them from Mainz for data analysis. JPL is very security conscious, sometimes to a fault, and an unfortunate aspect of their security policy is that computers that belong to foreigners cannot be connected to their network, even if those foreigners happen to be trusted partners who have provided scientific instruments for JPL spacecraft. So Daniel popped the file from the workstation onto a memory stick, shoved the memory stick into the back of the laptop and launched his data display program.

"Are you ready?" he said, looking up at Göstar.

Göstar took a deep breath. "Go ahead."

Daniel pushed the button and the data from the drive system popped up.

"It's perfect!" he shouted. And it was. We stared at it, not quite believing what we were seeing. The plot was telling us that the drive system wasn't simply as good now as it had been during cruise, but that the drive was better than it had been during cruise. In fact, it was looking just like it had back in Florida! Whatever had happened to the instrument during launch, the landing had somehow, inexplicably, fixed it. Our Mössbauer spectrometer was healthy, for the first time since June.

How this could have happened I cannot fathom. Something got bent out of place by the launch vibration, and it got bent back into place by the landing shock? It seems impossible to believe. It can't simply be that the drive behaves strangely in zero g, because the one on *Opportunity* never did this. I'm baffled. But as long as it stays this way, it doesn't matter. The thing works!

I hugged Göstar, who was almost speechless with relief. Down the hall, in the APXS and Microscopic Imager rooms, the same scene was being repeated with less apprehension and bafflement, but with just as much joy. The health check data from all three instruments showed that they're in good shape and ready to go as soon as we can get off the lander.

Along with the health checks, we've continued to go through the Navcam images of the scene around the lander that came down yesterday. We've named our first feature, a shallow dust-filled depression about twenty meters to the northwest of the lander. Everybody on the team is pretty burned out from lack of sleep now, so we're calling it Sleepy Hollow. It's probably a small "secondary" impact crater. When impact craters form they throw out debris, and when the biggest pieces of debris hit the ground they can create small, shallow secondary craters. We're thinking right now that Sleepy Hollow may be where we want to go right after we've gotten off the lander, the idea being that there may be bedrock exposed along the inner lip of the crater. Of course, all of this is based on just some pretty bad Navcam images. The plan could change dramatically once we get enough Pancam stuff down that we can really see what we're looking at.

SPIRIT SOL 3

I don't think we've landed on bare lakebed deposits.

The Pancam images are coming down in a big way now, and Jim Bell and his guys put together the first really good-looking color panorama today. It's a great picture, but it sure doesn't show anything that looks to me like sedimentary rocks. Most of the rocks that we see are dark gray, with smooth, almost sandblasted-looking surfaces. There aren't any obvious grains or crystals in them, and there definitely isn't anything that looks like sedimentary layering. Instead, these rocks look like nothing other than very hard, fine-grained pieces of broken up lava.

I'm trying to stay optimistic in the face of this discovery, but it's hard. The most damning evidence so far is that some of the rocks look like they actually have vesicles in them. Vesicles are little round voids in rocks, like the holes in Swiss cheese, that form when bubbles of gas develop in magma. And if these really are vesicles, then this stuff is definitely volcanic.

So if it's all lava, what happened? Gusev is a great big hole in the ground with a dried-up riverbed flowing into it, so the sediments we came here looking for must be around somewhere. But I'm beginning to fear that sometime after the sediments were deposited, the crater floor—or at least the part of it that we've landed on—got covered over with lava. Maybe the lava oozed up through cracks in the ground, or maybe it flowed in from somewhere else. Either way, though, it seems to have covered the sediments here, burying them from view. I've spent most of the day staring at these Pancam pictures, and I'll be damned if I can see anything in them that I can convince myself isn't a busted-up piece of lava.

We knew going in that this could happen to us here, of course. It was one of the risks we took when we picked this site. And we always figured that if it happened, the best way to deal with it would be to head for the nearest impact crater we can find. We don't know how thick the lava is, but if it's thin enough, maybe a crater will have dug through it, exposing the sediments that must lie somewhere down below. Sleepy Hollow could be a possibility, but it's so small that I'm starting to think that it's unlikely to be deep enough. Our best bet is probably the big crater off to the northeast. That thing's a quarter of a kilometer in diameter, and it probably dug down several tens of meters when it formed. So maybe we'll find sediments exposed inside it, or maybe we'll find sediments in chunks of debris that were thrown out onto the surrounding terrain.

Of course, we may not find any sediments there at all. But we won't know unless we go look.

SPIRIT SOL 5

The pace has slowed as we've struggled with how to deal with the blockage of the forward egress ramp. There seem to be two still-inflated "bubbles" of airbag fabric in front of us, and they're big enough to obstruct our intended route off the lander. We can turn on the lander deck and head off in a different direction if we have to, but it's an ugly maneuver, not risk-free, and the best thing to do before trying anything that tricky is to see if we can pull the airbags in. We've tried running the airbag retraction actuators over the past sol, tugging on the cords that are spun out through the bags, but nothing budged. We're going to try it once more before we give up, but nobody has high hopes of success. It's looking like at least Sol 10 before we'll be able to get off this thing, and it could be even later than that.

Mini-TES is working now, so the whole payload is checked out except for the arm and the RAT. We've gotten down our first Mini-TES spectra of the soil around the lander, and the stuff looks just like average martian soil seen from orbit. This spot where we've touched down is not exactly a geologically thrilling locale.

The more humdrum the stuff we've landed on appears, the more we think about what the features on the horizon may offer, especially those big hills to the southeast. I learned today that NASA Headquarters has decided that they'd like to name them the Columbia Hills, after the fallen space shuttle, with each summit named for one of the *Columbia* astronauts. I can't tell the team this yet, though, because NASA wants to okay it with the families of the astronauts before announcing it. But we should be ready to go public with it soon.

We also need names for the craters around us, and we had a big discussion about that today in the fourth-floor science assessment room. For individual rocks we're simply using whatever name pops into the head of the first person who decides that a certain rock is interesting. For big craters, though, it seemed to me that we ought to give it a little more thought. The sense on the team was that it would be a good idea to use some kind of theme for naming craters, rather than just picking names that are completely unrelated to one another. After a lot of talk, four different possible themes came up. One was to name them all after coins: Penny, Nickel, Dime, etc., because they're round and come in different sizes. A little unimaginative, but it would be a subtle way of thanking the DIMES team for getting us to Gusev. The second was to name them after American Indian tribes. That's maybe a little too America-centric for a mission with an international payload, though. The third was to name them after historic ships of exploration, and the fourth was to name them, with what could be unfounded optimism, after lakes.

I don't usually like to make decisions by voting, but this seemed like a good time to do it, so I told the team that I'd tape an envelope to my office door where they could each cast a vote for their preferred scheme for naming craters. I counted the votes this evening, and lakes beat out ships of exploration by a small margin, with the other two far behind. So we'll use lakes for craters, and we can use ships of exploration for hills if NASA's plan to name the hills falls through for some reason. I'm going to ask Jim Rice, who's a team member with a keener-than-usual interest in the history of exploration, to come up with lists of lake and ship names.

SPIRIT SOL 7

I'm so tired I can barely think. Writing is a strain. The switch to Mars time is taking its toll on the whole team now, and I'm not sure that adrenaline will carry us much longer. The problem is not that I'm spending too much time working. Some nights I spend ten or twelve hours in bed, but I sleep for three. The rest of the time I lie awake in the dark, my circadian rhythm so completely disrupted that sleep won't come.

We're making slow but steady progress toward getting off the lander. Rover standup is done, with everything fully deployed and latched in place. So *Spirit* looks like a real rover now. We've given up on the idea of trying to drive off straight ahead. Instead, we're going to turn 120 degrees to the right, and drive down the right rear egress ramp.

Tomorrow's main event is the release of the arm, also known as the Instrument Development Device, or IDD, firing one of the pyros that caused the most consternation before launch. If that goes well, then on Sol 9 we'll fire the last wheel release, cutting the middle wheels free, and on Sol 10 we'll sever the last electrical connection between the rover and the lander. At that point *Spirit* will be completely free and ready to move, though still standing on the lander deck. On Sol 11 we'll turn in place to face the chosen egress ramp. And then, if everything has gone perfectly, Sol 12 will be the big one: egress onto martian soil.

We're not going to go very far once we've got six wheels in the dirt, though. Nobody's ever looked at the soil of Mars with a microscope or a Mössbauer spectrometer before, so anything we do with the IDD once we get off the lander will be new. We're planning to slap the thing down on whatever dirt we've got in front of us once we egress and spend a couple of sols looking at that. Once we've got a good first soil measurement down on the ground, then it'll be time to go hunting for our first rock.

I don't expect much excitement from the rocks around here, though. We got the first good Mini-TES spectra of some rocks analyzed today, and to no one's surprise they look like basalt, the most common kind of lava.

SPIRIT SOL 12

I came in a few hours early today for the downlink that'd let us know whether or not egress went as planned. I was optimistic. The tricky part came over the past couple of sols, when we cut the cable and did the 120-degree turn in place on the deck to face one of the unobstructed egress ramps. That deck is littered with all sorts of junk, and turning on it was one of the nastiest maneuvers the rover will ever have to do. But we knew where all the junk was—after all, we had put it there—and we had rehearsed this move dozens of times in ISIL. So the probability of success was high. Still, it was good yesterday to see front Hazcam images, taken after the turn, showing a nice clear egress ramp straight in front of us.

Compared to that maneuver, today's egress drive looked easy. It's nothing but a three-meter straight-line drive, dead ahead. I was in the SMSA for the downlink. The first news, as usual, came from the Mechanisms guys: near-zero tilt, and just over three meters of new odometry on the wheels. There's no way that could have happened without a successful drive off the lander. Still, the thing that did it for me was the rear Hazcam image.

I have the image in front of me as I write this. The lander dominates the scene, warped across the frame by the Hazcam lens. The egress ramp is dead center, with its distinctive diagonal-plaid markings, and airbags are bunched on either side. The things that really strike you, though, are the wheel tracks, running straight and true back toward the lander. The soil is fine and cohesive, and the imprint of each tread is preserved in the soil almost perfectly. The wheels hardly sink at all into this stuff, at most just a centimeter or two. It's beautiful, and it's hard, visual proof that *Spirit* is on Mars, with six wheels in the dirt and ready to explore.

It's 2:10 A.M. Pacific time on Thursday, January 15, 2004. Today is one of the happiest days of my life.

SPIRIT SOL 13

I walked into the SMSA today and there was Steve Gorevan from Honeybee, staring at the screen in the front of the room, misty-eyed. There seem to be a lot of misty-eyed people around here these days. The IDD is deployed out in front of the rover now, and we just got down the first front Hazcam pictures of it. It's a beautiful thing to see. Steve was checking out the brushes on the RAT, which are in great shape. I was looking at the "Bodo-board" taped to the outside of the Mössbauer, a kludgy little electronics board that Bodo Bernhardt in Germany designed to fix a last-minute problem that we'd had with the instrument back in ATLO. Lori Shiraishi, an engineer who'd put years into building the IDD, was standing there with us, critiquing some of the work she'd done, wishing she'd had the time to make it look a little prettier. It all looked so familiar. The thing that really hit us as we looked at the picture wasn't that it was Mars in the background. It was just the fun of seeing our hardware again.

We also got the first really high-resolution Pancam images of the soil right in front of the rover down today, and one of the little rocks in the scene shows absolutely unambiguous vesicles. Damn. Once we've finished using the IDD on the soil in front of us we're going to plan a drive to our first rock, but I'm already pretty certain what we're going to find.

There's been a lot of discussion over which rock we should go to first, but the choice was simple once we really got down to it. Sleepy Hollow isn't looking like anything very special anymore; it's just rimmed with chunks of the same stuff that we see everywhere else. So the thing to do is simply to go after the closest conveniently shaped big rock that we can find. One possibility is a couple of rocks named Sashimi and Wasabi that are more or less dead ahead of us. Those aren't real big, though, and they have pretty strange shapes, so instead we've chosen a larger pyramid-shaped rock a bit to the left.

Whoever first looked at this thing had simply called it Pyramid, but once I realized, a couple of days ago, that it was likely to become a very important rock for us, I also realized that we ought to find a more imaginative name for it. I asked John Grotzinger, a geologist on the team from MIT, to come up with something, and after looking at some pictures of mountains, he convinced himself that it looked a little bit like Mount Marcy, in upstate New York. So he suggested the name Adirondack. I liked it, and that was what we used. Within twenty-four hours I was getting phone calls from newspapers and radio stations in the Adirondacks, wondering why we'd picked that name.

SPIRIT SOL 14

We got our first Mössbauer spectrum down today, from the soil in front of the rover. Göstar was asleep in his apartment when the data hit the ground. Daniel Rodionov called him to wake him up, and he drove in to JPL at autobahn speeds. It wasn't easy to resist the temptation to look at the data while we waited for him! We had gotten the files transferred to the laptop and ready to display, and we even took a quick peek at the drive system data to make sure that the thing was still behaving itself. But we couldn't look at our first real spectrum without Göstar.

Göstar arrived, slightly out of breath, and Daniel waved him over to the laptop and motioned for him to push the button. A spectrum came up on the screen, the first Mössbauer spectrum ever returned from another planet.

Göstar tilted his head to one side and stared at it for a moment.

"It's a doublet," he said, thinking like a spectroscopist. "It's not quite a doublet," as he looked a little closer. And then, as the realization that a decade of work had finally paid off suddenly hit him: "But it's a Mössbauer spectrum!"

There were hugs all around, and some tears: a replay of the scenes that had followed landing two weeks ago, but on a smaller scale and in German. Over the next six hours the Mössbauer guys, in the most impressive celebration I've seen since we landed, killed off two bottles of champagne and a quart of Glenlivet. I had to ask Ralf Gellert to get them back to their apartment in one piece.

With the success of our IDD work on the soil in front of us, the next step will be the drive to Adirondack. It's three meters away, which is pretty far, so there's a good chance we won't get there in one shot. Approaching rocks with one of these rovers is like playing golf. You've got a target off in the distance that you're trying to hit. Problem is, the rover doesn't always go exactly where you want it to go. So you approach the target with a drive. The drive gets you closer, but if the target is a distant one, it doesn't get you all the way there. So you drive again, coming a bit closer. Sooner or later, after you've done enough drives, you're there, able to reach out and touch it with the IDD. Some rocks are par threes: far enough away that it's probably going to take you three sols to get to it. Some are par fours or more. The longest approach that we're pretty confident we can do in a single sol is about two meters, so at three meters away, Adirondack is a par two. We're going to try to get to it in a single sol, and we'll see what happens. But the most likely outcome is that it'll take us two sols to get there.

SPIRIT SOL 15

Life has settled into a routine. I'm finally into the Gusev groove with my sleep cycle now, and the synthetic jet lag of the jump to Mars time is gone. It's a blessing having a martian alarm clock. Somebody up at NASA Ames, where I worked so many years ago, bought a bunch of clock radios a while back, and then went into them and hacked the electronics to make them run on Mars time. Try to find one of those at Wal-Mart. So I awake now each sol to an alarm that goes off at 2 P.M. Gusev Local Solar Time, no matter when that happens to be on Earth. Today it was at about 5:15 A.M. Pacific, but I've gotten to the point now where that really doesn't matter anymore.

My apartment's still barren and sterile; I've been meaning to put some pictures up on the walls, but I haven't had the time. I fall out of bed each martian morning to exercise my one significant culinary skill, brewing a pot of coffee. Even that's ugly. I broke the glass carafe of my coffeemaker the first week out here, so now my coffee drips every morning into a steel saucepan. I gulp down the first cup while I'm dressing, and I carry the second with me as I head down the elevator to the garage below the apartment building. The car right across from the elevator has California plates that read GON2MRZ. I have no idea who it belongs to, but the car hasn't moved once since I arrived, so maybe the owner really has. I drink the second cup on the 210 freeway as I make my way toward JPL. The freeway's clear in the predawn hours again this morning, but in a few more sols, as Mars time slides around, I'm going to start hitting rush hour. The coffee cups pile up in the backseat of my car, one per sol, until I run out. Every five sols I have to take the whole fistful of them back up into the apartment, wash them and start over again.

I'm on shift today as SOWG chair, a job I've been rotating since we landed with Ray Arvidson and Larry Soderblom. During the quick drive up the freeway I mentally run through the commands that we uplinked the sol before, and what we should expect to see on today's downlink.

The big issue today, of course, is whether or not we made it to Adiron-dack. After a few minutes I pull off the 210, past La Cañada High School and past the exclusive-looking riding club that's right next to JPL. The guard smiles as she waves me through the front gate.

The very worst part of every sol is the walk from the parking lot to building 264. I love working the uplink shift, because that's where we make the decisions about what will happen on Mars tomorrow. But there's a major downside to it: Sleeping through the downlink shift that precedes each uplink means that I'm not there when the data first come down. So I miss all the discoveries. Worse, if anything goes really wrong, I'm not going to be on shift when we first find out about it. So the long walk from the parking lot to 264 every sol is filled with apprehension. Maybe this will go away if we ever get enough good sols behind us. But right now, with everything so fragile and new, I arrive at JPL each sol anx-ious about what I might find.

Today brings no disasters, but for the first time since egress I step into the SMSA to find that we've gotten a bad break. Today's *Odyssey* pass was a bad one, with the orbiter low in the sky from where *Spirit* sat, and to-day's direct-to-Earth comm session had the high-gain antenna pointed straight at the PMA head. The result was that all we got down after the drive to Adirondack was one right Hazcam and just a fraction of the left Hazcam. The right image was good enough to prove we've reached Adirondack, but there wasn't enough stereo coverage to show us where we can place the IDD. So we birdied our first par two, but we can't take advantage of it.

SPIRIT SOL 16

We got the rest of that left Hazcam down today, so we'll be ready to put the IDD instruments on Adirondack tomorrow. Meanwhile, we've been going through the data from the IDD work on the soil. There's nothing terribly surprising about it. The MI images are beautifully focused, and they show fine-grained particles that seem to be clumped or aggregated together. The APXS results look very much like the soils that were seen at the old *Viking* and *Pathfinder* sites. They show that the soil here has a bit of sulfur and chlorine in it. Those are chemical elements that tend to be present in salts, maybe a clue that a little bit of some kind of salt is what's causing the clumps in the soil to hold together. But there's nothing very unusual in the chemistry.

The biggest surprise so far, such as it is, comes from the Mössbauer. There's a fair amount of a mineral called olivine in the soil. Olivine weathers away pretty easily if there's much water around, so the fact that we're seeing olivine in the soil with the Mössbauer says that there wasn't a whole lot of water involved in whatever processes made the soil here. Olivine is common in basalt, and the soil here is just looking like finely ground basalt with a tiny bit of salt sticking it together.

I've been listening carefully to the science discussions on the team, and I think we're getting a little too hung up on the rocks in the immediate vicinity of the lander. It's obvious why: They're the only rocks that we can see so far. But there's no reason to expect that there's anything special about the place where the lander happened to come to rest. We certainly should look at enough rocks around the lander to make sure that we've covered whatever geologic diversity the place has to offer. But after that, I really think that we need to get moving on toward that crater to the northeast. I looked over the list of lake names that Jim Rice came up with, and we both thought that "Bonneville" had the nicest ring to it. So Bonneville Crater it is.

It turns out that Bonneville is a place we could have come to grief if we hadn't had DIMES and the little sideways-pointed TIRS rocket motors. I had a talk today with Rob Manning about the EDL reconstruction work they've been doing to get ready for the *Opportunity* landing. DIMES worked on *Spirit*, giving the onboard computer a good solution for what the horizontal velocity was during the descent. Then, in the final seconds on the parachute, at about three hundred meters above the ground, we got hit by a gust of wind that was blowing at least ten meters a second sideways. The chute started to tilt over, heading east and picking up speed. The gyros sensed the tilt, and the computer used the DIMES-sensed horizontal velocity and the gyro-sensed tilt together to figure out what was going on, firing one TIRS motor to kick us back in the other direction. Our horizontal velocity at touchdown was ten meters a second. Without TIRS and DIMES, though, we would have whacked into the southern wall of Bonneville with a horizontal velocity between twenty-two and twenty-five meters a second, which is something like fifty miles an hour. Whether or not we could have survived a hit like that is a matter of conjecture.

SPIRIT SOL 17

The IDD is on Adirondack, and we're starting to collect data. Meanwhile, we've converged on a long-range plan. Once we've taken a good first look at the rocks right around the lander, we're going to head off to Bonneville. With any luck, we'll find something interesting enough there that'll make us want to spend a fair amount of time exploring it. After we're done, and assuming that the rover's still alive, the next thing to do will be to head toward the Columbia Hills. It's an awfully long way to the hills, something like two and a half kilometers from Bonneville, so there's a very good chance that we won't make it. But the plains around Bonneville look pretty much the same in every direction, so if we're going to drive we might as well at least drive toward something interesting.

What the Columbia Hills are made of is anybody's guess. But they seem to be older than the lava-covered plains we're on, sticking up through the lava like islands in a sea of basalt. Some people who've looked hard at the orbital images think they're seeing layering in the hills. I'm not convinced of this, but at least it may be something different from what we're sitting on now. I'm still holding out hope that Bonneville will reveal a lot. But if it doesn't, the hills are probably our best hope.

15 "WE'VE GOT NOTHING"

SPIRIT SOL 18, 1900 GUSEV LOCAL SOLAR TIME

I PULLED INTO THE JPL parking lot at my usual time this afternoon, and I made the walk from the parking lot to building 264 with the usual apprehension. Mark Powell, one of the ground data system software dudes, was on console as science data coordinator as I stepped off the elevator onto the fourth floor, and into the SMSA.

My first words to him were the same as they always are: "How're we doing?"

Mark grimaced. "We've got nothing," he replied.

What? "What do you mean, 'nothing'?"

"Nothing as in no data. We haven't heard anything from the spacecraft all day."

"Nothing?"

"Nothing."

I tensed. Hearing nothing from your spacecraft can mean a lot of different things, but none of them are good. I walked quickly to the flight director's console, where Jennifer Trosper was on shift as mission manager, and she spun the tale out for me. This morning the weather had been awful in Canberra, with heavy rain and lightning. The downlink from the spacecraft had been ragged, with lots of data dropouts, and the uplink had been bad, too. NASA built all three of the Deep Space Network complexes in places with very dry climates to try to get away from problems like this, but sometimes the weather gets you anyway. So it's very likely, Jennifer explained, that the rain had kept some of the commands that they had been trying to send up to the spacecraft from getting onboard. The spacecraft is supposed to recognize when it's gotten an incomplete command load and ignore it, so that shouldn't have caused a problem. But ten minutes before the communications pass was supposed to end they abruptly lost all telemetry from the spacecraft. Since then all they've heard has been two "beeps" that came down through the low-gain antenna—no data, just a little tone from *Spirit* to let us know she was still there.

With that news, I hunkered down in the SMSA for what it looked like could be a very long night. And while the loss of direct-to-Earth telemetry is bad enough, in the last few hours the story has taken a more ominous turn. When *Mars Odyssey* flew over Gusev and listened for a signal from *Spirit* a couple of hours ago, it heard nothing at all. So something bad definitely has happened on our rover.

I don't know how worried I should be at this point. I don't think the spacecraft is likely to be at any significant risk, at least just yet. The performance of the power and thermal systems has been outstanding so far, so we shouldn't be at risk of running down the batteries or freezing the vehicle at night, at least for a few sols. And the fault protection software on this machine is very good—she knows what to do to keep herself alive when something bad has happened. So we've probably just gotten ourselves into some kind of minor trouble here. Maybe we let the temperatures get a little too high in the WEB, and *Spirit* shut herself down to cool off. Or maybe the onboard list of communication windows has gotten scrambled somehow, so she's not talking to us when we expect her to.

Chances are that we'll get this all straightened out by sending commands up through the low-gain antenna, and we'll be back in business within a few sols. But we're certainly going to lose some science.

This is the first major in-flight anomaly we've had, and I have never seen this team more focused than they are right now. As I write this we are gathered in 875, the project conference room, and everybody's here: Pete, Richard, Matt, Jennifer Trosper, Mark Adler, Gentry Lee, all the mission managers and flight directors who are supposed to be awake and a bunch more who are supposed to be asleep. We're trying to work our way methodically through all the various possible fault scenarios.

It's a difficult discussion, in large part because we've got almost no data. Telemetry from the spacecraft is what you use to work out problems like this when they happen, and it's hard to make much headway on an anomaly when all you've got to go on is the fact that you have no telemetry. So the focus instead has been on listing all the things that might have gone wrong, and calling on the team's knowledge of how this fiendishly complicated beast is supposed to behave to figure out which ones might have caused the kinds of data dropouts that we've seen. Nobody seems scared, at least yet, but everybody is very intense. The meeting keeps breaking up into little side discussions, and Matt and Jennifer have to keep reining everybody in and getting them back on track.

There are several things that could have done this to us. One is what's called a low power fault, which is what happens when the batteries get too low. Nobody can think of a good reason why this would have happened, and it seems awfully unlikely. But if we did get into a low power fault condition for some reason, the rover simply would have shut down— consistent with the silence that we saw during this afternoon's *Odyssey* pass—and we won't wake up until we have sunlight solidly on the solar panels again tomorrow morning. If that's what happened, the way the vehicle is supposed to respond is by talking to us at ten bits per second over the low-gain antenna at 11 A.M. Mars time tomorrow. So if we're in low power mode we should also expect silence during the overnight *MGS* or *Odyssey* passes that are coming up in a few hours. The vehicle should just sleep through both of them, conserving whatever battery power is left. And then we should hear something tomorrow morning.

Another one that also seems pretty unlikely is called a WEB over-temperature fault, where things inside the vehicle get so hot during the afternoon that there's a danger of overheating something. Like low power, nobody can think of a reason why a WEB overtemp would have happened. We know how hot it's gotten inside the vehicle for each of the last seventeen sols, and we've never come anywhere close to the red line. Could some component inside the WEB suddenly have gotten stuck on, dissipating a lot more heat than it did before? It doesn't seem likely, but it could happen. WEB overtemp is probably the most benign of the failure scenarios, since it should make the vehicle go through a graceful, controlled shutdown once the temperature gets too warm. If it was a WEB overtemp fault that did this to us then we also would have been asleep during the afternoon *Odyssey* pass, but things should be cool enough by this evening to let the vehicle wake up normally and talk to us through both *MGS* and *Odyssey* tonight. So if we get a signal through either orbiter tonight, it'd be a pretty strong indication that it really was a WEB overtemp that got us.

So those are the unlikely scenarios. The likely scenario, at least in my mind, is that the flight software has gone off into the weeds somehow. The software now is a hell of a lot more stable than it was back in the dark days of thermal vac, but after seventeen sols of surface operations there's some chance that we're doing things to it that we've never fully tested yet. So there's a pretty decent chance that we've broken the software in some way that we didn't know it could be broken.

Could the list of communication windows that the vehicle keeps on-board have gotten corrupted somehow? Nobody can think of a way for that to happen, but it certainly could lead to the symptoms we're seeing if it had. Could the computer have gone through an unexpected reboot for some reason? If that happened it'd kill off everything that was happening onboard at the time we lost communications, but once the computer came back up again the comm window list should still have been there. It doesn't seem like it was.

Did we confuse the software by running some motors in a way we had never tried before? At the exact moment when we lost the signal we should have been doing a health check of one of the Mini-TES mirror motors, at the same time that we were steering the high-gain antenna to keep it pointed at Earth. Did this screw us up somehow? Maybe a motor problem would have killed off communication through the high-gain antenna, since we need motors to steer the antenna. But why would it have killed off communication with the orbiters? The UHF antenna is fixed, just like the low-gain antenna; we don't steer it, so it doesn't use any motors. This theory seems pretty far-fetched, but unlike some of the others it at least has the virtue that we can try it out on one of the rovers we have on the ground and see if anything bad happens. So a group is getting spun up to go off and do that test.

I can't help thinking that somehow the bad uplink caused by the rain in Australia must have had something to do with it. A partial load of commands gets in, the software gets confused and off we go into the swamps. But Glenn Reeves swears that the software should recognize and reject any kind of incomplete command load, and he's the architect for the whole software system. So the juxtaposition of the bad uplink with the anomaly is probably just a coincidence, but it sure seems like a weird one to me.

And, of course, with all the focus on whatever's gone wrong with *Spirit*, the final trajectory correction maneuver for *Opportunity*—which I had hoped would improve our chances of landing someplace in Meridiani where the geology looks interesting—has been canceled. We're on a trajectory toward a safe part of the ellipse, and with all the other problems we've got, we're just going to ride it out.

SPIRIT SOL 19, 0200 GUSEV LOCAL SOLAR TIME

I'm pacing in the SMSA with my headset on, waiting for the overnight *MGS* pass. I'm not really expecting that we'll hear anything on this one, since the most likely scenario seems to be that the onboard list of comm windows is messed up somehow. It's well into the evening now Pacific time, but everyone's here: Pete, Firouz Naderi, everybody up the JPL chain of command. The atmosphere is tense, and nobody's saying much.

Abruptly, the voice of UHF Telecom comes on the VOCA net: "We've got data!" I let out a whoop, and there's cheering and backslapping around the room. *Spirit's* still alive, and she's talking to us! I step out of the room to find someplace quiet, and pull out my cell phone to call Mary back in Ithaca. She had made me promise to call her when the pass came, no matter how late it was and no matter what the news was. A sweet, sleepy voice comes on at the other end of the line, and all I can manage to say is "We got it." Not until my voice cracks as those words come out do I realize how worried I really was about this.

We talk for a few moments and I ring off, letting her fall back to sleep. But as I step back into the SMSA, immediately I can tell that something's wrong. The hubbub has died down, and the smiles are gone. I jam my headset back on just in time to hear MDOT announcing that he didn't get any valid data during the pass, followed almost immediately by UHF Telecom with the news that the pass was much shorter than she expected it to be. Damn. *Spirit* woke up and started talking to *MGS* just when she was supposed to. But everything she sent was garbage, and the radio shut off abruptly only two minutes into what was supposed to be a thirteen-minute pass. After that there was only silence.

Suddenly, the situation has gone from worrisome to scary. What the hell is going on here? We obviously didn't trip a low power fault, since the rover wouldn't wake up in the middle of the night if that had happened. And the comm windows aren't messed up, since she talked to *MGS* right on schedule. Overheating in the WEB is looking pretty plausible right now, but why would that have happened? And why were the data garbage? And why did the pass end early? Nothing adds up, and without any valid data it's almost impossible to put a picture together. We're groping in the dark here.

SPIRIT SOL 19, 0600 GUSEV LOCAL SOLAR TIME

The overnight *Odyssey* pass is over now, and there was nothing on it. Not even garbage this time, just no signal at all. This is starting to look very bad.

At least we've got a theory for what happened during the *MGS* pass. There really aren't that many things that can make a transmission stop in the middle of a pass. The fact that the pass started at all says that the computer had to have been up and running, at least at first. So it looks like maybe we had a reboot during the pass. Glenn Reeves, our software architect, has pointed out that if we had some kind of fatal error right when the computer first turned on, it could have caused symptoms very much like what we saw. A fatal error on startup would have caused a shutdown fifteen minutes later, which, it turns out, is just about exactly two minutes into the comm pass. And that's when we lost the signal. And if the fatal error had affected the communications software, it would have meant that the computer wouldn't send any valid data to the radio. With no data from the computer, the radio would have just transmitted garbage, which is what we saw from it before the signal died. So I think Glenn's onto something.

But why would we have had a fatal error on startup? And why did we see nothing at all during the *Odyssey* pass? We still don't have the whole picture here.

I need to sleep soon, though I'm not sure if sleep will come. *Opportunity* hits the top of the atmosphere in less than seventy-two hours.

SPIRIT SOL 20, 0000 GUSEV LOCAL SOLAR TIME

There's been no improvement. I managed to sleep for a few hours, and then I came back in to the lab for a while during the afternoon, Gusev time. The mood in the SMSA remained tense and grim. Most of the day had been consumed by sending commands to try to get *Spirit* just to send back a beep and confirm that she was still alive. It's a bit of a guessing game, doing these beeps. The rover listens for signals from Earth in different ways depending on what kind of state she's in. If things are more or less normal, she sits there expecting to hear commands from Earth that were sent at a rate of 31.25 bits per second. But if something's gone really wrong and she's in some kind of fault condition, she listens instead only for commands that are sent at a slower rate of 7.8125 bits per second. We can only send commands at one rate at a time, so we have to guess what rate she's listening at and hope that she'll hear us and respond.

So if we command a beep and we don't hear anything back, what does it mean? That we sent the command at the wrong rate? Or that something's wrong with the computer onboard and it wasn't listening when we wanted it to? Or that something's wrong with the transmitter onboard and it can't beep? It's damned hard to infer anything useful from a complete lack of data.

The one piece of good news from the day is that just before Earth was about to drop below the horizon at Gusev, cutting off any chance of communication, we got back a beep. We could tell from the timing that it was one that had been commanded at 7.8125 bits per second, which is the rate used in a fault state. So *Spirit*'s still there. But she's a very sick rover.

I went back to my apartment and crashed for a few hours more, and then came back in again, driving up the 210 freeway at sunset. It's midnight now at Gusev. The sky is black there, and the temperature is sixty degrees below zero and falling. There's no energy flowing from the solar arrays into the batteries, and there hasn't been for hours. If *Spirit*'s asleep like she should be, conserving her energy for the morning, then she'll make it through the night just fine. But if she's awake, things could get rough.

Heading back to room 875 for the next round of guesswork, I run into Gentry Lee as I get off the elevator.

"Any new theories?" I ask hopefully.

"Yeah," he says, "but they're all bullshit. Nobody has come up with a theory that can fit all the observations with one failure. Either something is going on that we fundamentally don't understand, or more than one thing has gone wrong onboard."

This is bad. We need to come up with at least a plausible guess about what's gone wrong before anybody can start finding a way out of this mess. Theories that involve multiple failures are very unsatisfying. Either they require some underlying cause that may have been extremely damaging to the spacecraft, or they require a very unlikely coincidence. I'm holding out hope that it's something that nobody has thought of yet.

We're gathered now again in 875, in the biggest meeting our brand-new conference room has ever seen. The conference table, stretching the full length of the room, is surrounded elbow to elbow with grim-faced engineers. I've known every one of them for at least three years now, and I've never seen them looking like this. Every chair along both long walls is likewise filled, with more people spilling onto the floor, and two deep out into the hallway. I'm scrunched down with my laptop on the floor against the front wall of the room, looking almost vertically up at the projection screen. It's very hot in here, and I can tell that I'm not the only one who hasn't had a chance to take a shower in a while.

The meeting is idling in slow circles, with everyone looking at their watches and getting ready to head back downstairs for the next *MGS* pass.

Pete's having none of it: "Why is this meeting breaking up for an *MGS* pass? Is there anybody here who needs to be on console?" Sheepish looks are cast around the table; no, there's nothing that anybody here can do in the SMSA during an *MGS* pass but sit and worry.

"Okay, that's what I thought." Pete's keeping everyone's eye on the ball. "So let's forget about *MGS*, and let's focus on what really matters. What are we going to do with the spacecraft tomorrow?"

Matt Wallace steps in and tries to get the discussion back on track. "Okay, so what are the top theories?" he asks.

Daniel Limonadi, who looks like he hasn't slept in a very long time, sighs. "We don't have enough data to have any top theories." And he's right. Without some kind of real information from the spacecraft, it's going to be very difficult to work this thing out.

Daniel plugs in his laptop and starts taking us through a big table that he's put together listing the time that each event happened, what actually happened at that time and what inferences we can draw with bulletproof certainty from whatever happened. That third column of the table doesn't have much in it. We're stepping through the table line by line, with everybody jumping in to add details and challenge assertions.

There are a few things that we do know. On top of the weather problems at Canberra, we know now that the Aussies were also having problems pointing their antenna that day, and that they can't say for certain that they were pointed directly at Mars at the moment that we lost contact. So the fact that telemetry was cut off then doesn't necessarily mean that anything had gone wrong on the rover yet.

Abruptly, four or five pagers and cell phones go off around the room almost simultaneously. No joy: No data came down on the *MGS* pass, and in fact the orbiter didn't even detect a signal from *Spirit* at all this time. It's been a sol and a half now since we've gotten any data at all from our spacecraft.

The meeting lurches onward, but we're floundering. There's just not enough information for us to tell what the hell is going on. We absolutely must figure out how to reestablish communications well enough to start getting some real engineering telemetry.

In the absence of information, we focus on why the information is absent. There are two big mysteries. First, why didn't we hear anything over the high-gain antenna on the afternoon of the anomaly? Second, why didn't we hear anything during the *Odyssey* pass a couple hours later? There's a long list of possible explanations for both, but it's Richard Cook who zeroes in on the key question: "What do these two have in common? Is there any one failure that can account for both of them?" The only obvious one is WEB overtemp.

And here are two more questions: Why did the *MGS* pass last night end early, after just two minutes, and why did we get nothing but garbage data before it cut off? And again, what's common to the two problems? Glenn Reeves thinks he came up with the answer for that one last night: If we had a fatal error in the software that's responsible for sending data to the radio, it could hose the data and then also cause a reboot after two minutes of transmission. So that's a working theory, though nobody has any idea why it would happen.

The beeps aren't worth much, but they do tell us a little. The beep that we got yesterday afternoon was one that we know was commanded at an uplink rate of 7.8125 bits per second. The fact that the rover was listening for commands at that low bit rate means that she was in some kind of fault mode—that *Spirit* herself had realized that something was wrong.

But what was wrong? There aren't too many things that can make the vehicle switch to the low uplink rate. One is a low-power fault, which is what happens when the battery gets too low. That one is starting to seem alarmingly likely now, since we really seem to have lost control of the vehicle. If there's no way to control what's going on onboard, there's no way to assure that an overnight shutdown will take place like it should. And that can burn the batteries down pretty fast. The other fault that can send us to the low uplink rate is what's called an X-band fault. This is what happens when the vehicle senses that something is wrong with the direct-to-Earth communications. And that doesn't seem too unlikely, either, given that we're not getting any direct-to-Earth communications! So it's pretty likely that we're in either low-power fault mode or X-band fault mode, or both. And there could be a lot more wrong than that.

So now we finally get down to the nub of it: What do we do tomorrow? Battle lines are quickly drawn. How aggressive do we get about trying to reestablish communications? Should we try to get a beep first to confirm that *Spirit*'s alive and commandable? Or should we just start slinging up commands to set up new comm windows and hope for real data? Richard Cook, Art Thompson and Rob Manning are all in favor of doing the beeps first. Chris Salvo and Daniel Limonadi want to be more aggressive and try to get some data early in the day. Chris pleads, "Let's not squander the morning on beeps when what we really need is data." Richard settles on a compromise approach in which we'll try one beep, and then move on to setting up comm windows whether we hear the beep or not.

Now we turn to the issue of continuous reboots. This is a variant of the software-in-the-weeds theory that says that something that has gone wrong is making the computer sense a fatal error each time it boots up, responding with a reboot and another fatal error, in an endless cycle. Glenn came up with this one after he thought some more about what happened during the *MGS* pass last night, and I think he may be on the right track. It's a scary scenario, though, because a lot of the things that could cause it involve broken hardware. And if the hardware is broken it's broken; the best we could hope for in that case would be to figure out why we're dead.

One thing we can do to run down the repetitive reboot theory is go through every line of the flight software and look for anything that's capable of causing a fatal error and a reboot. So that's exactly what Glenn has spent his day doing. It's laborious work, and he's turned up a whole bunch of possibilities. None of them seem very likely, but it's as good a fit to the little information we have as anything anybody has come up with. And that frightens me.

One thing that's been on everybody's mind is what would happen if we had completely filled up the flash memory onboard. The flash is what we use on the rover to store all of our data products, and after eighteen sols of operations it's closer to being full than it's been at any time since

The scene when we took our first look around in Gusev Crater. The shallow, dust-filled depression nearby is Sleepy Hollow, and the dark splotches on the floor of the hollow are airbag bounce marks. It was a beautiful view, but those didn't look like sedimentary rocks scattered about.

Spirit's first look at the Columbia Hills. They seemed impossibly far away.

Six wheels in the dirt at Gusev... *Spirit*'s view back toward the lander.

Mark Boyles and me in the SMSA as we receive the data confirming *Spirit*'s successful deployment. One of the best moments of my life.

The first picture from *Spirit* after the Sol 18 anomaly, showing the arm still deployed on Adirondack. It was good to see she was still alive.

The end of *Spirit*'s instrument arm. We built the piece of the Rock Abrasion Tool that has the American flag on it using metal from the wreckage of the World Trade Center in New York.

The very first Navcam image from *Opportunity*, with layered bedrock just meters away. Unbelievable.

Left to right: Chris Voorhees, Kevin Burke, Randy Lindemann and Mark Boyles
checking out some 3-D images from *Spirit* in the SMSA.

NASA/JPL/CORNELL

Airbag bounce marks in Eagle Crater, so clear you can almost see the stitching.

AP/DAMIAN DOVARGANES

Joy and astonishment in the SMSA as the first *Opportunity* images hit the ground. From left to right: Matt Wallace, Chris Lewicki and Joel Krajewski.

NASA/JPL/CORNELL/USGS

Freaky little hematite balls. Our first close-up look at the "blueberries" of Meridiani Planum.

NASA/JPL/CORNELL

Enhanced color image of El Capitan, showing the "chicken scratches" that we quickly realized were crystal molds. This was the picture that finally convinced me that there had once been water at Meridiani.

Eagle Crater after we were done with it. We made a bit of a mess of the place.

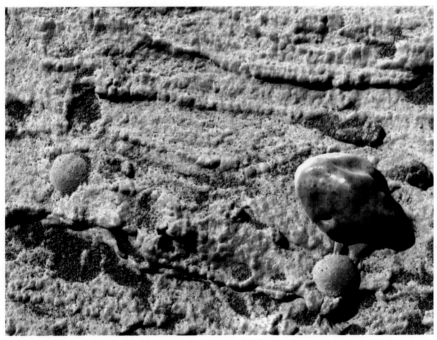

A Microscopic Imager image showing current ripples in the rocks of Eagle Crater, formed long ago by flowing water.

Wheel tracks in the rearview mirror as *Spirit* hits the road.

Bonneville Crater, the big disappointment.

Rover driver Chris Leger.

Wheel tracks behind *Opportunity* during the sprint to Endurance Crater.

Our first good view of Endurance Crater. It was a scary-looking place.

Endurance Crater from the southeast side.

NASA/JPL

Driving one of the test rovers on a test fixture at JPL, trying to decide whether or not it'd be safe to send *Opportunity* into Endurance Crater. The slope is so steep that the technician has to wear a rope and a safety harness. The rover, on the other hand, is on its own.

NASA/JPL

Opportunity sees her shadow as she heads over the lip and down into Endurance.

Enhanced color image of RAT holes in the layered sediments of Endurance Crater.

The dune field on the floor of Endurance Crater.

A panoramic view inside Endurance Crater.

Approaching the base of the Columbia Hills.

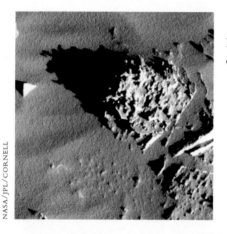

Pot of Gold...the rock that almost drove us crazy.

Spirit's view out onto the Gusev plains, from up in the Columbia Hills.

launch. Is there a way to fatal the software by filling up flash? We sure had problems with flash back during thermal vac. It's a good-sounding theory, but they tried filling up flash in one of the rovers here on the ground today, and nothing bad happened.

We straggle out of the room as the meeting sputters to a finish, tired and discouraged. I can't let myself believe that *Spirit*'s dying, but I don't see a way out of this yet, either.

We're really lost.

SPIRIT SOL 20, 0900 GUSEV LOCAL SOLAR TIME

It's a few hours later now, after a quick, restless flop on the futon on the floor of my office. If we don't start getting data again today, then chances are we're on a path to losing this thing. DSS-43, the big antenna in Tidbinbilla, is up and ready to start commanding. And the weather in Australia, thankfully, is good.

The first thing we're doing is commanding a beep at 31.25 bits per second. This is unlikely to do anything, since *Spirit*'s probably still in a fault state, listening at 7.8125 bps. But we're covering our bases. Next up will be another beep that we'll command at 7.8125. That's the beep that we ought to hear. If we hear either one of the beeps, we'll then command *Spirit* to open up a comm window on her low-gain antenna at whichever rate we find is working. If we don't get either beep, we could be in serious trouble.

It's 9:54 P.M. in Canberra and 12:54 P.M. in Madrid. Mars is setting in Australia and rising in Spain, so we're handing off from DSS-43 to DSS-63 to listen for the beeps. The first beep, if we get it, will come at 11:59:38 UTC, universal time, just two minutes after the handoff to Madrid. The second one will come at 12:06:55 UTC.

12:01 UTC: No news from Spain. . . . Looks like we didn't get the first beep. No surprise there. Now here comes the big one.

12:06:55 UTC: Listening.

12:08:55 UTC: Still listening, two minutes in. I'm watching the Doppler display on the screen behind me, and it's just flat.

12:12:55 UTC: Nothing. No beep. Maybe the beep command didn't get in. Or maybe we're dealing with some problem that nobody's thought of yet. Or maybe she died during the night.

This is the worst moment since the anomaly started, and I honestly don't know what to think. I used to have a lot of faith in this rover's ability to keep herself safe as long as the software is running properly. But my faith in the software itself is so shaken now that I don't know anymore. *Spirit* is gone, and *Opportunity* is closing in on Meridiani Planum. The atmosphere over Meridiani is still puffed up, and the EDL guys are still fiddling with chute deployment parameters. It's all looking very bad.

This whole thing could come out fine, of course. We've been through one landing already, so we know that *Opportunity*'s EDL design is fundamentally sound. And goodness knows spacecraft have seemed irretrievably lost and then been recovered before. But with *Spirit* silent and *Opportunity* about to hit the planet, I can't escape the thought that everything we've worked for may be gone seventy-two hours from now.

12:31 UTC: *Whoa . . . what???* DSS-63 just came back on the net and said that they think they may be seeing something! We didn't command this, but sure enough I'm staring at the Doppler plot and we've definitely got a signal. Telecom says it's a couple of dB above noise. It's *Spirit!*

Holy shit. Now we're actually getting data, coming down off of *Spirit*'s low-gain antenna at 10 bits per second! Mark Adler is on shift as mission manager, and he quickly comes on the net and explains that this is a comm window that the rover would have opened up on her own if we had had an X-band fault, or if we'd had a complete loss of mission clock. X-band fault is something we've thought about already, since it's one of the things that could have sent us into the 7.8125 bps uplink rate. And loss of mission clock would be something that would have happened if the battery had gone completely down to zero.

So which one of these did it? At this point, you almost have to ask,

Who cares? At least she's still alive! It's morning at Gusev, the sun's shining on the solar arrays and *Spirit* is talking to us.

And there's a way to figure it out. The length of the comm window should depend on which fault mode we're in. If it's mission clock the window should last for sixty minutes, while if it's X-band it should last for only thirty. All we have to do is wait and see how long it lasts. And of course there could be a lot more information than that in the data we're slowly getting down, once we can open it up and look at it.

12:42 UTC: Damn, we just lost the signal! That wasn't sixty minutes, and it wasn't thirty minutes, either. . . . It was eleven minutes, as if the computer crashed partway through the pass. Glenn Reeves's theory about the repetitive reboots is starting to sound pretty likely. But what's causing them? We all cluster around the flight software console, waiting to see what came down in that tiny eleven-minute fragment of data. And, of course, once Flight Software opens it up and looks at it, he discovers that it's a message saying that we've had a fatal error onboard. Big help.

We desperately need more data. It's still early in the sol, and the rover's awake, so it's time to go after this thing. Systems is recommending that we command a high-priority comm window at 120 bps, and Telecom says she's go. We radiate the command at 12:58 UTC, and settle in to wait for the two-way light time to pass.

Meanwhile, Fault Protection has had a little time to dig into the fatal error message that we got during the eleven-minute low-gain pass. She reports that it was a real-time fatal, which means that the computer detected a fault condition on startup, and then waited fifteen minutes before shutting down. The shutdown was probably what cut off the pass. So it looks like Glenn is right. The thing I really don't like is the news about the time tag on the error message. Either it's gotten corrupted somehow, or *Spirit* thinks that the year is now 2053. We may indeed have had a complete loss of mission clock.

13:22 UTC: We should be five minutes away from seeing data, if we haven't been clobbered by a reboot. That's a pretty big if.

13:27 UTC: All right, we've got two-way lock on the spacecraft, and we're getting data again. The spacecraft is commandable and talking to us. So now we'll see what we see.

13:33 UTC: Telemetry! We're getting event records, and the time tags on them are correct! We didn't lose the mission clock after all, which is a relief. So now we know that we must be in X-band fault mode, since it's the only possibility left that fits all the symptoms. We're actually making a tiny bit of progress here. But after a few minutes it becomes clear that we're getting the same few event records over and over again, which means that we're not going to learn a whole lot about what's happening on the spacecraft from this pass.

13:46 UTC: DSN reports loss of signal. This is considerably earlier than we commanded the pass to end. We definitely seem to be in a repetitive reboot cycle, with the reboots taking us down long before we can finish each pass.

14:17 UTC: We've commanded another high-priority comm window at 120 bps. We should get a signal back at 14:36. Maybe we'll see something different this time.

14:36 UTC: Two-way lock on the spacecraft, right on the money. And this pass is really working. Hot damn. We're getting currents, temperatures, voltages, records of what the software has been doing. This may be what we need to figure this thing out. And the fact that we're getting real data suggests that whatever is causing the reboots might be something other than seriously broken hardware.

A bunch of us—Glenn, Richard, Jennifer and many more—are clustered around the fault protection and flight software consoles as they plow through the data from this pass. We've got a very sick spacecraft. It's going to take a while to mine the data completely, but the key finding is that the battery state of charge is extremely low—so low that *Spirit* must have been awake for most of the last two nights. We've got to get her shut down hard and let the power system get healthy again. After a short discussion, the conclusion is to dust off the old SHUTDOWN_DAMMIT command that we used to use when things went to hell back during thermal vac. SHUTDOWN_DAMMIT should shut things down cold and should prevent any comm windows from opening up until the commanded wake-up time, which won't be until tomorrow morning. We're going to radiate it to the spacecraft and give the power system a break.

At this point it's been a very long time since I've had any sleep, and I decide to head home and try to get some real rest. I pick up a few odds and ends in my office on the sixth floor, then head down the elevator and start making my way to the parking lot. Just as I'm about to leave the lab I run into Pete, who's headed for building 180 at high speed. He looks unhappy.

"Did you hear?" he asks.

"Hear what?"

"It woke up again."

"What?" I ask. "You mean after the SHUTDOWN_DAMMIT command?"

"Yeah," Pete replied. "It woke up again."

This is awful. We still don't have control of this vehicle. If the SHUTDOWN_DAMMIT command didn't work, that means *Spirit* will probably be awake through the night again, crashing and rebooting over and over all night long. I don't know how much more of this she can take.

SPIRIT SOL 21, 0700 GUSEV LOCAL SOLAR TIME

I'm back in the SMSA after my first semidecent sleep since the anomaly hit. There's no new data since yesterday. The power system has to be getting very ragged by now, and it's about fifty-fifty whether the battery was low enough by sunrise this morning to trigger low-power fault mode. If low-power mode got triggered—and if it actually worked—it'd be a good thing, since the fault protection software then should have kicked in and made the vehicle take all the necessary steps—mostly shutting down systems and recharging the battery—to stay safe. But we've got no real confidence that any of the fault protection software works properly now, the way things are going. And if low-power mode doesn't work, and if we don't get control of this thing some other way, we could be done in another sol or two.

The one piece of good news is that Glenn now has a theory about the root cause of the problem. His theory is that the file system in the flash memory onboard the vehicle has somehow become corrupted. The file system is a bit like the filing system in a library: It's a way of organizing all the information that's stored in flash. And if the file system got corrupted somehow, that's definitely the sort of thing that could send the software off into the weeds.

During a brief quiet moment I ask Glenn why he thinks it's a problem with flash.

"It was that pass yesterday where we got nothing but real-time data," he says. I think back to it, and he's right: *Spirit* was telling us all about what was happening at the time she was transmitting, but nothing about what had gone before. She wasn't sending us any recorded data. It was as if she had no memory of whatever had happened to her.

"So what could cause us not to be able to get any archived data?" Glenn goes on. "We were getting stuff, but none of it was coming out of the flash. The software had to be working, because it was producing the real-time stuff. But where was the stuff from flash?"

So maybe he's right. And to my surprise, there's a way to test his theory. The flash file system has been one of the weakest parts of this software since the very beginning. It's part of what kept driving us nuts back during thermal vac. And because Glenn's been worried about flash from the beginning, it turns out that he built in a special trapdoor that should let us boot the computer up without using the flash file system at all. Using a command that is colorfully named INIT_CRIPPLED, we can initialize the system in a way that bypasses the flash file system completely, building a temporary file system in RAM, the random-access memory in the computer. If a corrupt flash file system is the problem, then booting with INIT_CRIPPLED should keep the software from crashing. Of course, the rover can't actually *do* much in this state—we need the flash to go about our normal business. But it might be a step toward regaining control of the vehicle. So the plan for today is to use INIT_CRIPPLED commands to try to regain control of the vehicle, and then another SHUTDOWN_DAMMIT to try to get her shut down for real, and start the battery charging.

12:34 UTC: Okay, here we go. We've been hammering the spacecraft with INIT_CRIPPLED commands for a while now, hoping that at least one of them will get in. At the same time, we're about to start listening for a comm window that the spacecraft would have initiated automatically at 0930 Gusev time if we didn't go into low-power fault mode overnight.

12:49 UTC: Now is when we would be hearing that 0930 comm window if it were happening, but we're not getting anything. This could mean that we indeed tripped low-power mode overnight, but it could also just mean that we're in a repetitive reboot cycle again, and the computer happened to be in the middle of a reboot. We're in the dark again.

The next thing to wait for is a comm window over the low-gain antenna at 14:16 UTC. This is one that we would see if we *had* tripped low-power overnight. And meanwhile, we're radiating a SYSTEM_RESET command, which should reset the system after all of these INIT_CRIP-PLED commands we've been sending. So if we do see the low-gain window at 14:16, it should end early because of the reset that we're commanding right now.

14:19 UTC: We've got a signal, which probably means that we indeed tripped low-power mode overnight. If everything's working right after the INIT_CRIPPLEDs, we should see the reset that we just commanded hit the rover at 14:26 UTC, shutting the comm window down prematurely.

14:26 UTC: Loss of signal, right on the money. Yowza! This reset is one that we commanded, which means that *Spirit* was actually receiving commands and responding to them normally for a short while. So here we go. Mark's mission manager again today, and he's just authorized the ACE to radiate a command that'll open a high-priority low-gain comm window at 120 bps, starting at 14:51. This is the moment of truth.

14:51 UTC: Got a signal. Now we'll see what we see. . . .

15:06 UTC: We're getting real data! Not garbage, and not the same event records over and over, but real data, for the first time since the anomaly. The INIT_CRIPPLED worked! God bless Glenn Reeves for the paranoia that made him put that command into the system and leave it there.

And the data are looking pretty good. Power is reporting that we did indeed trip low-power last night, but that's hardly a surprise. We've got one badly abused power system up there. Realizing that *Spirit* was finally responding to commands normally, we radiated a SHUTDOWN_ DAMMIT partway through the pass, and we saw it hit the spacecraft and cause a shutdown right on schedule. All we've got to do now is hope that she stays down through the rest of the day, and overnight. If that happens, she should come up tomorrow morning with reasonably healthy batteries.

16:22 UTC: A little while ago we commanded the vehicle to send us a beep. We waited twelve minutes for the command to get to the spacecraft, and another twelve minutes for the beep to get back, and we heard nothing. That's great news! Based on everything that has happened, it probably means that *Spirit* was sleeping peacefully when the beep command arrived, and never heard it.

I really feel like we've turned a corner today. *Spirit* has been mentally ill for the past three sols, and she was beyond our control. Today she regained a measure of sanity.

And on a whiteboard in the SMSA this evening, I saw these words: "The *Spirit* was willing, but the flash was weak."

I'm truly amazed by the intuition that enabled Glenn to realize that it was probably a bad flash file system causing the problem. A corrupt flash file system definitely would cause a reset every time the computer came on, sending us into a repetitive reboot cycle. And it turns out that repeated reboots would send us into X-band fault mode. We were using the high-gain antenna, moving it to keep it pointed at Earth, way back when the anomaly first happened. If the system goes down when we're moving the antenna, then the rover realizes after the reboot that she doesn't know exactly where the antenna is pointed. So then, if you tell her to use the high-gain antenna when she doesn't know where the thing is pointed—which is the first thing we did after the anomaly—she says, "I can't do that" and sends herself into X-band fault mode.

And, of course, the repetitive reboot cycle also explains why we went into low-power mode. The computer was rebooting and rebooting all night long, never going to sleep like it was supposed to. So naturally we took the battery down to almost zero. Everything fits.

It still could be a long road back. We don't know what caused the flash corruption in the first place yet. If it's damage to the flash hardware itself, then that could seriously curtail what we can do with the vehicle from here on out. But we'll see what we see. It could be something less sinister than that. At least a consistent story about the cause of the problem has emerged. And at least *Spirit* is behaving like a sane rover again.

The thing that strikes me most about all this is how crucial it was to have that INIT_CRIPPLED command in the system. It's not the kind of command that you'd *ever* expect to use under normal conditions on Mars. But back during the earliest days of the project Glenn realized that someday we might need the flexibility to deal with a broken flash file system, and he put INIT_CRIPPLED in the system and left it there. And when the anomaly hit, it saved the mission.

I've gotta buy that boy a beer.

16 EAGLE CRATER

I T'S LANDING NIGHT FOR *Opportunity*. The mood in the CMSA is relaxed and loose compared to last time. Partly that's because we know now that we have an EDL design that works, and partly it's relief that the *Spirit* anomaly is finally under control. *Spirit*'s going to be in rehab for a while, but it's looking increasingly likely that there's no hardware problem with the flash, and that once we get all the software issues cleared up she'll be good as new. So I've left *Spirit* in the hands of Mike Carr, John Grant and Ron Greeley, who'll be the SOWG chairs over at Gusev for the foreseeable future. Ray Arvidson and Larry Soderblom and I are done with *Spirit* for now, and we've made the jump tonight to Meridiani time. It's a big jump, and I'm feeling it already. Meridiani is halfway around the planet from Gusev, so the two martian time zones are a full twelve hours apart. It feels like getting on a plane and flying to India.

I'm standing again at my place in the back row of the CMSA, tonight between Richard Cook and Mark Adler. Pete's pacing again. As I look down the line I notice that O'Keefe, Elachi and Weiler all have headsets on this time; they weren't able to hear what was happening when *Spirit* landed, and that was a problem that had to be fixed. Another thing that stands out as I look around the room is how many people are wearing the same outfit that they wore for the first landing, myself included. We may be feeling looser than we did last time, but we still need all the good mojo we can get.

HRS venting, cruise stage sep, all the events leading up to atmospheric entry go as planned. Then, a few minutes from the top of the atmosphere, the room begins to go quiet. We're about seventy-five miles above the martian surface now, a little over four hundred miles from the landing site. As the chatter in the room subsides, the only voices on the net are those of Wayne Lee, ticking off the velocities and altitudes, and Rob Manning, calling out the accel tones. The tones climb through four, five, six and seven in rapid succession, and then stairstep back down again; we're past the acceleration peak and deep down into the atmosphere.

Polly's watching her screen alongside Rob: "There it is, Rob, parachute deploy!" And then, seconds later, "Heat shield! Heat shield!"

Everything's working, just like last time. We get radar lock on the ground, and then the tone indicating that the computer has a firing solution. Here we go.

"RAD firing!" yells Polly.

"RAD has fired," confirms Rob, and then, a few long seconds later: "Getting bouncing signal."

Each callout from Rob and Polly brings a brief burst of cheering, but now the room goes quiet as the airbags bounce and tumble across the surface. This is where we had the long, torturous silence last time.

But not this time. Polly calls out again: "We're seeing it on the LCP!" We're down, baby!

Again there are handshakes and again there are tears. Matt comes up and hugs me, almost lifting me off the floor. "We've got two!" he says, grinning. We sure do.

We've got roll stop now, and the signal from the vehicle is solid and strong. *Opportunity* is on Mars.

As soon as it's clear that everything's good, I walk over to von Kármán to watch the post-landing press conference.

Von Kármán is JPL's biggest auditorium, and tonight it's completely full. It's already past time for the press conference to begin when I find myself a place to stand in the back of the room. O'Keefe and Elachi and Weiler are up on the stage along with Richard and Pete and Rob, getting ready to tell the world that we've got two rovers on Mars. Two tiers of risers at the back of the room are packed with television cameras, the crews elbowing for position in the narrow space they've been allotted. It's a little before midnight Pacific time.

Just as the briefing's about to begin, there's a commotion from somewhere near the very back of the auditorium. I can't see what's happening, even standing on my toes, but the crowd is moving away from the doors to let someone or something through. A cheer goes up, starting first near the doors and spreading, and in come Adam Steltzner and Wayne Lee, arms in the air and fists clenched in triumph, with the entire EDL team behind them. When *Spirit* touched down three weeks ago they still had another rover to land, and things were pretty much strictly business. But tonight is it for them . . . their job is done.

The EDL guys begin to work their way slowly through the crowd, taking a long victory lap around the packed auditorium, high-fiving reporters and onlookers as they go. The cameras are rolling and this is going out live on national television, but all sense of order and decorum in the room is gone. O'Keefe and Elachi and Weiler are grinning up on the stage, shaking hands in turn with each of the EDL guys as they pass by.

And then the crowd begins to pick up a chant, scattered at first, then spreading: "Eee dee ell! Eee dee ell! Eee dee ell! Eee dee ell!" Adam and Wayne are waving their arms in the air, egging the crowd on. All I can think as I watch them in their moment of triumph is that this must be a space geek's most fantastic dream come true. How many engineers toil for long years in obscurity on a project with a doubtful chance of success, only to have their acronym turned into a victory chant on nation-wide TV?

It's two hours later now, and I'm back in the SMSA with Justin, waiting for the first pictures of Meridiani to come down. Chris Lewicki is Flight again; everything feels familiar. Chris is going around the room on VOCA, sub-system by subsystem, briefing everybody on what to look for in the data. Finally he gets to us.

"Imaging, Flight."

"Go Flight," replies Justin.

Chris looks over his own console at Justin, whose hands are poised at the keyboard. "How're your fingers doing?" Everybody laughs, impatient to see the pictures as soon as Justin's got them ready to go.

The SMSA is quieter and easier for Chris to control tonight than it was after the *Spirit* landing, and he appreciates it. "Thank you everyone for being well behaved," he says. "It's reasonably quiet tonight."

The reply from Flight Software is immediate: "That's because EDL isn't here!"

There's a brief flurry at the telecom console. Chris nods. "All subsys-tems this is Flight, we have a report that the data is flowing."

There's quick applause around the room before everyone gets down to business, the process much more orderly than it was last time. One by one the subsystem leads begin to call in to Chris with their data.

"Flight, Power."

"Go Power."

"Power's got thirty-two-point-six volts on the bus." There's sunshine on the arrays. Chris nods.

"Flight, Mechanical." I can hear the excitement in Joe Melko's voice at the mechanical console. He's the guy with the news that everybody wants to hear, and he knows it.

"Go Mechanical."

"We have confirmed solar-array deployment, MDD deployment,

ARA positions indicate Y petal down and the LPAs are in the cross position." All our critical deployments are complete.

"Mechanical, that's the best news I've heard so far today," says Chris.

"Here we go," says Justin to me quietly off-net, and then, on VOCA: "Flight, Imaging, full Hazcam coming down now." The room goes very quiet.

The first rear Hazcam image comes up on Justin's left screen, where only he and I can see it. I stare at it for a long moment, then another, and another.

What in God's name are we looking at?

It doesn't look like any picture of Mars I've ever seen. It doesn't look like any picture of anything that I've ever seen. There are no rocks. In every picture ever taken from the surface of Mars, the soil has been littered with rocks. Here there are none.

Instead there's soil, fine grained, dark and uniform. The only recognizable features in it at all are the last few airbag bounce marks, crisp and bright and cleanly outlined. A lumpy and strangely irregular horizon curves across the top of the image. And then, just below the skyline, off to the right, there's something different, something bright. I can't tell what it is.

Justin processes the image and drags it over to the right-hand screen, and there's a moment of cheering and applause. But the noise subsides quickly, as the reality of the bizarre scene on the screen begins to sink in.

"What the hell?" mutters Daniel Limonadi a couple of rows in front of me. The room seems oddly quiet.

Justin and I turn our attention back to his console. Packet watch tells us that the next picture is on the way, and Justin is ready. This one should be the full-frame Navcam that was taken before the PMA was deployed. It'll be pointed in the opposite direction, toward the front of the rover, and it should give us our first really decent look at whatever we've gotten ourselves into here.

The picture comes up on Justin's screen, and it's dark. There's something there, but it's underexposed. Justin applies a contrast stretch on the left screen while I wait impatiently.

The stretch hits, and instantly I realize what I'm seeing.

There's a layered outcrop of bedrock right in front of the rover.

This is impossible, it's too good to be true, it's too good to believe. I can't get any words out as I stare at the picture; it feels hard even to breathe. Justin drags the image to the right screen, and the SMSA goes crazy.

"Welcome to Meridiani," says Chris over the VOCA, his voice just audible above the din. "I hope you enjoy your stay."

Navcam images begin to stream down now, Justin throwing each one up on the screen as fast as he can process it. I stand with my mouth open as the images fly up, trying to take them in all at once. The outcrop is brighter than the soil, jutting out of a slope that seems to be facing toward us in every direction. It's impossible to tell how far away it is yet, or how big it is. But there's clear, obvious, in-your-face horizontal layering throughout it. Could this thing be made of sedimentary rock? It can't be; that kind of luck would be too much to hope for. It must be some kind of layered volcanic material.

The mood in the room is giddy, incredulous. This is nothing like the *Spirit* landing; this is nothing like anything any of us has ever seen. People are laughing, cheering, crying with each new picture that comes up. There's wonder in our faces as we stare at the screens. We're seeing magic.

Lewicki comes on the VOCA: "Now would be a good time for our PI's assessment of what we're looking at."

Oh Lord. I step to the front of the room, awkwardly thumbing my push-to-talk button.

"Flight, PI, voice check."

"Five by five, PI," says Chris, smiling. My headset's fine; it's my voice that's shaky.

I take a deep breath.

"I will attempt no science analysis, because this looks like nothing I've ever seen before in my life.

"We knew going into this that at a fine scale the texture of Meridiani Planum is unlike almost anything else on Mars. As we had expected . . ."

Another image comes up, stopping me in mid-sentence.

"Holy smokes. I'm sorry, I'm just, I'm blown away by this." I can't get anything out of my mouth that makes any sense.★

"That outcrop in the distance is just out of this world. I can't wait to get there. I've got nothing else to say, I just want to look."

Somebody calls out: "Did we hit the sweet spot?"

"This is the sweetest spot I've ever seen," is all I can manage in reply.

Once the comm pass is over Justin stitches together the first full Nav-cam panorama, and suddenly I realize that the horizon really does look awfully close in every direction. We're in a hole in the ground. Did we come to rest in a little impact crater? There are so few craters at Meridi-ani that it seems hard to believe, but there's no other good explanation for it. And suddenly the outcrop makes sense: It's exposed on the inner wall of an impact crater, and we're sitting in the middle of the crater, looking straight at it. We've just scored a three-hundred-million-mile interplane-tary hole in one.

How can this have happened? After so many years, and so many disas-ters, how can we suddenly have gotten this lucky? *Spirit*'s alive, *Opportu-nity*'s down, and we've got real bedrock in front of us at Meridiani, *in the first damn pictures to hit the ground!* Everyplace else on Mars that anyone's ever seen, there's been nothing but busted up chunks of rock that could

---

★ What I didn't realize was that the TV cameras were rolling and that my comments, such as they were, were going out live to the world. They showed up later in some strange ways. A few days after the landing, the following short piece ran in the *Korea Herald,* an English-language newspaper in Seoul:

"The day after the Mars rover, *Opportunity,* landed on the red planet and sent the first batch of photographs last week, the country's major afternoon daily, the *Munhwa Ilbo,* translated ar-ticles from the *New York Times* and other American press reports and published them. In the re-ports, a NASA expert, Dr. Steven W. Squyres, looking at an amazing picture just transmitted from Mars, was quoted as exclaiming: 'Holy smoke . . . I'm just blown away by this.'

"Thereupon, the *Munhwa Ilbo* ran the headline: 'The second Mars rover lands, sees mys-terious smoke' . . . It was fortunate for the *Munhwa Ilbo* that Dr. Squyres didn't shout: 'Holy cow.' "

have come from anywhere. Even if you can figure out what rubble like that is made of, it's out of context—there's no sure way to relate it to the place where you found it. But bedrock is geologic truth: It's the stuff that can tell you what happened, right here in this exact place, long ago. And that's what we've got.

And on top of that, this stuff is *layered*. We've got no idea how far away it is, so we don't know how thick the layers are yet. We'll figure that out soon enough. But whatever it is, this stuff is a martian history book, laid out right in front of us. We don't know how many pages there are yet, and we sure don't know what's written on them. But if we can figure this out, we'll have done what we came here to do.

OPPORTUNITY SOL 2

Pancam, MI, APXS and Mössbauer have all passed their post-landing health checks. The first three aren't a surprise, but I'm still baffled by the Mössbauer. Of all the instruments on both rovers, the *Opportunity* Mössbauer was the one that had me the most scared during cruise. Imagine being parked in the hematite capital of the solar system without a working Mössbauer spectrometer. But it's okay now, and the calibration data we got down today were perfect, nothing like the mess we were getting before we landed. Maybe something really was vibrating on *Opportunity* during cruise, smearing out the spectra. We'll probably never know.

But there's terrible news today, too, and it overshadows everything else. As the power subsystem guys were going through their data from the first night on the surface, they noticed that something in the rover was drawing a half-amp current that wasn't supposed to be there. It kicked in about an hour before midnight Mars time, and it didn't go away until about 10:00 the next morning. And whatever was responsible, it sucked more than 170 watt-hours of energy out of the batteries overnight.

There are a bunch of theories, but there's only one that makes any real sense, and that's that the heater that warms the IDD shoulder joint, the first joint on the arm, is stuck on. The amount of current is just right for that heater, and the turn-on and turn-off times are just what they should be for the thermostat that regulates it. Tests over the next sol or two should confirm the theory, but the power and thermal guys are all convinced already.

If they're right, we're in very big trouble. One hundred and seventy watt-hours is something like 20 percent of the total energy that the solar arrays produce in a sol, and that's now when they're as clean as they'll ever be. Later on in the mission, when the arrays start getting dusty, the percentage will be even higher. Put another way, after all the work that Randy Lindemann and the rest of the mechanical team did to cram those big arrays onto the rover, we'll be squandering a huge fraction of the energy that the arrays put out, every single sol. It's not a problem now, when we have so much power, and it shouldn't hurt us much for the next month or two. But if we can't find a way to fix this, eventually *Opportunity* is going to be crippled by it.

Part of this is my fault. Back when we first designed the IDD, I insisted that we had to be able to use it any time of the martian day or night. Mars gets really cold at night, too cold to move the arm unless all the joints have heaters on them that we can switch on when we need them. And if we really want to be able to move the IDD during the coldest hours, the heaters on each joint have to put out a lot of heat. I made a big fuss about this a couple of years ago, and the thermal engineers obligingly put big beefy heaters on all the IDD joints. So here we are on Mars now with a beefy heater that's sucking power all night long, and no way to turn the damn thing off.

OPPORTUNITY SOL 3

I drove into the lab this morning still pondering over the images that came down the first night. All I can figure is that we must have landed in a volcanic ash deposit. Maybe the layered bedrock outcrop off in the distance is lava, or maybe it's ash that was welded together as it fell, still hot from an eruption. But the dark stuff that's lying around everywhere else looks for all the world to me like an eroded heap of loose volcanic ash. And if that's what it is, then the big question is going to be whether or not it ever had water percolating through it, creating a hydrothermal system.

I walked into the fifth-floor Science Assessment room a few hours ago to start the sol. What I found was that today's downlink has made the story a bit more interesting.

One surprise is that the outcrop is tiny. It looked so imposing when we first saw it that the Pancam guys had started calling it the Great Wall. But earlier today Mike Malin sat down with some stereo images and worked out the real size of the thing. The outcrop is just a few tens of centimeters high at most; much of it is no higher than the curb on a street corner. The rover would be able to drive right over it. And what that means, of course, is that the layering that we see in it has to be thin. The individual layers can't be more than a few centimeters thick, which is too fine for it to be lava. It's got to be some kind of welded-together volcanic ash.

The other thing we realized today is that there are two different sizes of grains in the soil. One is very fine, but there are also a lot of granules a few millimeters in size, like gravel, scattered everywhere. I have no idea what this means yet.

The weirdest thing about the gravel is that there isn't any of it in the airbag bounce marks. You see the stuff everywhere else, but inside the bounce marks the surface is startlingly smooth, as if the bags had been pressed gently into talcum powder. Even the inverted impressions of the seams that were stitched by the sewing machine ladies in Dover are beautifully preserved. But the granules that surely must have been there when the airbags hit are gone.

We've been arguing all day about what this means. Some people think that the granules must be weak aggregates of fine grains, crushed into powder by the airbags when they hit. Others think that they're strong, like real gravel, and that when the airbags hit they were pushed down into the soil around them, out of sight.

Whichever it is, we should know soon enough. But I think it's going to be a while before we have this place figured out.

OPPORTUNITY SOL 6

Mini-TES is alive and working, and we've got the first data down. We've found the hematite! Phil Christensen proudly displayed the first Mini-TES map of the terrain around the lander today, and it's stunning.

According to Mini-TES, there's a little bit of hematite in the outcrop, and a lot of it in the soil. But the soil isn't the same everywhere; some has more hematite and some less. The most interesting thing is that the bounce marks in the soil have essentially no hematite at all in them. So somehow, when you bounce onto hematite-rich soil with airbags, the hematite vanishes.

The only explanation for this that makes any sense to me is that the hematite is in the gravel, and that when we bounced on it with the airbags it got pushed out of sight down into fine-grained soil that doesn't have any hematite in it. We still need to look at individual gravel grains with the Mössbauer once we get off the lander to confirm this. But if the coarse grains are what carries the hematite, it fits.

What's up with the outcrop, though, is a mystery. Mini-TES hasn't told us what it is yet, but there are a bunch of things that we now know it isn't. In particular, it's not basalt, and it's not limestone. I'm still guessing that it's welded ash, but we'll see.

The problem with the stuck-on heater is real, and it looks like it'll be with us permanently. Other than that, though, *Opportunity* is healthy and beautiful, standing on her own now and ready to be cut free from the lander. With a clear path down the front ramp, she could be ready for egress as soon as tomorrow.

OPPORTUNITY SOL 7

We're off. Matt is on shift as mission manager today, and when I got in he asked me if it'd be okay for us to egress early. We've always had a rule that we wouldn't egress until we got a full Mini-TES panorama down on the ground, the idea being that egress is dangerous and that we wouldn't want to risk such valuable data by doing something that had a chance of killing the rover. But that's a nice gentle ramp onto very smooth soil in front of us, and nobody's voiced any concerns about egress here not being safe. So with *Opportunity* ready, I eagerly gave Matt the go-ahead. I heard later that he'd already played "Born to Run" as today's rover wake-up song, so I think he knew what my answer would be.

The egress drive was perfect, and we now have twelve wheels in the dirt on two sides of the planet. The quote of the sol came from Joel Krajewski, who has been the architect of the whole complex process of unfolding these rovers and getting them off their landers. When the rear Hazcam that showed *Opportunity*'s first wheel tracks came up on the big screen in the SMSA, Joel turned to me, handed me a set of keys, and said "Here are the keys, Steve. Drive 'em like you stole 'em."

And we know where we might drive *Opportunity* now. The EDL analysis has been done, and it shows that we bounced and rolled for almost half a kilometer, coming to rest in a little crater about 20 meters in diameter. Around our crater lie flat, almost featureless plains stretching off in every direction. But to the east, about 600 or 700 hundred meters away, is a deep, fresh-looking crater about 150 meters in diameter.

We're going to be in this sweet little hole in the ground for a while, since we've got to figure out what's going on with that outcrop before we go anywhere. But after we've climbed out onto the plains, that crater to the east seems like the obvious next place to head for.

OPPORTUNITY SOL 8

Things have taken a peculiar turn.

We've gotten down the first Pancam images of the soil right in front of the rover, and the "gravel" that's all around us seems to be made of little spheres a few millimeters in diameter, strewn across the soil like beads fallen off a necklace. It's the strangest-looking thing I've ever seen on Mars.

It's hard to tell exactly how spherical they really are, since they're so small and hard to see. But there's no denying that they look almost perfectly round in the images we've got so far. We've been calling them "freaky little hematite balls," though the evidence that they're really made of hematite is still sketchy. We're going to try to get some images of them tomorrow with the MI, and those images may reveal much.

OPPORTUNITY SOL 10

Yesterday wasn't our sol. The morning uplink was mistakenly transmitted at the wrong data rate, so *Opportunity* didn't see it, and everything dominoed from there. Once they realized in the SMSA what had happened they did the best they could, but most of the sol was lost. The MI images were taken, but they happened too late in the sol to get them back down to Earth yesterday. So it made for a grumpy team and a long wait overnight.

Today, though, the images are down, and they're stunning. These suckers are round—I mean *really* round. We're calling them spherules now, short for "spherical granules," and there are two of them in the MI images that came down today. They are as close to being perfect spheres as I can imagine any geologic object being. They're way too spherical simply to be grains that have been tumbled over time by wind or water to make a rounded shape.

So what the hell are these things?

We've got three theories, all of them based on geologic processes that we know can make little round things on Earth.

One theory is that they're something called "accretionary lapilli." *Lapilli* is Italian for "little stones," and accretionary lapilli are small round stones that can form during volcanic ash eruptions. They're like volcanic hailstones, made by the millions as ash coagulates together into little balls while it's falling out of the sky after a violent eruption. Pompeii is buried in the stuff.

The second theory is that they're beads of glass. Violent geologic events like lava eruptions or meteorite impacts can throw droplets of molten rock into the air. The droplets cool quickly, and by the time they hit the ground they can solidify to form little spherical glass beads. They're found abundantly around volcanic "fire fountains" in places like Hawaii, and also around some meteorite impact craters.

The third theory is that they're concretions. When water percolates through rocks it can have all kinds of stuff dissolved in it, and if there's enough dissolved stuff, it may want to precipitate, or solidify, from the liquid. Once precipitation starts somewhere in the pore space of the rock, more of the same material can precipitate around this initial point of nucleation. Layer upon layer is added, like the way an oyster builds a pearl, with the precipitated minerals cementing the grains of the rock together to create a hard spherical mass. Concretions are found in a bunch of different kinds of rock on Earth, and it wouldn't be crazy for martian concretions to be made of hematite.

We're going to need to know a lot more about these things— especially what they're made of—before we'll be able to say which theory is right. But whatever they are, there's no question that we've stumbled onto something very strange here.

OPPORTUNITY SOL 12

We got a good Mössbauer measurement on the soil down today, and we followed it up with an MI image. The Mössbauer measurement was a disappointment, but the MI image was revealing.

I had hoped that we'd get a spherule into the field of view of the Mössbauer, but we missed. Instead, all we got was fine-grained sand, and the sand looks like basalt to the Mössbauer. So whatever happened here, one of the things that it made was a lot of basaltic sand. Not much there to get excited about.

The MI image, though, solved one of our mysteries. The Mössbauer has a contact plate on the front of the instrument, with a small hole in the center that the instrument looks through. We pushed the contact plate lightly into the soil when we were setting up to make the Mössbauer measurement. And after taking the Mössbauer away, the imprint of the contact plate was clearly visible in the MI image, like a tiny version of the imprints of the airbags we can see in the Pancam images.

The difference here, though, is that we have both "before" and "after" pictures: two MI images of the very same patch of soil, one taken before we put the Mössbauer down, and another after we took it away. And when you look at those two images together, it's immediately obvious that as the contact plate hit the soil, it pushed granules down into the sand and out of sight. Some are buried completely, and others are sticking partway out. So that puzzle, at least, is solved.

At the moment, the favorite theory among most of the team is that the spherules are accretionary lapilli, since that's the most obvious way to make so many little round things. But if they're really lapilli, where's the volcano? We don't have a story that hangs together yet.

Our next step is going to be to go take a first look at the outcrop.

We'll have to work it over thoroughly, of course, but before we even plan that I just want to drive to the nearest piece and take a quick taste, so that we'll get some idea of what we're dealing with. The outcrop forms an arc partway around the crater, and the right-hand end of that arc lies dead ahead of us, just meters away. So that's where we're going.

The first name that anybody came up with for the rock at the end of the outcrop, unfortunately, was Snout. Like Pyramid back at Gusev, this seemed like a bad name for what could be an important rock, and again I asked John Grotzinger to come up with something better. So Snout is now Stone Mountain, named after the landmark in Georgia.

We should be there soon.

OPPORTUNITY SOL 15

I'm relaxing in Huntington Gardens in Pasadena on a rare afternoon off, sitting in the same bamboo grove where I wrote so much of the proposal seven years ago. The place hasn't changed much.

Today has been the best day of the mission since we landed. I'm still working at Meridiani, and I may be for a while. Over at Gusev, they completed the first RAT grinding today, putting a perfect hole 2.7 millimeters deep into Adirondack. Steve Gorevan was the only Honeybee still standing when I got in, and even he looked pretty beat. But it sounds like they were a very happy bunch of 'bees. The RAT works, in real martian rock.

The news from Meridiani is even better. I awoke last night at 1 A.M. Pacific, exhausted from the day's work but still buzzed. I lay there in the darkness, thinking how badly I needed to get back to sleep, but at the same time realizing that the first close-up Pancam images of the outcrop—Stone Mountain in particular—must be down. I told myself that they'd be there in the morning, and that I should go back to sleep.

Then, barely, I heard my pager vibrating in the living room where I keep it when I don't want to be awakened. I flipped open my cell phone and speed-dialed in the dark. The message was from Justin Maki: "Hey Steve, it's Justin. I don't know if you've seen those Pancam images that came down on *MER-B,* but if you haven't, you should check 'em out. . . . It's pretty amazing. The little spheres seem to be embedded in the rock and they're getting weathered out, according to Ray. Go to a computer near you as fast as you can, because they're pretty cool."

Oh jeez. I gave up, walked into the next room, opened up my laptop and connected to JPL. And just as Justin had promised, the images were astounding. Indeed, the little spherules, which are now variously called "blue balls," "grape nuts," and "blueberries" by different people on the team, are unquestionably weathering out of the outcrop.

The images show that the outcrop is layered on scales far finer than we had ever imagined. Indeed, it's so fine that you'd really call it laminated rather than layered, with individual laminae that are just a millimeter or so thick. The spherules, which themselves are a few millimeters in diameter, are embedded in this stuff, like blueberries in a muffin. And as the outcrop erodes away the blueberries drop out and roll down the slope, where they sit forever.

There's a growing body of evidence that the blueberries may be the hematite-bearing grains. One clue, of course, we already had: The blueberries sit on top of the soil, and they were pushed down into it, out of sight, where the airbags hit. Mini-TES sees hematite in the blueberry-covered soil, but it doesn't see hematite in the bounce marks. Another clue is that the latest Pancam images prove there's a very obvious color difference between the blueberries and the muffin: The spherules are distinctly more gray than the matrix that they're embedded in, and hematite can be a gray mineral.

The most interesting thing, though, comes from a little MI-Mössbauer-MI sequence that we did before the drive yesterday. This was just a quickie: one MI frame on the soil, a Mössbauer measurement less than an hour long, and another MI image. The great thing about the post-Mössbauer MI images is that they clearly show the "noseprint" that the

Mössbauer contact plate makes in the soil, so we can tell exactly where the Mössbauer was placed. The one really good Mössbauer measurement we have on soil so far, back on Sol 12, was pretty much all in sand, and it just looked like basalt. But in the quickie we did yesterday, the post-Mössbauer MI image showed that we got lucky and had a nice blueberry dead-nuts in the center of the Mössbauer field of view. And even with just a forty-seven-minute measurement, Göstar sees a very nice six-peaked "sextet" in the spectrum that looks like hematite.

If the blueberries are made of hematite, then it's hard to see how they could be lapilli. In fact, if the blueberries are very different at all in composition from the stuff they're embedded in, it's hard to see how they could be lapilli. Lapilli form in airborne volcanic ash, and the lapilli and the ash fall out of the sky together, so lapilli tend to be made from the same stuff as the ash around them.

So what should we do next? We don't know, since we really haven't had a good look yet at most of the outcrop. What we're going to do for the next few sols is just what a geologist would do on the scene: walk the outcrop. We're going to start at the right-hand end, where we are now, and spend several sols driving along it and shooting hundreds of megabits of high-resolution Pancam pictures and Mini-TES spectra. The *Opportunity* long-term planning group, which is led these days by Andy Knoll and Hap McSween, has been designing this over the past couple of sols. They've been calling it a "drive-by shooting," though I think I'm going to have to clean that up for the press briefings. Once we've got all the data down, we'll pick the one or two best spots we see, and go back and hit them with everything we've got.

OPPORTUNITY SOL 17

Oh my Lord . . . what is going on here?

The APXS and Mössbauer measurements on Stone Mountain came

down today, and they're astonishing. I caught Ralf Gellert in the fifth-floor APXS room just as I was heading off to a press briefing.

"Ralf, have you got anything yet?"

"Yes, I just finished analyzing the spectrum from Robert E," which is the name that somebody gave to the target on Stone Mountain.

"And?"

"And there is a lot of sulfur in this stuff."

"How much is 'a lot'?" I asked.

"Something like eight percent."

"Eight percent SO_3?" He was right, that was very high for a rock.

"No, no, eight percent sulfur."

"What?!?" That couldn't be right. I did the numbers quickly in my head; if there was that much sulfur in this rock, something like a third of it would have to be made of sulfate salts.

"Are you sure?" I asked.

"Yes, certainly. You can see it for yourself, look." Ralf pointed on his screen to the spectrum of Stone Mountain that had just come down, overlaid on the spectrum of soil that we got a few sols ago. The two spectra matched up pretty well, but there was one enormous peak in the outcrop spectrum rising far above its counterpart in the soil spectrum. It was sulfur.

I walked off to the press briefing with my head spinning.

The Mössbauer spectrum from Stone Mountain came down next, and it's even more bizarre. There's a very strong signal in it from something that we've never seen before. Göstar and Dick Morris think it might be jarosite.

Finding jarosite would be tremendously important. It's not a very common mineral on Earth, but it has been talked about in the past as something that might turn up on Mars. Jarosite is a sulfate salt, and it's one that actually has water in its crystal structure. You need to have water around to make it. If there's jarosite in this stuff, then water has to have been involved.

But was it? There's so much that's uncertain right now. The APXS measurement on Stone Mountain was just on the surface of the rock. We haven't RATted the outcrop yet, and without using the RAT on it first we can't be sure that the rock doesn't just have some kind of thin sulfur-rich coating or stain on the outside. And the jarosite is even more shaky. The Mössbauer measurement on Stone Mountain was just a short one, and the shorter the duration, the less certain the results.

So if you sit back and look at things dispassionately, we really don't have much yet. We've got finely laminated rock with spherules embedded in it. We think the spherules may be rich in hematite, but we haven't proven it. We think that the rock itself may be astonishingly rich in sulfur, but we haven't proven that the sulfur isn't just at the surface. And we think that the rock may have jarosite in it, though we haven't proven that yet, either.

So right now there are a lot more perplexing questions than there are definitive answers. If there really is that much sulfate salt in this rock, and especially if there's jarosite, I don't know how it could have been made without water. But the whole thing just seems so hard to believe. I know we came here looking for evidence of water. I know we picked this site over a hundred others because we thought it was likely to have it. But in my heart I never really believed or dared to hope that we would actually find what we were looking for—not like this. And now, with tenuous evidence starting to pile up in front of us, I find myself somehow resisting the idea. I feel as if I ought to be embracing this stuff, arguing that we've found evidence for water, leading the charge. Instead, I'm hanging back, almost afraid to believe it. It's strange, and I don't understand why I feel this way.

There's still a lot to be done. We'll see what we see.

OPPORTUNITY SOL 26

I'm back in Pasadena after a delightful six-sol break in Ithaca for Valentine's Day, the first time I've seen Mary and the girls since we landed. I left

Opportunity in Arvidson's and Soderblom's care while I was gone, and they did well. The drive-by shooting, which we renamed the "shoot and scoot," is done. The pictures are down, and they've been mosaicked together to make one giant image, which we've printed on a roll of photographic paper almost twenty feet long. It's stretched across a table in the fifth-floor science room, covered already with marks and scribbles that are the scars of the debate over what to do next.

The consensus, after much discussion, is that we should go to El Capitan. The name is another Grotzinger special. It's not derived from the famous El Cap in Yosemite, but instead from the one in Guadalupe National Park in west Texas. John is convinced that we're looking at sedimentary rocks here, not at volcanic ash, and Guadalupe is a place where he's seen sediments like what we may have at Meridiani. Our El Capitan is the biggest rock in the outcrop, pretty much dead center. It's all of about forty centimeters tall.

Grotzinger is an interesting guy. He's smart as hell, for one thing—a member of the National Academy of Sciences at some ridiculously young age, and acknowledged to be one of the best sedimentary geologists in the world. He's also a lot of fun to have around. He's tall and athletic, a former lacrosse player. He got his degree from Hobart College, not far from Cornell, and he actually did his thesis work in and around Cayuga Lake at the same time I was a student in Ithaca. We never ran into each other in those days, but Mars has brought us together now.

John's about to move over to Meridiani permanently, and he's very happy about it. I originally had him as one of the long-term planning leads over at Gusev, since that's the place where I thought we were most likely to run into sedimentary rocks. But with a bunch of basalt at Gusev and what looks increasingly like sediments at Meridiani, John starts as long-term planning lead for *Opportunity* tomorrow.

Over the past sol or two, John and I have had some very lively discussions about what we're seeing in the shoot-and-scoot mosaic. He thinks we might be seeing ripple cross-bedding.

Cross-bedding is something that can happen when particles are laid down in a flowing fluid. The particles can be sand, or silt, or ash; the fluid can be air or water. When particles settle out from a motionless fluid, they tend to form nice flat, parallel layers. But when the fluid is moving, you can get some layers—that is, some beds—that are tilted with respect to the others around them: cross-beds.

There's no doubt that we're seeing some cross-bedding in the out-crop. Even the first images suggested it, and the shoot-and-scoot mosaic proves it. That doesn't mean a lot by itself, though, since cross-bedding can develop in so many different ways. But in a few places, John thinks we may be seeing the kind of tiny cross-beds that form when liquid water, not air, is what's transporting the grains. I'm not convinced that he's right, but I'm not convinced that he's wrong, either.

Recognizing these things is a subtle business, and it's the kind of thing that sedimentary geologists spend years developing a knack for. If it were anybody other than John I'd be skeptical, and even John doesn't feel very certain about it. But he's got a good eye, and I do have to admit that there are a few places in the mosaic where we're seeing things that look tanta-lizingly like the remains of ancient current ripples.

If these really are ripple cross-beds, they're a crucial find. They'd be evidence not just that water once flowed through the rocks—which is an idea I'm slowly becoming more comfortable with—but that the rocks themselves were laid down in water. It's the difference between water you can draw from a well and water you can wade in. Problem is, there's not much more we can do to learn about them with Pancam. The ripple fea-tures, if that's what they are, are tiny, just a few centimeters across. They're right near the limit of what we can see with Pancam. The cameras have 20/20 vision, but they're bolted to the top of the mast. If you were there with your own eyes, you'd just get down on your hands and knees and look, but we've got no way to move our cameras closer to the ground. The pictures we have now are ambiguous, even though they're the best that Pancam can provide.

Tonight, though, I had another idea. The IDDs on these vehicles have been performing beautifully so far. Could we actually use the IDD to take a big mosaic of images with the Microscopic Imager? We've never given this kind of thing much thought, and we've certainly never tried it. The MI field of view is only three centimeters across, so covering a big rock with it would be like wallpapering with postage stamps. But it's the only way I can think of to get the kind of resolution we need to solve this problem.

John doesn't know it yet, but his first job as LTP lead for *Opportunity* is going to be to design a campaign to search for ancient martian current ripples with a microscope.

OPPORTUNITY SOL 27

Today we made our approach to El Capitan, with Pancam imaging at the end of the drive. I stepped into the fifth-floor science room just as the pictures had started to come down.

"Holy snipeshit!"

What? It had to be Grotzinger. I walked over to the LTP lead workstation, where he was processing the first good Pancam picture of El Cap.

It was bizarre. The rock's surface is crisscrossed by tiny gashes, looking like little chicken scratches.

"They've gotta be crystal molds," said John.

"You think?" I replied. But I had to agree, they sure looked like it.

Crystal molds form when little flat, tabular-shaped crystals of some mineral precipitate from water that saturates a rock. Gypsum, another sulfate salt, is one of the minerals that does this on Earth. The crystals grow in place, either forcing aside the stuff that's already there or replacing it completely.

Then, something happens. Maybe the water chemistry changes and the crystals dissolve away. Or maybe conditions dry up and the crystals, which are softer than the rock around them, simply erode away. Either way, you wind up with little tabular-shaped voids in the rocks where the missing mineral used to be, the molds of crystals that formed and then were destroyed, long ago.

And that's what I think we're seeing in El Cap.

This discovery, for me, is a landmark. Ever since we landed, I've been resisting the idea that there was water here at Meridiani. Maybe I just want to be very, very certain before declaring victory. But whatever the reason, I've been fighting the idea in my own mind for weeks. Today that changed.

It's not that little gashes that might be crystal molds provide some kind of bulletproof evidence that water was here. You have to be very careful about drawing conclusions from features that, after all, are just holes left behind by something that isn't there anymore.

But still, these gashes have pushed me over the edge. It's the preponderance of the evidence.

I'm not the only one who has been wrestling with this. It's been fascinating to watch the whole team as these clues have been revealed to us, sol by sol over the past few weeks, like some kind of weird martian mystery novel. All of us have come to this remarkable experience with our own backgrounds, with our own set of prejudices. And each of us has reacted to the mounting evidence for water here differently. Some people leaped joyfully off the cliff when we got our first hint of the outcrop's composition at Stone Mountain. Grotzinger was one of the early true believers, and Andy Knoll was another. Others still aren't convinced yet, Mike Malin most notably among them. But today did it for me. I simply can't see how you can make rock like this without a lot of water being involved.

Feeling in my gut that water was here is one thing; proving it is another. We've still got a lot of work to do. We've got to show that the rock is rich in sulfur all the way down, and we've got to confirm the jarosite. We've got to see if the blueberries, which I'm starting to think have to be concretions, really are rich in hematite. And we've got to go after those current ripples, if that's what they are, with everything we have.

I have no idea how long we're going to be in this crater, but it's starting to look like it could be quite a while.

OPPORTUNITY SOL 28

A bunch of MI images of El Cap just came down, and they clearly show that the gashes are little planar things, shot through the rock at crazy angles. Each one's a centimeter or so long and a millimeter or two wide, many of them fat in the middle and tapered to points at the ends. These things simply have to be crystal molds; I can't think of any other explanation for them that makes sense.

And along with the gashes, we're starting to see enough blueberries in the new MI images now to get a sense of how varied they are in shape and size. Most of them are really round, but every now and then we see one that's a doublet: a blueberry with another blueberry growing out of its side. And while the single ones usually don't show much structure, one or two of them have stripes, like the stripe in a croquet ball, running across them, parallel to the laminations in the surrounding rock. Everything about this makes sense if the blueberries are concretions, but we still don't have a compelling story yet. Somehow we've got to figure out what they're made of.

OPPORTUNITY SOL 29

Another sol, another revelation. Today's came in the form of new APXS data from El Cap. The surface of El Cap is rich in sulfur, just like Stone Mountain, but it's also got *bromine* in it. Bromine is a weird thing to find in a martian rock. Bromine shows up in some rocks on Earth, usually as bromide salts. And the best way I know of to make bromide salts is by evaporating seawater or something like it. I'm a true believer in the water idea now, but I sure didn't see this one coming.

Of course, just like at Stone Mountain, this is still just a surface measurement. We've got to figure out what this stuff is *really* made of, and the only way to do it is to punch a RAT hole into El Cap. This is why we built the RAT, and it's time to put it to work.

Steve Gorevan is nervous, though. We built the RAT to handle every reasonable kind of rock we could think of, but we never imagined anything like sulfate-laden sediments shot through with hard little blueberries. The 'bees have been sweating ever since they saw Stone Mountain, and with the day of reckoning now upon us they're haunted by visions of blueberries popping out of El Cap and rattling around inside their creation like steel balls in a pinball machine. But it's time to do it, and I've told them just to give it their best shot and that we'll deal with whatever happens. The first target we've picked for them has been given the Dickensian name McKittrick MiddleRAT, and tomorrow's the day we're going to go after it.

OPPORTUNITY SOL 30

Oh yeah! We hit McKittrick MiddleRAT with the best we had, and it worked. I got into the SMSA today to find Gorevan and the rest of the 'bees marveling at some post-RAT MI pictures that had come down just moments before. Near as we can tell, we got four millimeters into the rock and then stalled, probably when we hit a blueberry. But that's deep enough. And if the engineering data from the RAT can be believed, the rock is really soft and the berries in it are really hard . . . it's like grinding into cottage cheese with marbles in it. I'm amazed that the thing worked, frankly, and the 'bees, behind all their pride at this accomplishment, seem almost as surprised as I am. Still, it did work, and now we've got a window into the rock that the APXS and Mössbauer can look through. We've got the APXS in the hole already, and tonight it'll collect the data that should resolve the sulfur question one way or the other.

After that, we'll go after the jarosite. Usually when we do a Mössbauer measurement, we run the instrument for something like eight or ten hours and that's good enough. But I want to kill this problem. So as soon as the APXS is done, we're going to spend an entire sol doing a twenty-four-hour Mössbauer measurement in the RAT hole. Then, just to really be certain we've nailed it, we're going to narrow down the velocity range on the instrument and take another twenty-four-hour spectrum that's tuned especially to pick up jarosite if it's there.

And then we'll put in another RAT hole, and do the same thing all over again.

It's the best we can do. In a few more sols we should know.

OPPORTUNITY SOL 32

Today I was in my office, waiting for the El Cap RAT hole data to come down, when Grotzinger walked in.

"You got a minute?" he asked.

"Sure."

"I think I've got this cross-bed campaign worked out," he said as he settled his long frame into the chair opposite me. He had a careful sol-by-sol plan with him, sketched out on a sheet of paper in his hand.

I reached out and took it from him, squinting at his careful writing. It took only about five seconds for me to see that it wouldn't work. John's plan had us driving from one rock to another to another, up and down the outcrop, spending a couple of sols at each one to take what looked like hundreds of MI images.

"John, we can't do this! This is ten sols!!"

"But that's how long it takes, Steve, if we're really going to do it right."

"John, ten sols is more than ten percent of the entire mission! We can't spend that much time on a fishing expedition. You have to cut it down."

John looked crestfallen. He'd been putting a lot of thought and effort into this thing, I knew, and here I was, shooting him down before we'd even gotten to the details. But that stuck heater has me spooked, sucking all of those watt-hours out of our power system every day. I simply don't know how long *Opportunity* is going to last. And on top of that, there's a lot of skepticism on the team about these supposed ripples. If we're going to convince everyone that we should do this, we're going to have to get it down to no more than a week, tops. Even then it could be a tough sell.

I worked with John through the details of his plan, pointing out a couple of places where we might be able to shave a sol or two off it. He's promised to come back in a few days with something that'll be a little easier to swallow.

OPPORTUNITY SOL 33

The APXS and Mössbauer data from El Cap are down.

There's still a lot of sulfur. In fact, the amount of sulfur in the rock went *up* after we RATted into it, not down. So this rock doesn't just have a sulfate coating on the outside . . . it's sulfate-rich all the way through.

We've also got some great Mössbauer spectra, and together they make a slam-dunk case for jarosite.

We've still got work to do, but these measurements nail it. There was water here.

Yowza!

There's another story brewing, too, emerging when we look at all our APXS data together. We're seeing bromine everywhere, but in very different amounts in different parts of the outcrop. In fact, the bromine concentration seems to be all over the place in this stuff, varying hugely from one spot to another.

Rudi and Ralf have been thinking about it, and they've got an idea.

Along with bromine, we also see chlorine in the rock. The two elements are probably present in bromide and chloride salts, respectively. Bromide and chloride salts are alike in a lot of ways, yet we're finding that the amount of bromides, relative to the chlorides, is enormously variable. Why?

This kind of variation, it turns out, is something you see in some deposits on Earth called evaporites. Evaporites are salt-laden rocks that form when seawater evaporates away, leaving its salts behind. And while bromide salts and chloride salts are similar in many respects, one way that they're very different is how easily they dissolve in water.

So here's how it goes: Suppose you've got a big puddle of seawater, with bromide and chloride salts dissolved in it, and it starts to evaporate. The first salt crystals to form are chlorides, and they begin to solidify and settle out in layers. The bromides, though, stay dissolved, because they can. Then, as more and more water evaporates, the brine that's left behind becomes more and more concentrated. Eventually it gets so concentrated that even bromide salts start to form, settling out on top of the chlorides that formed earlier. So once it's all over, you've got a stack of salts that's mostly chlorides at the bottom, with more and more bromides as you get toward the top. The amount of bromine in the rock varies relative to the chlorine, just like what we're seeing.

So Rudi and Ralf think that our bromine variations may be telling us that these rocks are martian evaporites. And I think they may be right.

As I watched Ralf's quiet pride today while he showed his latest spectra to the team, I thought back to Florida. The APXS on *Opportunity* is his "Lazarus" instrument, the one that he brought back to life after it died during the thermal test. On a day like today, that seems like a very long time ago.

So now we've got a compelling case that there was water in these rocks. We still need to see if the blueberries are made of hematite, and we still need to figure out the supposed ripples. But there's a lesson from El Cap. We killed the sulfur and jarosite problems, and we got our bromine surprise, because we spent all the sols here that we needed. When Grotzinger comes back with his new ripple plan, I think I'm going to go for it no matter how long it is.

OPPORTUNITY SOL 35

John's got a new plan, and it's down to six sols. That's good enough for me, and it'll be good enough for the team.

His focus is on just two rocks, which he's named the Dells and Last Chance. We'll do Last Chance first, actually, and then go on to the Dells, blanketing them both with MI images. If the ripples are there, this is the best chance we've got at seeing them.

OPPORTUNITY SOL 39

Today was like nothing I ever imagined back when we designed this payload. The most MI pictures I ever figured we'd take on a single target was about five, since that's how many you need to be sure to get something in focus. But we're parked in front of Last Chance now, and we just finished taking *120* pictures of it with the MI. All in one day, all on one rock that's about the size of a football. Unbelievable. I sure hope it all worked.

We also got a cool little surprise today. Even after we've driven away from it, El Cap continues to delight. We just got down our first color Pancam shots of the RAT holes we put in El Cap, and they each show a wonderful brick-red stain in the "RAT droppings"—the haloes of fine-grained rock dust left behind by the RAT as it grinds. The blueberries are gray, but we know that when you take gray hematite and grind it up finely, it turns brick red. So this is one more clue that the berries are made of hematite. We've got to get a good Mössbauer spectrum of these things!

OPPORTUNITY SOL 44

We're done shooting the Dells, and the pictures from both Last Chance and the Dells are dribbling down bit by bit. We don't have enough of them down yet to say anything, and the wait is maddening. Every day I go and see Grotzinger, and every day he's hopeful but uncertain. So we keep waiting. I love these rovers, but damn, they're slow.

Meanwhile, with the ripple campaign done, we're now finished with the outcrop except for one thing. We've still got to prove beyond any doubt that the berries are made of hematite.

The Mössbauer is the tool for the job, but it's not an easy thing to do. The berries are only a few millimeters across, and the field of view of the Mössbauer is four or five times that. So if we try to use the Mössbauer to look at a single berry embedded in the outcrop, what we'll actually see is a whole lot of outcrop and a little bit of berry.

But we've got a plan. In the shoot-and-scoot mosaic, Andy Knoll spotted a place where there happens to be a little bowl-shaped depression in the outcrop rock. A bunch of blueberries have rolled down into it, where they're now all clustered together. Andy's calling it the Berry Bowl. We're going to drive over to the Berry Bowl while we wait for the last of the maybe-they're-ripples images to come down, and try to take a Mössbauer spectrum on a bunch of berries at once.

And with that, we should be done with this crater. The crater has been our entire world for the last month and a half. But all of Meridiani Planum is out there, beckoning, including that huge crater off to the east. Once the outcrop science is wrapped up, we'll need to get moving.

The strange thing, though, is that hardly anybody wants to leave.

I've been listening carefully to the conversations on the team lately, and they're a little disheartening. All everybody seems to talk about now is this or that observation they still want to make inside the crater, mostly looking at the soil. The soil is interesting stuff, I guess, but it's not what we came here for. What we came here for was the water story, and once we've nailed this blueberry thing we'll have done what we can in the crater about water.

It was getting obvious that I needed to do something about this, and today I did. We had a little come-to-Jesus meeting for the whole team this afternoon, and after hearing everyone's plans and schemes I laid it out for them: We have to be on our way out of this crater by Sol 60, a little over two weeks from now. There was a lot of grumbling, and some people still aren't very happy about it. But I'm convinced it's the right thing to do. There's a big world out there, and *Opportunity* isn't going to last forever.

OPPORTUNITY SOL 46

We're at the Berry Bowl, with the Mössbauer squarely on berries and taking data. And meanwhile, we've finally got the full MI mosaics down and assembled for both Last Chance and the Dells. There are ripples in this stuff that were made by flowing water! All along I was worried that maybe we were seeing "eolian" ripples, the kind made by wind, not by water. But the new MI mosaics have convinced Grotzinger that our ripples really did form in water, and they've convinced me, too. John sent the pictures to John Southard at MIT, who's probably the world's leading authority on ripples in sedimentary rocks. Here's the reply he got:

> When I looked at the photo I took a breath and said 'Wow!' To me it looks clearly like ripple trough cross-lamination . . . It does not even remotely look like any eolian cross-stratification I have ever seen.

John's e-mail to me after he heard back from Southard was succinct: "We've nailed this thing." There's no doubt in my mind now that some of these rocks were laid down, long ago, in liquid water.

So how do we top this? How do we make an observation with these rovers that means more than finding evidence for an ancient body of salty water on Mars? I'm not sure we can.

With news like this, a celebration of sorts was in order, and it seemed utterly logical that it should be at John's expense. So this afternoon we got some JPL letterhead stationery and drafted up a very official-looking invoice for $24 million—six sols of work at $4 million a sol, carefully itemized and charged to the Massachusetts Institute of Technology—and slipped it into Grotzinger's mailbox. He opened it while sitting in the back of the room during the *Opportunity* science assessment meeting. It didn't take him long to figure out that somebody was putting him on, but for about five seconds the look on his face was priceless. We're a very happy team tonight.

And we've finally picked a name for our little hole in the ground. We're going to call it Eagle Crater, after the *Apollo 11* Lunar Module, using the ships-of-exploration theme that didn't quite make the cut back at Gusev. And it's also got a nice little double meaning. If you're golfing on a par three hole, an eagle is a hole in one.

OPPORTUNITY SOL 48

We've done it. The Mössbauer spectrum of the Berry Bowl is down, and it's got by far the strongest hematite signal we've ever seen. The best explanation we've got for the blueberries is that they're hematite-rich concretions. So that wraps it up. It's time to get moving, as soon as we finish burning up twelve more sols doing the odds and ends I promised the team we'd do before we leave.

What comes next, once we're out of here? The plains outside the crater could be interesting, and they'll certainly be different. It'll be fun to get the rover out there and let her unwind over open territory, just to see what she can do. But I don't think there's going to be a whole lot of science on the plains.

I guess that what we should do next really depends on how we feel about what we've accomplished so far. Do our discoveries in Eagle Crater amount to Mission Accomplished? I want them to, but the more I think about it, the more it doesn't feel like they do. We came to Mars to learn whether or not the place ever could have supported life. And now we've got an answer, sort of. At this particular spot, long ago (we don't know when), for a little while (we don't know how long) the answer is yes: Mars could have supported life. But how long was it wet? How much water was there? What was the climate like? We really don't know. This may be Mission Accomplished, but it's not giving me the kind of satisfaction that I feel like it ought to.

What we really need is more bedrock, deeper down in the ground. The rocks in Eagle Crater are great, but they're only about a foot thick. It's a history book all right, but every page is missing except the last one. We don't have any idea what came before the water. And if the answer's out there, it lies deeper down in the ground.

The closest thing that could give us access to deeper rocks is that big crater off to the east, which we've now named Endurance, after Shackleton's ship. So that's our next target. It would be easy for us to get bogged down on the plains, just as we have in Eagle, and we need some kind of goal to push us toward Endurance. So I'm going to tell the team that I want us to be there by Sol 90, the end of the nominal mission. I think we can do it—those plains look like they should be pretty easy driving. But it's going to be a long six weeks. I wish I had a better idea of how much rover we're going to have left once we get there.

17 ENDURANCE

W HILE *OPPORTUNITY* BEGINS THE long drive toward Endurance Crater, I'm about to jump over to *Spirit* for a couple of weeks to help get us moving toward the Columbia Hills.

Ever since we landed, the big hope for getting at the sediments that must be somewhere underneath all that basalt at Gusev was Bonneville Crater. Bonneville is surrounded by a low rim, so it was impossible for us to see what was in it from a distance. We had to drive to the rim and look inside.

The traverse to the crater was long and arduous, struggling through the basalt boulders at twenty or thirty meters a sol. As we grew closer the terrain became rockier and rougher, the rover entering the blanket of debris that was thrown out of the crater by the impact that formed it. The ejected rocks that we passed by all looked like chunks of the same basalt we'd been seeing since we left the lander, but hopes were still high on Sol 65 as we made the final climb up the rim.

The first look into the crater dashed those hopes. It was a pretty view, the most scenic landscape that we've seen at either site since we landed. But there were no layered rocks exposed anywhere in the crater. Instead, Bonneville looks like a big version of Sleepy Hollow. It's shallow, it's filled with sand and dust and the inner wall is lined with the same lumps of shattered basalt that we've seen everywhere else at Gusev. Our hoped-for window into the sediments below the basalt revealed nothing.

So after two months of work to get there, Bonneville Crater was a bitter disappointment. For the scientists who've been over on *Opportunity*, reading about *Spirit* in the daily reports, the failure at Bonneville was a brief diversion from the glories of Eagle Crater. But for those who've persevered with *Spirit* from the start, it's left them discouraged and tired.

They also seem to be losing focus. A source of distraction since we landed at Gusev has been the so-called white rocks. They're not really white; they're more like a light tan in color. But they look different in Pancam pictures from all the gray basalt blocks, and that gave the team some hope.

No one dared say it, but at first the hope was that the white rocks might be carbonates. The idea that Mars had a warmer and wetter past has always raised the question of what could have allowed a warmer climate. The simplest answer is that long ago Mars may have had a dense atmosphere of carbon dioxide, leading to a "greenhouse effect" that warmed the surface. But if that happened, where did all the CO_2 go? The simple answer would be that it's been locked up in carbonate rocks on the surface: limestones. Problem is, no one has ever found carbonate rocks on Mars.

There wasn't much other than their light color to suggest that the white rocks might be carbonates. They certainly didn't look anything like carbonates to Mini-TES. Instead, they just looked like martian dust. And another clue was that all of the white rocks seemed to be low and flat-lying, an ideal geometry for accumulating a dust coating.

Still, we had to look, and back around Sol 80 we spent eight sols thoroughly working over a white rock near the rim of Bonneville named Mazatzal. We brushed it with the RAT. We drilled into it with the RAT. We hit it with everything we had, and in the end the answer was clear: It was a chunk of basalt with a coating of dust on the outside.

But now, some people want to keep chasing other white rocks. Maybe Mazatzal was unusual, they say, maybe the Big One is still out there. I don't believe it, and I think it's time to give up on the Gusev plains and move on.

So today I sent this e-mail to Matt Wallace and Jennifer Trosper:

Hi guys: In order to keep things from bogging down too much at Gusev ("ooh . . . another white rock!"), I'm about to do the same thing there that I have done now twice at Meridiani: Give the team a date by which I expect them to reach a certain milestone. Both times I have done this at Meridiani (out of Eagle Crater by Sol 60 and at the rim of Endurance Crater by Sol 90) it has had the effect of enabling the really high-priority science while still keeping things moving forward. The same is needed at Gusev.

So I'm planning on giving them a target date by which I expect us to reach the base of the Columbia Hills. . . . The thing I'm still wrestling with is what I should give them as a date. What this depends on, of course, is what a realistic drive capability per sol will be. . . .

What do you think would be a reasonable *sustained* average driving speed (meters per sol) at Gusev? In other words, if all we did every day was drive and acquire the remote sensing necessary to support driving, what could our long-term average speed be between here and the Columbia Hills? I'll then take this number and margin it to enable what I feel is an appropriate amount of science along the way.

One thing that I know will help is that the mobility guys have been working since January on improvements to the software that the rovers use to find their way around obstacles. Until now, *Spirit* and *Opportunity* have been timid, easily frightened into immobility by small rocks and hillocks that pose no real threat. That's why the drive to Bonneville was such an ordeal. But the new software should make them smarter and more courageous, and we started uploading it to both vehicles a few days ago.

With this in mind, Jennifer's reply was encouraging:

"I'd like for us to see how the new software works with the autonav fixes before we say for certain what we could expect. . . . My guess/hope is we should be able to get ~75 meters on a good sol."

If she's right then I see two different ways of doing the math, and they both lead me to the same answer.

Jennifer thinks we can go 75 meters on a good sol, so let's call it 60 meters on an average sol. Today is Sol 100, and we're 2.2 kilometers from the base of the hills as the crow flies. Put in all the wiggles and squiggles that we need to go around rocks and other obstacles, and that's more like 2.7 kilometers as the rover drives.

At 60 meters a sol, 2.7 kilometers would take forty-five sols. I don't want to spend the entire time driving; we need some time to stop and do science along the way. Figure three quarters of the sols are mostly spent driving, and one quarter of them mostly doing science. That makes it sixty sols to do the whole thing, putting us at the base of the hills by Sol 160.

Do it another way. Right now *Spirit* has plenty of electrical power, but that won't last. Dust is building up on the solar arrays, and the seasons are beginning to change. The calculations are sketchy, but right now they say that sometime around Sol 220 *Spirit* is going to be pretty much out of gas. And based on our experiences at Eagle Crater, I figure I'd like to have at least a couple of months in the Columbia Hills before that happens. Count back two months from Sol 220, and you're at Sol 160.

So Sol 160 looks to me like the target. I've just got to convince the team of it.

APRIL 17, 2004

Today I had a heart-to-heart with the *Spirit* team, laying out the situation at Gusev. I didn't give them the Sol 160 number yet. Instead, I just gave them the facts: Jennifer's estimate of 75 meters on a good sol. The 2.7 kilometers to the hills. The difficulty of keeping *Spirit* doing science much beyond Sol 220. They're smart people, and with the facts in front of them they're likely to come to the right conclusion on their own. So I'm going to give them a day to let it simmer.

Complicating everything, though, is that we've just made the switch from living on Mars time to living on Earth time.

Mars time has worked well for the science team. Nearly all of us are far away from our homes, living in barren little apartments and focused exclusively on operating rovers that are more than a hundred million miles away. Except for the rare and inconvenient moments when the real world intervenes, life on Mars time is simple and tolerable.

For the JPL engineers, though, it's different. They're living at home, where they've always lived, most with spouses and many with kids. They have lawns to mow and soccer games to go to. The JPLers follow a schedule of four sols on and three sols off, and for most of them that means four sols on Mars time followed by three days on Earth time, week after week. The whipsawing of this weekly jet lag is killing people, and it's been clear since we landed we wouldn't be able to stay on Mars time indefinitely.

So about a week ago, we changed. We're not really living strictly on Earth time, though. It's more like Mars time with a hole in it. The rovers' schedules still slip thirty-nine minutes every day with respect to Earth time; that can't change, because it's dictated by how fast the two planets rotate. But we've established a keep-out zone between 22:00 and 07:00 Pacific, when no operations activities on Earth are allowed, giving people a chance to sleep when it's actually nighttime.

The thing that makes this even somewhat possible is that we've gotten a whole lot better at our jobs. Back when we first landed, it took us seventeen hours every sol between when we saw a downlink and when we had the next uplink ready to go. We've got that down to more like seven hours now, and sometimes it's even less.

So for a sol when the uplink time is, say, 14:00 Pacific, we can start working at 07:00 Pacific and still make it. Next sol the uplink is thirty-nine minutes later and it's even easier. The uplink time slips around by thirty-nine minutes each day, and as long as we're in the part of the cycle where Earth time is synched up reasonably well with Mars time, things are good.

Where it gets tricky is when we're out of synch, when adhering to a strict Mars-time schedule would force us to stay awake and plan during the overnight keep-out zone. That's when we run into what we're calling "restricted sols," and we're in restricted sols right now with *Spirit*.

Restricted sols are no fun. We have to plan them before we're really ready, during the daytime hours on Earth when we haven't seen the downlink from Mars yet. And with no new pictures on the ground to show us where we are, we can't drive the rover or use the arm. Each sol has to be planned two sols in advance, which means that for a while, until Earth and Mars synch back up again, we can drive only every other sol.

It's frustrating and confusing, and it's going to slow us down on the way to the hills. The fraction of sols that are restricted will drop as we get more used to Earth time, and I'm hoping that the new navigation software will give us something close to the sixty-meter average that we're counting on. But it's going to be difficult. With *Spirit*, it always seems to be difficult.

Meanwhile, over at Meridiani, where rocks are scarce and driving is easy, they did 140 meters in one sol yesterday, an all-time record.

APRIL 18, 2004

Dave Des Marais from Ames is the long-term planning lead for *Spirit* now, and after yesterday's little heart-to-heart he also came to the conclusion that Sol 160 is when we need to reach the hills. So in today's end-of-sol discussion he gave that number to the team. There was a lot of talk and no outright objection, but as I listened to the discussion I didn't get the feeling that everybody was happy about it.

Indeed, after the meeting several people came to me privately to complain. But their complaints only reinforced my opinion that 160 is the right number. Some thought 160 was too aggressive, and that we should allow more time for science along the way, arriving at the hills around Sol 170. Others, Jim Rice most vocally among them, think that we're wasting our time on the plains, and that we should try to reach the hills by 150. With the opinions bracketing 160, I got together with Dave, and we finalized it. Sol 160 is the target. It's time to get moving.

APRIL 20, 2004

Dave has a plan for getting us to the hills. He calls it the sol quartet.

The idea is simple. As we work our way across the plains, there are some things we want to do on a regular basis. Take pictures and Mini-TES spectra of the stuff around us. Reach out with the arm and find out what rocks are made of. Look up at the atmosphere. For the most part, this isn't the kind of science that'd make us go out of our way to drive to a particular rock or patch of soil. It's just part of the business of being careful observers, keeping an eye open for anything really interesting as we work our way across the plains.

So Dave's idea is to use a pattern of four sols, the sol quartet, that we'll repeat over and over as we move toward the hills. Each sol in the quartet has a good long drive in it. And each one also has something else. One sol of the quartet uses the arm to look briefly at whatever's in front of the rover before driving. Another makes particularly thorough observations of the atmosphere, and so on. The idea is to have something in there for everybody, with each key observation being repeated at least once every four sols.

Dave put a lot of work into making the sols in the quartet fair and balanced, knowing that looking after everyone's science is the key to keeping morale up during the long forced march to the hills. And when he presented it to the team today, they went for it.

There's going to be some tweaking of Dave's plan. The amount of power from the solar arrays will drop daily over the next sixty sols, and I've got a nasty feeling that they're going to have to start carving some science out of each of the quartet sols in order to keep the rover moving. But it's a start, and the mood among the *Spirit* team today is better than I've seen it in a long time. There's a new challenge in front of them now, and most of them seem eager to take it on.

At Meridiani, *Opportunity* has made it to Fram Crater. This is the major landmark on the way from Eagle to Endurance, a tiny impact crater nine meters in diameter.

From what I hear, the rocks at Fram seem to be made of the same sulfate-rich, blueberry-laden stuff that we saw back at Eagle. The strangest thing is the way that some of them have eroded. Maybe it's because of how the rocks at Fram are perched up on the lip of the crater, exposed to the wind. Whatever it is, some of the blueberries there are suspended out at the ends of thin stalks of sulfate-rich rock, looking like the slightest touch would snap them off. The *Opportunity* guys think it's wind erosion, the hard blueberries protecting the stalks immediately downwind from them while the sandblasting eats away at everything else. They're calling them blueberries-on-a-stick.

Fram Crater, of course, is named for Amundsen's ship. The *Fram* itself is still around, well cared for as a floating museum in Oslo, and I spent an hour tonight trying without success to track down an e-mail address for the museum, wanting to let them know that we've named a crater on Mars after their vessel. I hope they'll read about it in the news.

Opportunity's work at Fram won't take long. Once it's done, it'll be time for her to move on to Endurance Crater, and for me to move back to Meridiani.

.ᗧ. 2004

ᗐack on *Opportunity*. We're only a couple of sols out from Endurance now, and each sol the crater grows visibly larger on the horizon.

We can see a little bit of the interior, enough to tantalize, but it's still impossible to tell what we're dealing with. We're coming in from the east, and the crater wall is higher on the far side than it is on the near side. So we see the top part of the western wall, facing toward us and glowing gold in the morning sunlight in Pancam pictures. From this distance it looks almost clifflike, imposing and actually a little frightening. But it's impossible to tell yet how much of that is real and how much is just fore-shortening. We should find out soon enough.

The question that we'll face once we get to the crater, of course, is whether or not we can go down inside it.

The orbital images of Endurance are enticing. They seem to show layering in the walls, maybe well down into the crater. Everyone is mind-ful of the disappointment at Bonneville, of course, and nobody wants to get our hopes up too high. But that was *Spirit*, the hard-luck rover, and here at Meridiani things could be different. If there really are layers of rock exposed in the walls of Endurance, then they're deeper down and older than the rocks we saw back at Eagle. And if that's the case, they could tell us what came before all the water.

But can we get to them? If everything is as steep as the far wall looks right now, it may be flat-out impossible. Mike Malin, whose camera on *Mars Global Surveyor* took the orbital images, thinks the whole thing is go-ing to be so steep that all we'll be able to do is look in from the lip, say "Oh how pretty," and move on.

It all boils down to how good *Opportunity* is at going up and down hills, and that's not something we know much about.

As near as we can figure, the only thing that limits how steep a slope we can drive down is being sure that we don't flip the vehicle over. That happens at about 45 degrees, which means that we should safely be able to go down any slope that's significantly gentler than that. And I've got to think that somewhere around the rim of Endurance there has to be a slope that's gentler than 45 degrees.

Getting *up* again, though, is another matter.

Our experience so far in driving uphill at Meridiani isn't encouraging. Back on Sol 56, the first time we tried to climb out of Eagle Crater, we failed. The slope there was only about 17 degrees, and we tried to do a fifteen-meter drive, straight up it and out. Instead we ended up spinning our wheels in place for the equivalent of fifteen meters, digging six nice deep holes in the sand.

We got out safely the next sol, driving diagonally across the slope and over the lip, but the experience spooked everybody pretty badly, both at JPL and at NASA Headquarters. I've talked to a bunch of the mobility engineers, and they all think that we would have done better if we'd tried to climb out over the rocky outcrop, rather than doing it up loose sand. But even if we can climb up a 20 degree slope on rock, which is just conjecture, that's still nowhere close to the 40-ish degree slope we think we can go down.

So here's the problem: If the gentlest slope we can find anywhere around the rim of Endurance is between 20 and 40 degrees, then this thing could be a permanent rover trap. We'd go in just fine, and as long as we kept going downhill we'd be okay. But as soon as we tried to climb back out, we'd stop in our tracks. Endurance Crater would be the place that *Opportunity* dies.

That sounds bad, of course. But she's got to die somewhere, and as I look at Endurance, I can't help thinking that maybe it's the right place to do it. If we just peer into this crater and then leave, like we did with *Spirit* at Bonneville, the closest interesting thing in any direction is some "etched terrain" that lies almost three kilometers to the south. This stuff looks intriguing in the orbital pictures: rough and hummocky, different from the smooth plains we've been driving around on. The best guess, though, is that it's made of rocks very much like what we're hoping to find down inside Endurance, eroded into irregular shapes by the wind. So if we can find what we're looking for by climbing down into the crater, why pass it by in hopes of surviving a three-kilometer slog to the south in search of the same thing?

The slog to the south is not without appeal. Endurance could get old after a while, no matter how interesting it turns out to be at first. But I don't know if we actually can make it all the way to the etched terrain. *Opportunity*'s stuck-on heater is still sucking 170 watt-hours out of the rover every sol, and we're starting to feel it badly.

Back when we landed, when the arrays were putting out more than 800 watt-hours every sol, losing 170 watt-hours to the heater wasn't a big deal. But now, with the array output below 600 watt-hours and falling, it's really starting to hurt. The minimum output that we need just to keep *Opportunity* alive is something around 280 watt-hours, and if you stack the wasted 170 watt-hours on top of that you get 450 just to do nothing all day. An array output of 600, which is where we've been lately, leaves just 150 watt-hours to work with for a sol. That's not much. And an output of 500, which is where we'll be in another several weeks, leaves 50. That's almost nothing.

So we've got to find a way to fix this. The thing we're going to try is called Deep Sleep.

Deep Sleep is clever, and possibly dangerous. It was Leo Bister who came up with it. The concept is simple: Just disconnect the battery overnight. Sometime after the sun goes down, we'd execute a command to take the battery completely off the power bus. That means that *no* heaters could be operated on the vehicle, whether their switches are on or not, because there wouldn't be any energy flowing to the switches. No power would get wasted overnight, and when the sun hit the arrays the next morning the rover would feel the current start to flow and would automatically put the batteries back online.

Of course, that's all theory right now. We've never tried it on a flight vehicle, and if the batteries don't come back on line the way they're supposed to when the sun comes up, it'll be a very bad day. But if it works it'll save *Opportunity* more than a hundred watt-hours every sol, which is a ton. So it's probably worth the risk.

There's a price to be paid even if it works, though, and it's pretty high.

First, if the batteries are off-line during the night, that means we can't do any nighttime science. Nighttime is prime time for doing long Mössbauer and APXS measurements, and on a night when we use Deep Sleep we can't run the instruments because there's no current coming from the batteries to run them.

Next, no batteries at night means no nighttime comm passes. *Odyssey* and *MGS* each fly over *Opportunity* twice per sol: once each in the afternoon and once each in the predawn hours of morning. Any night that we use Deep Sleep will be a night when we can't have our A.M. orbiter comm passes, because there's no battery online to power the rover during the pass. So if we use Deep Sleep a lot we'll cut the amount of data we can send back to Earth almost in half.

And the worst part about Deep Sleep is that it's likely to kill Mini-TES.

Mini-TES has a lot of complicated optical components inside it, and one of them is called a beamsplitter. Most of the optical elements we used inside Mini-TES are mirrors, but the beamsplitter isn't. The beamsplitter has to let infrared light pass through it, like a window, which means that it has to be transparent. And normal optical materials like glass aren't transparent in that part of the infrared spectrum. So instead of glass, we used a kind of salt crystal called potassium bromide.

The beamsplitter inside each Mini-TES is made of a disc of potassium bromide about the size of a quarter, glued into an aluminum holder. The problem is that potassium bromide and aluminum don't shrink by the same amount when they get cold. The potassium bromide shrinks more than the aluminum does, and if things get cold enough the potassium bromide will shrink far enough back from the aluminum around it that it'll crack. If the beamsplitter cracks, Mini-TES has had it.

We've been worried about this problem from the start, of course, and to keep Mini-TES safe we put a "survival heater"—a heater that'll kick in automatically whenever the instrument starts to get dangerously cold during the night—on it. But if the battery is off-line because of Deep Sleep, the Mini-TES survival heater can't kick in, no matter how cold it gets.

We don't know how cold the beamsplitter can get and still survive. A few years ago we did a quick test of a few beamsplitters at low temperatures. At minus 50 degrees C they survived, and at minus 60 degrees C they cracked. We didn't try any temperatures in between. So somewhere between minus 50 degrees and minus 60 degrees is the death zone, and we don't know where.

The *Opportunity* thermal engineers also don't know exactly how cold Mini-TES is going to get once we start using Deep Sleep. But they've done some calculations, and their calculations say that it'll probably be a few degrees below minus 50, getting slowly colder as the weeks and months wear on.

Before too long, I know we're going to have to start using Deep Sleep on *Opportunity*. But sooner or later it's going to kill Mini-TES.

APRIL 30, 2004

We've arrived at Endurance Crater. Like coming upon the Grand Canyon on Earth, it happened suddenly.

The sol we planned yesterday was simple: a twenty-meter drive toward the crater, and five Navcam images at the end of the drive to see what we see. I stepped in to the SMSA just as the data started to flow today, and the early word from Mobility was that we had covered seventeen of the twenty meters, stopping short because the rover had sensed a hazard. It had to have been the lip of the crater.

Today's *Odyssey* pass was agonizingly slow, so I decided to begin the science meeting without the images, getting a jump on the planning process. We were just getting started when Tim Parker interrupted.

"First Navcam frame is down. . . . Anybody want a look?"

There was a passionate chorus of approval around the room. It was obvious that we couldn't get work done while the first pictures of Endurance were hitting the ground, so I told Tim to route the image to the big screen.

It was startling. Our planning simply stopped as we looked at it, overwhelmed. One by one over the next hour and a half the other four images came down, each one more spectacular than the one before. By the end of the pass we had a five-frame Navcam mosaic on the ground that revealed all of Endurance Crater.

The crater is deep. It's too early to be sure, but the vertical drop from the top of the far wall to the floor looks close to twenty meters. It's also steep. In fact, it's damn dangerous-looking.

The far, western wall dominates the scene. The wall is a precipice along much of its crest, and if the images don't lie there are segments, meters high, that are so steep that they overhang. Crags and pinnacles dot the crest, and long ribs of rubble extend down from it toward the crater floor. The lower walls of the crater are steep and forbidding, wide swaths of featureless sand studded at irregular intervals with rounded, oddly perched boulders. On the floor of the crater is a spectacular dune field, the undulating crests like frozen waves in the images. This thing is no Eagle Crater, and it's no Bonneville. It's like nothing we've encountered before.

We're going to have to be very careful here. Those cliffs are the kind of thing that *Opportunity* could fall off and die if we screw up. We're feeling utterly overwhelmed, and at the moment I have no idea how we're going to deal with it.

The slope immediately in front of us doesn't look particularly intimidating, though that may be deceptive. It's only 18 degrees near the top, but it steepens below that to some higher angle that we haven't been able to measure yet. And there's nothing that looks like intact bedrock on it.

Instead, it's what Grotzinger likes to call "a dog's breakfast," just a jumbled-up mess of pavement-like slabs with no organization or structure to them. There's rock down there for sure, but it'd be impossible to tell exactly where it came from. A bunch of it probably slid down from right where we're standing now.

The far side of the crater, though, is a different story. The tallest cliffs there are at least several meters high, and even in Navcam images from this distance it's obvious that they're made of dramatically layered bedrock. There's no way of telling yet what it's made of, but the thickness of the stratigraphy exposed here has got to be at least ten times what it was at Eagle Crater. That's ten times more geologic record than we've been able to see until now.

Better yet, most of the rocks across the crater don't seem to be busted up and jumbled like the ones at Eagle were. If that's true, then the real principles of stratigraphy apply here: The youngest rocks are on top, the older rocks lie below, and everything in the whole stack can be put into strict and unambiguous time order. That means we can do real history here, not just say that there was a moment in the past when something in-teresting happened. If we can work our way down into Endurance Crater, it'll be like traveling back in time.

Of course, in order to do it we're going to have to find a way of get-ting at this stuff.

I don't think we can get at it by driving down from where we are, and we certainly can't drive down those cliffs on the far side. What's caught our eye, though, is what's at the far left and right ends of the Navcam mo-saic, on the northern and southern walls of the crater. Here we see what looks like the same layered rock that makes up the cliffs on the far side, but wrapping around onto much gentler slopes. The slopes look easy enough to descend, though I have no idea whether we'd ever be able to make it back up them again. Larry Soderblom has already decided that we ought to go down, and that the slope at the left end of the Navcam pan, on the north side of the crater, is the place to do it. I'm calling it Larry's Leap.

MAY 2, 2004

We're partway through taking a huge Pancam panorama of Endurance crater, one that surely will become one of the signature images from this mission. It'll be hundreds of megabits, so it's going to take several sols to get it on the ground. We've got some of it down now, though, and the part we have shows Larry's Leap in detail.

The slopes look reasonable, though I say that without really knowing what "reasonable" means. I'm worried, though, by a broad band of what looks like sand separating two of the rock layers. If that's anything like the sand that bogged us down on our way out of Eagle, then Larry's Leap could be a one-way trip.

The Pancam pictures of the other place where we might be able to dip our toes into the crater, down on the south side, haven't come down yet. I asked Grotzinger to come up with a name for that area, and he suggested Karatepe, which is an archaeological site in Turkey. We should get a look at it sometime in the next sol or two.

I've been stewing about this whole issue of whether to drive into Endurance for weeks now, and last night I was up a good part of the night thinking about it. What I'd love to do is simply say "let's go on in" and deal with the consequences later. I really do think we'll do better on rock than sand, and even if we try to get back out and fail then we'll die in a good place.

What I'm worried about, though, are issues of perception and blame. If we get stuck in this thing by accident it's going to look like a failure whether it was the right thing scientifically or not. And with *Spirit* still not turning up the science we've been hoping for at Gusev, it'd be a damn shame for something to cast a pall over the success that we've had so far with *Opportunity*. So, much as I hated to do it, today I fired off an e-mail to Richard Cook and Matt Wallace requesting that the *MER* project do a formal assessment of our ability to climb out of Endurance Crater.

I know what my e-mail will stir up. We never do anything the easy

is project. There will be meetings. There will be PowerPoint
ations. There will be tests and tests and more tests. We'll brief the
ts up the line to every manager imaginable, both at JPL and at NASA
Headquarters. And there will be formal decision meetings with everyone
up to and including Elachi and Weiler getting their licks in. It's just the
way we do these things.

In the end I bet we'll get approval to go into the crater, because I
think it's the right thing to do whether we can get out or not. But it's all
about shared accountability. If there's any chance that we're going to get
stuck in this thing, I want *everybody* who might want to complain about it
later to have been in on the decision.

MAY 4, 2004

The Pancam images of the biggest cliff on the far side are down, and it's
spectacular. We've decided to name it Burns Cliff, after Roger Burns.
Roger was at MIT for years, and he was one of the first scientists to suggest that sulfur chemistry could be important on Mars. He even predicted
that jarosite might be found there someday. Roger died a few years ago,
too early to learn that he was right. He would have loved this place.

What Burns Cliff is actually made of is anybody's guess at this point.
But it sure has some dramatic layering in it. The most striking feature is
some spectacular cross-bedding near the base of the cliff, with steep-
sloping beds below truncated plainly by flat-lying beds above. This cross-
bedding is huge, visible from more than a hundred meters away, and it's
nothing like the tiny ripple cross-beds that we saw back at Eagle. Instead,
this looks like the kind of cross-bedding that you get inside sand dunes.
So it's looking right now like what came before our evaporites at Eagle
Crater was probably a dune field. The current guess is that the dunes were
made of volcanic sand, and that Burns Cliff is basaltic sandstone, something that would be a real rarity on Earth. But we'll see.

Richard and Matt reacted to my e-mail with enthusiasm; indeed, Matt had already concluded that we needed to do the same thing. But it's going to take a long time. Matt drew up a rough schedule today, and it's probably going to take something like a month before we'll have a decent story to take to JPL management and NASA Headquarters. The big hang-up is that we'll need to build some kind of steep test fixture that we can drive a rover on, and then take one of our Earth-bound rovers and do a bunch of tests on it. So we're trying to come up with productive ways to keep *Opportunity* busy for a whole month while the slope testing gets done.

We've been arguing back and forth on Karatepe vs. Larry's Leap as the ingress point. The slopes look pretty much equivalent, the steepest sections at both places somewhere between 25 and 30 degrees. Both of them look scary, to be honest about it. Most of us are leaning toward Karatepe, with Grotzinger leading the charge, the reason being the direction that the slope faces. We've been on Mars long enough now that the seasons are changing, and winter is coming on in the southern hemisphere. Each sol the sun moves a little farther north in the sky, which is part of why our power situation is deteriorating.

But if we can tilt the solar arrays toward the sun—to the north—we should be able to improve the situation. And at Karatepe, which is on the southern rim of the crater, the slope inside the crater faces north. If we go in there, the arrays should give us something like eighty watt-hours more each sol than we'd get at Larry's Leap. Even Soderblom was convinced when he saw that number.

MAY 5, 2004

The pictures of Karatepe are down now, and it's just about perfect. It's rockier than Larry's Leap, and the layers exposed there connect right up with the ones in Burns Cliff. If we're going to go into Endurance Crater, Karatepe is the place to do it.

The initial shock of what Endurance looks like is finally wearing off. Grotzinger led a great end-of-sol discussion today, and I think we've got a handle now on how we're going to attack this thing. The crater's round, of course, so the obvious thing to do is circumnavigate it. The question is whether to go clockwise or counterclockwise. I was initially for going clockwise, since that'd put us at Karatepe something like a month from now, about when Matt thinks that the hill-climbing study will be done. But Grotzinger argued for counterclockwise, and he won me over. That'll put us close to Karatepe early, and if the climbing study goes better than we expect it to, we'll be ready.

As we work our way around the crater, we're going to stop two more times, shooting two more big panoramas like the one we just got. It's hard to see how we could do much better. I'm worried, though, about how much time it'll take, with the solar array output dropping the way it is. We're down to less than three hours of operations a sol now, and it's just going to get worse.

We're planning to do a first test of Deep Sleep two sols from now. We don't expect to break Mini-TES when it happens. Before this sleep we're going to dump a bunch of power into the rover, making it warm and toasty inside. That's hugely wasteful, though, and we can't do it every time we sleep. We'll do it for this one-time test, to keep Mini-TES safe, but it's not a viable long-term strategy.

I hope to hell that Deep Sleep works. Taking the battery off-line sounds simple in principle, but if things don't go right when the sun comes up in the morning, we've had it. It shouldn't be a problem. Putting the battery back online uses only the rover's lower brain functions, and the main computer doesn't have to be on. So even a hopelessly confused vehicle should be able to handle it. Still, I'm going to breathe a lot easier once this test is behind us.

MAY 7, 2004

Chris Salvo's mission manager report for today sums it up:

> At approximately 12:51 PDT on Friday, May 07, 2004, we received data from a 9:30 LST Sol 102 downlink session. Woo-hoo! The spacecraft performed very well. Deep Sleep executed as planned and we saved a great deal of energy through the night as a result. Mini-TES temperatures appear to have dipped to -46 degrees C at around 8:00 LST, the coldest we had allowed the instrument to get so far in this mission, but warmer than the -50 degrees C which is the lowest proven temperature for this instrument.

I walked into the SMSA just as the downlink confirming that we had survived Deep Sleep was coming in. Chris's musical choices for the sol, played as usual over VOCA, were perfect. The first, blasting as I walked into the room, was Handel's Hallelujah Chorus. It was followed immediately by Led Zeppelin's "Dazed and Confused."

MAY 20, 2004

For two weeks we've been working our way, with excruciating slowness, around toward the second panorama position on the southeast side of Endurance Crater. The ground here slopes gently to the south, and the southward tilt of the solar arrays is killing us.

As I write this I'm in Montreal, where I've been presenting our results to the American Geophysical Union. Today's mission manager report from Rick Welch back at JPL, which I downloaded minutes ago, was typical:

Opportunity spent Sol 113 sleeping and partially recharging her batteries. On Sol 114 she successfully executed a 16 m drive to approximately 5 m from the southern rim of Endurance Crater. *Opportunity* is healthy but continues to have an adverse southern tilt of solar array acquiring less than 500 W-hr of energy and limiting her to approximately one hour of activity and one UHF communication pass to stay energy neutral.

We've got to do something.

I tracked down Phil Christensen, who's also here in Montreal, and we got together on a bench in a hallway outside one of the meeting rooms. Phil looked tired and a little grim; these months of flight operations are starting to wear on everybody. I wasn't sure how he would react to what I had to say.

"I know what you want to talk about," he began.

"Yeah, I'm sure you do," I replied. "It's time to pull the trigger on Deep Sleep."

"I know, and I agree," he said to my relief. "I just have one request."

"What's that?"

"I'd like to finish up a good Mini-TES panorama from the second pan position before we do it."

That made sense to me. In fact, it made sense to get the Mini-TES pan done and to get it down onto the ground and analyzed before we did anything. The Mini-TES data we had gotten from the first pan were suggesting that some of the cliff faces might have sulfates in them, instead of basalt, which was a surprise. I didn't want to take a chance of breaking Mini-TES until we'd had one more look.

So that was it. We talked through a few more details. Then, as Phil got up to leave, I asked him a question that I knew he couldn't know the answer to.

"Phil, do you think the instrument will survive?"

"Yeah, I do," he said slowly. "We'll see, but I really do."

I hope he's right.

MAY 22, 2004

We're finally at the second pan position, three weeks after we pulled up to the first one. We're going to set up camp here for a while, taking another huge Pancam panorama and getting as much Mini-TES data as Phil wants. After that, and with Deep Sleep at the ready, we'll continue on counterclockwise around the crater, with what I hope will be much greater speed toward pan position three.

MAY 24, 2004

I got back to JPL from Montreal last night, and today I hiked up to the Mars Yard with Matt Wallace to see the test fixture that we're planning to do the hill-climbing tests on. It looks awful.

The thing is a giant steel sandbox, a couple of feet deep and maybe thirty feet on a side. We've got the biggest yellow forklift that I've ever seen in my life up there, rigged to go under one edge of the sandbox and lift it to whatever angle we want it at. Once it's up, thick steel stanchions get propped underneath, perched on squares of one-inch plywood that are stacked to get the angle right.

What's awful is what's inside the sandbox. Matt's guys went down to the local Home Depot, picked up a bunch of garden paving stones, and glued them onto a big plywood slab in the sandbox. They filled in the gaps between the stones with sand, which is the one thing that's Karatepe-like about it. But the paving stones themselves are as smooth as slate, and they offer none of the rough projections that the real rocks at Karatepe do.

"We can't use that!" I protested to Matt when I first saw it. "It'd be a ridiculous overtest."

"You mean because it's so smooth?"

"Of course I mean because it's so smooth! It's *way* too smooth. That stuff looks nothing like the rocks at Karatepe."

We took Pancam pictures looking down at Karatepe during our drive by it a couple of weeks ago, and they show the slope there in nice detail. It's rocky, with irregularly shaped stones that indeed lie flat and pavement-like on the ground. But the surfaces of the rocks aren't smooth at all. They're roughed up and lumpy, with embedded blueberries sticking out of them. And it's those rough, irregular surfaces that I'm counting on to give our cleats something to grab onto.

"Well, we're planning on putting some quick-setting concrete and some BB's or something onto it over the next few days," Matt explained, "to give it a rougher texture. But that'll take three or four days. For now, I think this is a good test to do."

He was seriously talking about driving the rover on these slick paving stones. I could see why, but the problem was that he was thinking logically.

"Look," I said, "if this weren't such an emotionally charged issue, I'd agree with you completely. If we try to drive up this stuff as it is and the rover slips, then we've learned nothing, because it's an overtest. No harm done. And if we try to drive up it and somehow we actually make it, then we can declare victory.

"What I'm worried about," I continued, "is the psychological damage if we fail our first test. We had problems in the sand back at Eagle, and that got most of the managers here and back east convinced that we can't climb hills. The first test we do on this thing is going to be incredibly high profile, and whatever happens, the word is going to get around immediately. So I'd rather not have us fail it. Even if we do succeed on a reasonable surface later, the failed first one will always call the results of the successful ones into question."

Matt thought about it for a moment. I knew he could see my point, but I also knew that he wanted to get on with it.

"Okay, look, how about this," he countered. "How about if we bring the rover up here tomorrow and try to run it up the slope, but we just call it an 'engineering checkout'? It's the first time we've driven the vehicle on this fixture with this kind of surface. There are all kinds of problems we could run into. We'll keep it low profile, and we'll just tell anybody who comes that it's a checkout of the test setup, the computers, the cables, that sort of thing. Something to give us something productive to do while we wait for the guys to finish roughing up the rocks."

"Okay," I said, casting another uneasy glance at the paving stones. "I guess that's a reasonable compromise."

And I suppose it is. But I don't think there's any way the rover's going to be able to climb that thing.

MAY 26, 2004

I'm in Florida, where I just gave a *MER* talk, and as I was heading for the airport this afternoon my pager went off. It was Matt. I pulled out my cell phone and dialed.

"Hey Matt, it's Steve, what's up?"

"Well, I thought you'd like to hear the results of our little 'engineering checkout.' We did it today."

"And?"

"And the thing scampered up the slope."

"Really? How steep?"

"Twenty degrees for the first test, but we think there's plenty of margin above that."

Oh my goodness. "How much slip was there?"

"We didn't actually measure it, but there was almost none. It couldn't have been more than ten percent. This thing's a hill climber," he added proudly.

"How do you know there's margin above the twenty?"

"Well, they did what they call a 'fish-scale test.' You know those little spring-loaded scales they use to weigh things with, like in a fish market? They used something like that, a force gauge, to tug on the rover while it was on the slope, to see how much force was needed to make it slip. It was a lot, and when they did the numbers they found that we have margin all the way up to about thirty-two and a half degrees."

Thirty-two point five degrees! Karatepe here we come.

"And this was on the slick paving stones?"

"Yep."

"Man, we've got us a robot mountain goat. Thanks for letting me know."

I shook my head, amazed, as I hung up the phone. We'll still need to do a complete test program so that we'll have our story straight for Elachi and Weiler, but it's already obvious what the answer is going to be.

So do we keep going around the crater like we'd planned? Or do we just sprint back to Karatepe and do it?

MAY 31, 2004

We've talked it over on the team, and we're going to go straight back to Karatepe and set up for the entry. *Opportunity* is a wasting asset, and the sooner we get on with it, the more likely we are to get it done successfully.

And now Mini-TES is a wasting asset, too. We started using Deep Sleep regularly a few nights ago, and so far the instrument hasn't broken. The temperatures have been going to minus 52 degrees C. It'll get colder, though, as winter comes on, and one morning sooner or later Mini-TES will be dead. So if we're going to go in, now's the time to do it.

Of course, we still need to brief this whole thing up the food chain on both coasts, but after the tests in the Mars Yard I don't expect problems. The results are compelling, and even if they weren't the argument to go in would still be solid.

The real question now is what we do once we get in there.

When you look at the Pancam pictures we took from the top of Karatepe a couple of weeks ago, the upper part of the slope is a jumbled dog's breakfast, made up of blueberry-laden rocks that look like the same stuff we've been seeing for months. A few meters down the slope, though, everything changes. The rock becomes contiguous and regularly banded, the jumble giving way to intact stratigraphy. We don't know what the rock down there is made of, though we're starting to suspect from the Mini-TES data that it may be sulfates instead of basaltic sandstone. And we don't know how deep it goes before running into sand. But it sure looks good.

So I've worked out a plan with Matt, and it goes like this: We'll move down slowly, in several steps. The first few steps, while we're still on the jumble, will be all about engineering. The first sol, we'll drive in one rover length and then drive all the way back out again, just to prove we can do it. The next sol, we'll drive in two or three rover lengths, and then back uphill one or two, so that we're just into the crater.

And then, if we haven't been slipping too much, we'll be ready to go. The slopes at Karatepe run anywhere between twenty and thirty degrees, most of it on pretty solid-looking rock. Twenty degrees is fine, and thirty is close to the hairy edge of what we think we can do. So we'll work our way slowly downward, sampling with all of the instruments on the IDD as we go. It'll take time, and the RAT should get a real workout. We'll measure the composition and texture of the rocks in a systematic way, working our way backward through martian time.

How will we know when to quit? We'll quit if we run out of rock. Sooner or later the rock will give way to sand, and there'll be no reason to go on.

We'll also quit if we reach a point of no return . . . a place where the slope is so steep that if we go down we won't be able to get back up. Either way, we'll stop, take a last look around and then leave. The rover climbs backward better than it does forward, so we'll just put it into reverse and back straight up the slope and out, all in one shot.

I am *really* looking forward to this. We put so much into these rovers, and it feels very good now to be pushing them so hard, both here and over at Gusev. But it's not just for the adventure of it. Our foray into Endurance will be down the first true stratigraphic section ever studied on another planet.

JUNE 4, 2004

We're going in. We briefed it to Weiler today, and he went for it with gusto. He started off by asking what he calls one of his "Homer Simpson questions": What would we lose if we got stuck in the crater? As soon as he realized that the main thing we'd lose would be the chance to do a three-kilometer drive in search of the same kinds of rocks we'll find in the crater, the deal was sealed. I briefed the science story and Richard Cook briefed the engineering story, and in the end, Ed just said, "Sounds like a good idea. Let's do it."

JUNE 8, 2004

Here we go.

Yesterday we commanded what should be the first ingress/egress cycle into Endurance: six wheels into the crater, take some quick pictures, and six wheels out again. We're gathered now in the SMSA, waiting for the data through the P.M. *Odyssey* pass. As usual, it'll be engineering telemetry first, followed later by the images, so we're clustered around the mechanisms and mobility consoles, a big crowd.

The first data should have been here a while ago, actually. It's hard to believe that anything really bad could have happened from such a simple drive, but the silence is getting a little unnerving.

And then, ten minutes behind schedule, the data start to hit.

Chris Voorhees sees it first: "Tilt is nine degrees. That means we're back out!"

"Or that we didn't go in at all," I reply. All nine degrees tells us, near as I can see, is that we're not on the slope inside the crater, which is a lot steeper than that.

Chris gives me his "yeah, I guess you're right" look, but seconds later Jeff Beisadecki, the rover planner who designed the drive, looks up from his console.

"No drive errors," says Jeff with some pride.

Jeff's callout proves that Chris is right. If the drive completed with no errors, and if we're on a gentle slope, that means we've gone in and back out again. There are handshakes around the room.

Moments later, beautiful Hazcam images confirm it. The wheel tracks are crisp and clear, with no rocks out of place. The drive ended within centimeters of where we wanted it; the slippage can't have been more than 5 percent. The total yaw during the whole drive was only two degrees. Everything is on the money.

And then, best of all, the Pancam images start to come down. The rock below us is stunning, beautiful, spectacular. There's got to be five meters of intact bedrock exposed down there, with half a dozen horizontal color bands running through it. What is this stuff? Sandstone? Sulfates? Some kind of mix? It's impossible to be sure from up here, but it's all laid out there down below us. All we have to do now is drive in and find out.

It feels like the mission has started all over again.

18 POT OF GOLD

*S*PIRIT HAS REACHED THE Columbia Hills.

I've been immersed for weeks in planning *Opportunity*'s attack on Endurance Crater, so I've pretty much left the job of leading *Spirit*'s forced march across the Gusev plains to Ray Arvidson. His task hasn't been an easy one. The sol quartet has worked, but *Spirit* is starting to show serious signs of fatigue, even without a bad heater, and the *Spirit* team is flagging, too.

This whole mission has been hard for the *Spirit* guys. For the first three months after we landed, while we were all living on Mars time, hardly anybody switched from one rover to the other, deterred by the pain of going through that wicked twelve-hour time change. So during those early days the team split in two: a distinct *Spirit* team and an equally distinct *Opportunity* team.

The interesting thing is that even though we've gone to Earth time now, most people have stayed with the rover they started with. Maybe it's been loyalty, or maybe it's just been force of habit. Either way, though, it's the *Spirit* guys who've consistently felt shortchanged by events ever since we landed. We rejoice when *Spirit* touches down safely, and then *Opportunity* rolls to a stop right in front of a stack of sedimentary rocks. We rejoice when *Spirit* drives 50 meters in a sol, and then *Opportunity* reels off 140. We rejoice when *Spirit* gets to Bonneville and finds a pretty view, and then *Opportunity* gets to Endurance and finds layered cliffs of sediments. The *Spirit* guys love their rover as much as the *Opportunity* guys love theirs, and with good reason. *Spirit* has been a very tough machine. But there's no question that *Opportunity* has been the glamour rover of the mission, stuck heater and all, and the *Spirit* guys know it better than anybody.

Now, just as we've finally hit the hills, the power situation on *Spirit* is getting bad. With each passing sol, of course, the dust buildup on the solar arrays has gotten worse. And now on top of that, winter is really starting to set in at Gusev. *Spirit*'s landing site is at about 15 degrees south latitude, which is much farther from the equator than *Opportunity*'s. Compounded by the dust buildup, the low angle to the sun in the northern sky is starting to hurt *Spirit* badly. The quartet sols were each rich with science back when we first set sail from Bonneville, but as the power situation has deteriorated Ray and the team have had to prune more and more science out of them. By the time we got to the hills, there wasn't much science left.

Power isn't the only problem. Several weeks ago, the mechanisms guys who look after *Spirit* started to notice that the drive motor on the rover's right front wheel is drawing a lot more electrical current than it used to. That's not healthy, and it may mean that the motor is on the verge of failing. *Spirit*'s got 3.3 kilometers on the odometer now, which is twice what *Opportunity* has, and five times the distance that either rover was ever expected to go. It's hardly surprising, then, that signs of wear are starting to appear, but it's serious cause for concern. It'd be agony to have struggled all the way to the hills and then have the mobility system give out just when we need it most.

So *Spirit* is hurting, but her traverse across the Gusev plains has been a remarkable accomplishment. For weeks, even as the odometry numbers climbed, the hills didn't seem to get any bigger. Every morning, before I'd head up to the fifth floor to plan the attack on Endurance, I'd drop into the SMSA and see how many meters *Spirit* had done the sol before. Most sols it was a solid 60 or 70 meters, double or triple what we were doing back during the drive to Bonneville. One sol it was more than 120. But every day the pictures looked the same.

Then, just a couple of weeks ago, the view began to change. It was a subtle thing at first, just a little more detail on this hillslope or in that valley. But sol by sol the hills kept getting bigger, and new details began to appear.

The summit that's been closest to us since we left Bonneville is Husband Hill, named after Rick Husband, the astronaut who commanded *Columbia*'s crew. Husband Hill has a small buttress, which we've simply called the West Spur, jutting out directly toward us. It's our point of first contact; in this sea of basalt it feels like our first landfall. And now, as we look directly up at the West Spur, it's clear that *Spirit*'s long search for bedrock at Gusev is finally almost over. Most of the rock in the hills is a rubbly mess, but there are also obvious outcrops in clear view, just tens of meters up the hill, made of who knows what. For all I know it's the same damn basalt we've been seeing for kilometers, but at least it's bedrock. We'll know soon enough . . . if our wheel doesn't give out, and if we're actually able to climb the steep slopes that now loom above us.

Our official date of arrival at the Columbia Hills was Sol 156, four sols ahead of schedule. The *Spirit* team takes considerable pride in that.

JUNE 11, 2004

Today I'm back in Ithaca for a too-brief visit with Mary and the girls, watching the *Spirit* SOWG meeting from the fourth floor of the Space Sciences building over a blurry videocon link. Now that we've arrived at the West Spur, the immediate plan is to look up the spur with Pancam and Mini-TES, pick the closest outcrop that looks halfway interesting, and take a short climb up to find out what it's made of. After that's done there's been talk about continuing on to the south, along the base of Husband Hill, in search of we're not quite sure what.

Mike Carr is chairing the *Spirit* SOWG today, and he's leading the group through his agenda, planning remote sensing, when out of turn and a bit abruptly, Justin Maki breaks in from the back of the room. Justin has a bad habit of browsing through images during SOWG meetings, but he rarely says anything about them unless he finds something he thinks is really exciting.

"Hey, a picture of a pretty interesting-looking rock just came down."

"Okay, put it up," says Mike, sounding annoyed at the interruption. He's almost done, and he wants to get on with today's plan.

Justin routes the image to a screen at the front of the room and even over the videocon link I can hear the gasps around the table at JPL. I zoom the SOWG room video camera in tight on the screen, trying to make out what they're looking at.

It's hard to tell what it is, but there's no question that Justin has spotted something odd. In fact, it's different from anything I've seen before, even at this resolution. On my screen the thing looks like it has dozens of strange little lumps sticking out of it, leaning this way and that in crazy directions. I don't know what to make of it.

There's murmuring and chattering around the SOWG room. Mike's meeting has gone completely off the rails. Amid the hubbub, Larry Soderblom, who's the SOWG chair for *Opportunity* today, walks in. In typical Larry fashion, he sizes up the situation quickly.

"I'm thinking you're probably going to want to get some more pictures of that thing," he says with fine understatement. "That could be the end of the rainbow."

Justin instantly places a Pancam imaging target on the rock. The name he's given it is Pot of Gold.

JUNE 13, 2004

I'm back out at JPL now, and puzzling over Pot of Gold. It's one of the strangest things we've found on Mars.

It's not very big, maybe twelve or so centimeters across, small enough that you could hold it in your hand. It's not bedrock; the thing probably rolled down here from someplace up above. From a distance, it looks for all the world like some of the rocks we saw with *Opportunity* back at Fram Crater, with little blueberry-like things dangling out on the ends of stalks. They can't possibly be blueberries, though, that would be just too weird. But it sure looks like this rock is full of some kind of little hard blueberry-sized lumps.

There are also some very strange planar structures running through the rock. These look hard, too, but instead of being little lumps they're thin planes, just a millimeter or two thick, cutting through Pot of Gold like jagged knife blades. The blades aren't parallel to one another; in fact, one of them runs at practically right angles to the others. So these aren't bedding planes in a sedimentary rock. In fact, there's nothing about this thing that looks remotely sedimentary to me, though I'll be damned if I can say what kind of rock it does look like. I've never seen anything like it, on Mars or on Earth.

One of the most interesting things about Pot of Gold is how abruptly we stumbled onto it. The Columbia Hills go on from here for kilometers, yet we literally hadn't moved more than *one meter* into them before this thing brought us up short. Is the whole hill complex made of stuff like this? Why was the very first thing we looked at upon arriving here so weird? The original plan when we got to the hills was to take one quick taste to find out what they're made of, and then to move on quickly to the south in search of something new. A lot of people on the *Spirit* team still seem to feel that that's the right thing to do. But I'm starting to wonder, especially with a filthy solar array, a gimpy wheel and the strangest rock we've ever seen practically right in front of us. It's hard to say what I expected to find in these hills, but it sure wasn't this.

The big question, of course, is whether or not Pot of Gold has anything to do with water. After today's *Spirit* SOWG meeting I had a talk with Steve Ruff, one of the Mini-TES guys who was hooked in over the videocon link from Arizona State. Steve has been one of the true *Spirit* faithful, spending every sol since January in Gusev Crater.

"I want this thing to be the end of the rainbow," he said with a weariness in his voice. "I really do. But Gusev has crushed my hopes so many times that I'm not going to count on it being anything."

I had to laugh; it was such a *Spirit* team thing to say. "You sound like you ought to spend a little time over at Meridiani, Steve. . . . It'd do you good." But secretly, I'm also hoping that this is finally going to be the Big One at Gusev. At Meridiani, everything has been served up to us as if we'd ordered it, with the Big One in the very first Navcam image we took on the night we landed. At Gusev, though, we've had to work for everything, with three kilometers of rover tracks behind us. How cool would it be if, after all that driving over all those months, we found what we came for here, too?

Why Pot of Gold looks the way it does is still a mystery, but in the rocks right around it I'm starting to think that I see something that I recognize: cavernous weathering.

Cavernous weathering is something I've seen in the Dry Valleys of Antarctica. It can happen when a rock sits out exposed to the elements for a very long time. Something happens to modify the outer skin of the rock. In Antarctica you can get a tiny amount of moisture seeping into the rock's surface, dissolving silica from the rock grains and then re-precipitating that same silica in between the grains. The silica glues the grains together, giving the rock a tough outer shell.

Then something changes. Maybe the climate gets drier. Sandblasting starts to eat slowly away at the outer surface of the rock. Finally, somewhere, the hard shell is breached, exposing the rock's softer interior. No longer protected by the shell, the interior weathers away quickly, eventually leaving just the shell behind. I've seen cavernously weathered boulders in Antarctica that were so hollowed out you could climb right inside them. When it gets really extreme the roof of the rock falls in completely, leaving just the broken fragments of the original walls standing.

Pot of Gold isn't obviously weathered this way, but some of the rocks around it are. Right next to Pot of Gold is a cavernously weathered rock with just a trace of the walls left, the rock's interior rotted away, leaving just a pile of crumbly crud inside. That doesn't mean that whatever happened here is the same as what goes on in Antarctica in all its details. But it does mean that some kind of process has made the outer shells of the rocks here harder than their interiors.

So what if Pot of Gold is cavernous weathering carried to an extreme? What if the knife blades are all that's left of the rock's old walls, some of them toppled over at crazy angles? What if the little lumps are just the crumbled remnants of a caved-in roof? It's too hard to tell from this distance; we need to move in closer and get a better look at the thing. A lot will hinge on whether it's stuck together well or whether the whole thing is a weak and crumbly mess. But Pot of Gold might make sense if it's a particularly funky-looking manifestation of the same thing that rotted the rocks next to it.

We're going to have quite a time trying to figure this out, if we get the chance. But I'm really getting worried that time is running out for *Spirit*. Since I've gotten back to JPL I've started checking into the power situation more carefully, and we're seriously running out of steam. If things keep going the way they have been, and if we don't do anything about it, we've got only a few more weeks before *Spirit* simply stops.

I don't think that *Spirit* is at serious risk of dying completely anytime soon. Even with a solar array output much less than what it is now we still should be able to nurse her along for months, going into a kind of hibernation during the darkest, coldest months of winter. And in fact, once spring starts to come on, sometime around December or January on Earth, the power projections say that we might be able to start doing some science again if she's still alive.

But that's for a future that may or may not come. For now it's a race to figure out these hills before winter shuts us down.

There are a couple of things we might be able to do to squeeze out a few more weeks before it'll be time to hibernate. One, of course, is Deep Sleep. I'm real nervous about doing that on *Spirit*, though, because it's so much colder at Gusev than at Meridiani. The thermal calculations say that *Spirit*'s Mini-TES will go down to minus 54 or minus 55 degrees C the first time we use Deep Sleep at Gusev, and that's way into the death zone.

So what I think we'll do instead—what I'm counting on to keep *Spirit* alive through the winter—is find slopes that face north.

This is something we never counted on when we planned this mission. In fact, it's something we never really thought about much at all. We picked flat landing sites for our rovers because flat means safe, and with only ninety sols and six hundred meters to work with, it never occurred to us that we'd have much of a chance of driving the rovers into steep, rugged terrain after we landed. But now it's happened. *Opportunity* is down in Endurance Crater on a sun-drenched slope at Karatepe, and all that sunshine is working wonders for her power system. If we can get *Spirit* up onto some north-facing slopes in the Columbia Hills it could do the trick there, too. In fact, with steep enough slopes and some luck, we might even be able to keep her doing a little bit of science through the winter.

Finding steep slopes in the Columbia Hills would mean some serious climbing, though, and I don't know how good a climber *Spirit* is going to be. *Opportunity*'s doing fine at Karatepe, but that's with all six wheels on solid bedrock, not on the crumbly stuff that the Columbia Hills look like they're made of. And even at her best, *Spirit* may have only five good wheels to climb with before long. So unless this rover turns out to be a much better mountaineer than we think she is, the best guess is that we have five or six weeks left to do science at Gusev before hibernation or worse. And whatever the story is with Pot of Gold, I want to try to nail it well before then.

JUNE 14, 2004

We drove closer to Pot of Gold yesterday, and now it's within reach of the arm. The Pancam picture from just short of the final position is revealing. I had thought that the little hard lumps in the rock might be just crumbly bits of "roof" that had fallen in, sitting in a loose pile, but now I can see that that's not the case.

The little lumps inside Pot of Gold are not simply loose bits of crumbled shell. Instead, they're solidly embedded in the rock, and weathering out of it, startlingly like the blueberries back at Meridiani. So the rocks here seem to have a hard, crunchy outer shell, a soft, chewy interior and little hard crunchy bits sprinkled in among the chewy stuff. What are the crunchy bits on the inside? Big crystals in an igneous rock? Pebbles in a sedimentary rock? Blueberries? We're going to take some MI images today, so tomorrow we may know more.

This is starting to feel like being back at Eagle Crater with *Opportunity*, every sol bringing some new clue. And it's wonderful to see how it has energized the *Spirit* team. I just wish I had some confidence that we'll figure it out before winter.

JUNE 15, 2004

I'm trying to split my time between *Spirit* and *Opportunity* now. I'm the *Opportunity* SOWG chair this week, so that's where my real responsibilities lie, but things are going so smoothly over at Meridiani that it doesn't take up all of my time. We're well down into the crater at Karatepe now, and today we wrapped up the science in the familiar dog's breakfast of jumbled rock that lies high on the slope. Next we'll move down into whatever lies below it. The power situation looks great, there's plenty of room in flash for new data, and we've got layered bedrock ahead. Business as usual in the land of *Opportunity*.

For *Spirit*, though, things aren't going as well. We had an IDD fault today because of a bad command, the result being that all we got down was one bizarre, perplexing MI image of Pot of Gold. None of the hard lumps are visible, and it's just a strange-looking mess. So we've still got a lot of work ahead of us at Gusev.

Down on Earth, though, we had a bit of a breakthrough today among the *Spirit* team. Most days, once the SOWG meetings are over, we try to have a get-together of the science team to hash over whatever issues we're dealing with at a more relaxed pace than the hurried tactical planning process allows. And today's discussion was all about what *Spirit* should do next.

For two and a half months now, the *Spirit* crew has been in forced-march mode, motoring across the Gusev plains just as fast as their wheels would take them. The goal was to get to the hills by Sol 160, and against the odds they did it. To do it, they had to put themselves into a pure drive-drive-drive mindset, and it's been a little hard for them to downshift back out of it. So for the past several weeks, even as the hills have loomed larger in front of us, I've been seeing plans for *Spirit* that would have us go bombing at high speed along the base of the hills, trying to get to a place called Lookout Point, half a kilometer south of where we are now, by winter. That would make good sense if the hills were as boring as the rest of Gusev has been; at least we'd have a nice view for the winter. But it makes a lot less sense now that the hills seem to have turned out to be interesting.

What I reminded everybody at today's gathering was that the reason we set the goal of hitting the hills by Sol 160 was so we'd arrive while we still had a little gas in the tank. And they did it. The race is won, we're at the hills and it's time to start doing serious science again.

So what I told them today is that there are no target dates anymore.

There's no date by which we have to leave the area around Pot of Gold. There's no date, ever, by which we have to get to Lookout Point. The only thing we have to do is to use every capability that *Spirit* has left to figure out what the hell is going on here. And after we talked there seems to be a developing sense that zooming off to Lookout Point is not the right thing to do. Instead, we're talking about spending several weeks working along the base of the West Spur and then just climbing the thing, searching for bedrock on north-facing slopes and wintering over somewhere high on the slopes of Husband Hill.

JUNE 16, 2004

Today at Gusev was one of the unusual sols that includes two good *Odyssey* comm passes in a single martian afternoon. The conditions have to be just right for this to happen. *Odyssey* flies above our site, low in the sky as seen from *Spirit*, on one orbit, and then it flies over again, low in the opposite part of the sky, just one orbit later.

I was expecting a lot from today's passes. The IDD fault the day before yesterday cost us a bunch of data, and today was planned to be a full recovery of everything we lost. We were expecting seven MI images of Pot of Gold, plus our first Mössbauer spectrum on it.

The first downlink was a disappointment. There was no Mössbauer data, and just three and a half of the seven MI images, all of them so out of focus that it was pretty clear that the remaining ones would be out of focus, too. This is a fiendishly difficult target to take pictures of with the MI, with all that weird topography on the surface. It's a problem we're going to have to figure out how to solve.

The second downlink was starting up just as I passed by the *Spirit* Mössbauer room on the fourth floor. Dick Morris was inside.

"Anything yet?"

"Still waiting."

I went next door to the SMSA and had a chat with MDOT, who checked her packet watch display and assured me that there were Mössbauer packets on the ground.

"Probably just haven't gotten through the pipeline yet," I told Dick. "I bet they'll be here soon."

"I'll be ready."

I did a lap or two around the floor, talking to people, seeing how things were going, killing time, giving Dick some peace. Then I went back to the Mössbauer room.

"Got anything?"

"Yep. . . . I'm just working on it now."

"Will it take a while, or can I hover?"

"Should only take another second."

I settled myself in a chair behind Dick as he brought the data file into the Mössbauer viewing program and got it set up to display the spectrum.

Dick floated his cursor over the PLOT button. "Wanna go out for a cup of coffee?" he asked, grinning. It took me back to our first Mössbauer spectrum, months ago, with Daniel and Göstar at the keyboard.

I smiled. "Plot it."

Dick hit the button, and the spectrum came up.

"Jesus," I muttered, "there's a sextet in there."

"Pretty strong one," Dick replied.

"You think it's hematite?"

"Could be." Dick is always cautious. "I'll need a little time to work on it."

I left so Dick could analyze the spectrum without distraction. An hour or so later I stopped back in again, and he was done.

Pot of Gold has no olivine in it—it's the first olivine-free rock that we've found at Gusev.

There's something else in there that Dick isn't quite sure of yet, though he thinks it may be pyroxene, another mineral that's common in martian rocks.

And, no doubt about it, there's hematite. Lots of it.

We're still a long way from having this thing figured out yet. All we've got so far is a couple of decent Pancam images, a single in-focus MI image that makes no sense to anybody and one quick Mössbauer measurement. No Mini-TES, no APXS, no RAT hole. Still, this is the biggest thing we've found at Gusev, by far.

I have to keep reminding myself that there are lots of ways to make hematite, and not all of them involve water. The keys will be to find out how the hematite is distributed, to learn what other minerals are there and especially to see what the chemistry tells us. Just like Eagle Crater.

Is the hematite concentrated mostly in the knife blades and the lumps? If so, maybe it was deposited from water that flowed through fractures and voids in the rock. Why isn't there any olivine? Is it because there was none to begin with, or is it because it's all been dissolved away? Olivine can be easily destroyed by water, and when a rock that has olivine in it is soaked in water, the olivine can be the first thing to go. If there really was water here was it cold, or was there a hydrothermal system, a hotspring?

Dick presented the new Mössbauer data a few hours later at today's end-of-sol discussion, with Göstar joining in proudly on the line from Germany, and there were whoops of victory around the room when the hematite spectrum went up on the screen. It's the best moment we've had at Gusev in months, and maybe the best since the night we landed.

Indeed, with this news today a sudden sense of joy and relief has swept over the *Spirit* team. I just pray that it's not misplaced. I think we've got something, but it's impossible to be sure yet. Suppose we get a contradictory story when the APXS data come down? So much hangs on the hematite's being related to water, and it may not be. If this thing really saw water, then the grim trudge across the Gusev plains will have paid off. But it could be a while before we're certain.

Still, at least it's not the same damn basalt, and just that is something to celebrate. To go all those kilometers across all that basalt simply to find yet more basalt in the hills would have been crushing. Pot of Gold certainly vindicates the decision to sprint for the Columbia Hills after Bonneville, and it may even be a step toward redeeming our choice of Gusev as a landing site.

So it's been a good day for the *Spirit* team. The people I've been thinking about most since the Mössbauer data came down, though, are the EDL guys, so many of them now moved on to other projects. They worked miracles to get us to Gusev, and for months it looked like we'd have little to show for their efforts. I saw Wayne Lee in the cafeteria at lunch today, just fifteen minutes after I'd seen the Mössbauer data, and I went over to tell him the news and shake his hand. I've got to give Adam a call, too.

JUNE 19, 2004

I love this rock, but it's starting to drive me nuts.

For one thing, it's just a really tough target to deal with. The shape and size are roughly what you'd get if you took a big potato, stuck a bunch of toothpicks into it, and then jammed jelly beans on the ends of the toothpicks. The damn thing's got tentacles.

We've finally gotten a bunch of good MI images of it down. The trick to getting them was simple: Just take a ridiculous number of pictures and a few of them are bound to be in focus. Up close, this thing is even crazier looking than it was in the Pancam images. The tentacles have little bulbous ends to them (the jelly beans); we're calling them "nuggets" at the moment, for lack of a term that actually means something. They're not spherical, and it's hard to convince yourself that they ever were. So I don't think they're concretions like the ones at Meridiani, hematite or no hematite. And the skin of the tentacles—indeed, everywhere on the rock that isn't coated with dust—has a weird pitted texture that's unlike anything we've ever seen. I see those pits, I think of the fact that there's no olivine in the rock and I wonder if the pits didn't form when olivine crystals just dissolved away.

But then there's the APXS data, which also just came down. It turns out that the mix of chemical elements in this rock looks pretty similar to what you'd expect for a basalt. It's certainly nothing exotic.

There are some patterns to it. The main one is that the rock is higher than usual in sulfur and chlorine, both of which are elements that tend to be moved around easily by liquid water. So that could be a clue. But just like at Stone Mountain back at Eagle Crater, we don't know whether we're simply dealing with a coating of salt, dust and other crud on the outside of the rock. And more discouraging, we don't see any of the big chemical changes that we'd expect if water had dissolved away some major minerals from the rock. That makes the theory about olivine going away look pretty weak.

What we badly need to do is RAT the thing and see what it's like on the inside. Is the hematite just a coating on the outside, or is the rock shot all the way through with the stuff? Does the pitted surface mean that it's porous on the inside, like a sponge? Is there any olivine on the inside? And what are the *real* element concentrations inside this thing? The APXS just sees the surface, and the surface of the rock looks like it's pretty contaminated with dust.

Of course, it's hard to imagine a much more difficult target for the RAT than this one. It's small, the hematite could make it very strong, and it's got an impossibly rough surface. We may have a real struggle on our hands trying to put a RAT hole into this.

As fascinated as I am by what's going on at Gusev, I'm mostly a spectator in this little adventure. I'm still the SOWG chair for *Opportunity* right now, and Ray is running the show at Gusev. I've got to keep reminding myself to give him room and let him lead his team without interference. It's not always easy.

JUNE 20, 2004

The frustration is growing at Gusev. I've been trying to keep my distance, and from that distance it has taken me a little while to piece together what happened there over the past few days. But I've got it now, and it's not good. We're going to end up having blown at least four precious sols screwing around with this rock simply because people didn't tell Ray what they really thought.

The problem has been the RAT. Pot of Gold is a ridiculously difficult RAT target, bizarre in shape and barely bigger than the RAT itself. So a couple of sols ago the Honeybees decided that they were queasy with the south-facing heading from which *Spirit* was approaching it. RATting Pot of Gold from that direction would work, they thought, but other things being equal they decided they'd be happier swinging around and coming at it from the west instead.

They mentioned this at an SOWG meeting, but none of the subtleties of their thinking came through in their words. Instead of the nuanced reality, the message that Ray got from them was black and white: In order to RAT Pot of Gold, the rover was going to have to be pointed east instead of south. And that would mean driving away from the rock and coming back at it from a different direction.

This news from the 'bees put the rover drivers on the spot. For days they'd been feeling anxious about the terrain around Pot of Gold, though they too hadn't conveyed their thoughts well. The ground is lumpy and irregular there, with steep little slopes and nasty pockets of loose sand and dust scattered about. It's the kind of stuff that makes precision driving difficult, and the truth is that we were probably lucky to get to Pot of Gold in the first place as painlessly as we did.

But just like the Honeybees, the rover drivers didn't give Ray a clear enough picture of what they thought. What they were thinking among themselves was that a drive to reapproach Pot of Gold from a different direction could turn into a real mess. The message as they conveyed it to

Ray in the heat of a hurried SOWG meeting was simply that this could be a tougher drive than usual.

So now we're stuck. Faced with incomplete information, Ray planned what seemed like a simple loopdey-loop to come at Pot of Gold again from the west. Instead, *Spirit* has been mired for three sols now, tilted to the west almost twenty degrees and slipping half a meter or more for every meter she tries to drive. The soil in the rear Hazcam images is churned into an ugly pile. Ray is grumpy and exasperated, and I don't blame him.

So today I went to see the Honeybees in their fourth-floor RAT room. Gorevan and the rest of the senior 'bees are back in New York now, with just the younger guys out at JPL keeping things together. The expressions around the room were cheerless; with the sun heading north daily, they know that this is costing us sols that we can't afford.

"Look," I said once we all got settled, "I know this thing is a bad target. Nobody is going to blame you guys if we try to RAT it and your sequence faults out. This thing is so far outside the design envelope for the RAT that I never would have even considered going at it during the nominal mission."

They listened without saying a word.

"But we're into extra time now, and we've got to start looking at it that way. Any science we get at this stage is gravy, and if we try something that doesn't work, it's no big deal as long as nothing breaks. I know you guys want every RATting to go perfectly, but let's just accept that this one won't, and see if we can get anything out of it at all."

They nodded; they obviously hadn't looked at it that way before.

"So I'm no longer interested in knowing what you think is the best angle to approach this thing from. That's an academic question that no longer applies. What I want you to do now is tell Ray what is the complete range of *acceptable* angles to approach this thing from. And then we'll see what we can do."

It took them an hour, but they finally emerged from their room and gave the numbers to Ray, written in blue ink on a little pink sticky note. When it came right down to it, there was quite a range of angles that they thought would work, and those angles included the south-facing heading that we were on back three sols ago before the churning began.

Live and learn.

JUNE 21, 2004

Another good day at Meridiani, and another bad one at Gusev. At Meridiani, we're working our way systematically downward through the Karatepe stratigraphy. So far it's all rich in sulfates, with not a trace of basaltic sandstone in sight. It took a lot more water to make these rocks than we originally thought. At Gusev, we got about a meter closer to Pot of Gold, with one more meter still to go. This is frustrating, difficult stuff to drive in.

The strategic issue at the moment deals with what happens after we finally finish up with Pot of Gold. The climb up the West Spur of Husband Hill is going to be an exercise in route-finding that has no precedent in the history of mountaineering. Some of what we'll have to do is familiar to any climber. We'll have to find a route that isn't too steep for our abilities. We'll have to find a route that's on good solid rock, instead of on crumbly stuff that's going to avalanche away from under us. But another part of it, which no climber I know has ever had to deal with, is that we'll have to stay on a slope that'll keep our solar arrays tilted toward the sun. That means north-facing slopes, and right now the slope in front of us faces west.

So how do we attack this thing? There's no point in trying to force our way up a slope that points in the wrong direction. It may be the most direct route, but the solar power will be so bad that we'll never make it. We're better off taking a circuitous route up a north-facing slope, even if it's a much longer one.

This is a problem tailor-made for Larry Soderblom, and today I appointed him chief route-finder for the first Martian Mountaineering Expedition. He's got a beautiful digital topographic map of the West Spur, made at the U.S. Geological Survey in Flagstaff using orbital images from *Mars Global Surveyor*. So now Larry's analyzing slopes, calculating solar array outputs and computing optimal routes from where we are to the best outcrops that we can see on the spur. Whatever he comes up with is what we're going to try.

Of course, the bum right front wheel could be a spoiler if it craps out on us. We can try to drag it if we need to, limping backward up the hill. But up how steep an incline? And for how long before something else breaks?

JUNE 23, 2004

Glory hallelujah, we're back to Pot of Gold! It's been almost two weeks since we first spotted this accursed rock. I'll be glad to be done with it.

JUNE 24, 2004

Another good day at Meridiani, as they all seem to be. We're working our way down through these layered rocks, and we've got it down to a routine. Each time we get into what looks like something new in the Pancam images, we first take some pictures of it with the MI to see what it looks like close up. Next we put in a RAT hole, and then we stick in the Mössbauer and the APXS to figure out what it's made of. Nothing to it. The very same RAT guys who caused all the commotion at Gusev have been doing an incredible job at Meridiani, punching in one clean RAT hole after another with the rover perched on a 28-degree slope. *Opportunity* continues to be the charmed machine.

Meanwhile, on the other side of the planet, this damn rock is kicking our butts. Yesterday was the big RAT attempt, and as expected it didn't go very well. The RAT errored out; worse, the rock moved. This thing really is too small and lumpy for us to put a clean hole into it, and now it's taking evasive action. But if you look real close at the post-RAT Pancam images of the rock you can convince yourself that we probably lopped the tops off some of the stalks. More important, it looks like the RAT brushes cleaned the surface off a bit, and maybe that'll be enough. We're doing overnight APXS now on whatever we got, and the sol that Ray and the guys planned today will include a mosaic of MI to survey the damage. After that we'll have done all we can do.

The strangest news from Gusev today has nothing directly to do with Pot of Gold. We've gotten down some recent Pancam images of the torn up terrain around the rover, and there's something in them that we've never seen before. With all that thrashing around we did trying to get back to Pot of Gold, we crunched over a lot of the shells of the cavernously weathered rocks around it. And everywhere we drove over those shells, the crunched-up remains are now really shiny. I'm not just talking about a little gleam here. Wherever the wheels contacted the shells, the Pancam pictures are so bright that they're completely saturated. This stuff is blazingly reflective. It's hard to be sure why, but I think there's a decent chance that it's because it's rich in hematite.

The sweet irony, of course, is that if we hadn't spent all those ugly sols churning our way back to where we should have been in the first place, we never would have seen this sparkly stuff. I hope the Honeybee guys are feeling better.

JUNE 28, 2004

After spending the better part of a month futzing with one little rock, we've finally had the breakthrough we were hoping for.

Two things have happened. The first was that we got down a beautiful six-frame MI mosaic on Pot of Gold after the RATting attempt of a few sols ago. We really did nick this thing a bit. We chopped off some stalks, we cut through some nuggets and we did quite a bit of brushing. The MI images also show that where we cut through the nuggets, the surface became shiny and sparkly, just like where we crunched the other rocks with the wheels.

The second thing is that we've gotten down the APXS data from the cleaned-off area, and it's really high in sulfur, chlorine and phosphorous. In fact, the levels are higher than we've ever seen in soil, and much higher than we've seen in Gusev rock. So it has to be that the rock is enriched in these elements, not just that they're in soil that's coated on the outside.

With these results, I'm convinced now that Pot of Gold was wet once, the first water-altered rock we've seen since we got to Gusev Crater. At long last, we've got evidence for water at both landing sites. *Spirit*'s luck has finally changed.

Still, there's so much here that we don't know. Yes, the Columbia Hills really are dramatically different from the plains around them. But what are they made of? How did they form? All we've got is an incomplete picture of one peculiar rock, with a limping rover and winter coming on. But our path now is clear. The only thing that makes sense is to start up the Soderblom route on the West Spur of Husband Hill, searching on north-facing slopes for new clues and hoping that the martian winter doesn't shut us down.

SEPTEMBER 12, 2004, TWO AND A HALF MONTHS LATER

It's been three days since we've heard from either *Spirit* or *Opportunity*. Mars is out of sight from Earth now, on the far side of the Sun, making it impossible to communicate with the rovers. This alignment of the planets is called superior conjunction, and we expect to survive it. We've been getting ready for it for weeks, and both vehicles have commands on board that should keep them safe for the ten sols or so that we'll be out of touch. In fact, they should both be busy doing science the whole time if all goes well.

Opportunity is still inside Endurance Crater. The descent down the crater wall took us all summer here on Earth. Our plan was to turn around as soon as we ran out of rock, but every time we thought we were almost done, we'd spot an outcrop farther down the slope, and we'd keep going. We're deep inside the crater now, almost to the dune field.

On the way down we put in a total of eleven RAT holes, each revealing the properties of a different layer in this incredible stack of sedimentary rocks at Meridiani. It was a trip back in time, with subtle and fascinating changes along the way in the chemistry and texture of the rocks. But our main discovery was simply that the rocks in Endurance Crater are blueberry-laden sulfates all the way down. Combine this with what we're seeing from orbit, and we're starting to suspect that there could be tens of meters of sulfate-rich sediments here at Meridiani. It took a *lot* of water to make these rocks.

Over at Gusev, the Columbia Hills continue to thrill us at every turn. After we left Pot of Gold, the mechanical guys came up with a way to try to fix our bum wheel. We ran the heater on it up to the highest temperature we could, and then moved the wheel through several rotations in both directions, the idea being to spread the lubricant in the gearbox around a bit, making the thing easier to turn. It didn't fix the problem, but since we did it there's been no more degradation of the wheel's performance; the situation has stabilized. We drag the wheel to prolong its life when we're on gentle ground, and we use it at full power to climb when the going is steep.

Spirit has turned out to be a better climber than anybody could have guessed. The West Spur is a rugged and difficult place, but the Soderblom Route up it has worked, and with steep northward tilts much of the way we've had the energy we needed to gain some serious altitude.

And there's bedrock in the Columbia Hills. We've found it, we've RATted it and now we know what it's like.

The rocks are indeed rich in sulfur, chlorine and phosphorous, like Pot of Gold was, all the way through. Many have hematite in them, and we're starting to suspect that some also contain goethite. If we're right about the goethite it's a big deal, much like the jarosite at Meridiani. Goethite is an iron oxy*hydroxide*—an iron mineral that, like jarosite, actually has water in its crystalline structure. So if goethite is there, and we're becoming pretty confident that it is, then that's one more solid piece of evidence that water once flowed through these rocks.

Some of the rocks in the Columbia Hills are also layered. We haven't reached any of them yet, mostly because so far they've been on south-facing slopes that we can't approach safely during the winter. But it's only a matter of time, and if we come out of superior conjunction safely there are some nice north-facing layered targets not far up the hill from where we are now.

We're still working out what happened here. But the best bet today is that these are "volcaniclastic" rocks—rocks formed when a violent volcanic explosion caused particles of ash, sand and gravel to fall from the sky and settle out in layers. Then water flowed through them and altered them. The water could have come much later, after the rocks were cold, or it could have been at the same time, in a burst of hydrothermal activity. But the evidence for long-ago water in the Columbia Hills is now inarguable.

What I'm still puzzling over is what, if anything, the water that soaked these hills had to do with the lake that once occupied Gusev Crater. It may have had nothing to do with it at all.

The Columbia Hills are old, certainly older than the basaltic plains that surround them. They rise like islands out of that basalt, and they were there long before the basalt was laid down.

Somewhere beneath all that basalt must be the sediments we came looking for. And obviously, those sediments must be older than the basalt is. But that doesn't mean that the rocks of the Columbia Hills aren't older still. In fact, the water that once soaked the hills may date from some truly ancient epoch that has nothing whatsoever to do with the lake that brought us to Gusev Crater. We've found a water story at Gusev, but it may be unrelated to the one we came looking for. Exploration is like that, I guess.

At Meridiani, there's no question that we found what we came looking for, and more. The rocks there were laid down in liquid water, in an environment that surely must have been suitable for some primitive forms of life.

Of course, being habitable and being inhabited are two very different things, and we have no way of knowing from our data whether life ever took hold at Meridiani. At least two things might have made it difficult. One is the chemistry of the water. There are places on Earth today that may have much in common with Meridiani back when it was wet. In a river called the Rio Tinto, in Spain, minerals like the ones at Meridiani, including jarosite, are being deposited. What makes the waters of the Rio Tinto remarkable is their extreme acidity. *Tinto* is a red Spanish wine, and the Rio Tinto derives its name from the color of its waters. As acidic groundwater percolates through rocks, it dissolves away iron compounds, emerging to darken the water of the river to a deep ruby red. As I think of the jarosite in the rocks at Meridiani, I have to wonder if long ago they were laid down in a ruby red sea under a pink martian sky.

It's an appealing image, but the kind of acidity that Meridiani's minerals may imply could have posed a real challenge to the origin of life. The acid waters of the Rio Tinto are teeming with microbial life, but those are organisms that evolved *into* that environment, descended from more conventional microbes that got their start elsewhere under more favorable conditions. Whether life ever could have first begun in the acid waters that may have prevailed at Meridiani is a question we cannot answer.

Another challenge to life at Meridiani is that despite all the evidence for water, it probably was dry there most of the time. The rocks at Meridiani don't show signs of deposition in deep water. Instead, they're probably dirty evaporites, formed when shallow, muddy, salty water dried up and left its residue behind. Grotzinger's cross-beds in Last Chance and the Dells were made by flowing water, but there are other cross-beds at Meridiani that look like the ones that form when wind piles grains up into dunes. We don't have evidence for a deep ocean at Meridiani. More likely it was like a salt flat or playa, wet occasionally but dry much of the time. Salt flats on Earth are also rich with life, but again, it's life that developed first elsewhere and then evolved into that harsh ecological niche.

So Meridiani was no evolutionary paradise. Still, conditions there were more suitable for life than they were anywhere else we can point to on Mars, making Meridiani a tempting target for future exploration. Not only do the rocks there tell of wet, habitable conditions long ago. They're also rich in minerals that precipitated from liquid water, and that could still preserve within them evidence for whatever was once in that water, trapped forever like bugs in amber. The thing I'd love to do with the rocks at Meridiani is get some of them into laboratories back on Earth, where they can be picked apart molecule by molecule by the world's best scientists to reveal whatever secrets they hide.

As *Spirit* and *Opportunity* continue to explore, my wonder at their longevity grows. Our worries over being able to live for ninety sols and drive for six hundred meters seem almost laughable now as the rovers settle in for their ninth month on Mars. I always understood that Sol 90 was when the warranty would expire for each rover, and that an expired warranty didn't mean that the wheels would fall off when the Sun came up on Sol 91. I thought even before we launched that we'd probably make it beyond ninety sols and six hundred meters, and in my secret heart I was confident that if we landed safely, at least one of the rovers would make it to Sol 120 or even 150, traveling as much as a kilometer. But multiple kilometers, hundreds of sols, and rock climbing in mountain ranges and impact craters were never in my thoughts. Even as we stood at the rim of Bonneville Crater on Sol 90, with the Columbia Hills looking impossibly far away, that kind of performance was difficult to believe in. I'm glad that we did.

The longevity we've had is testimony, more than anything else, to the skills and the guile of the engineers who built the rovers. Engineers take their requirements very seriously, and no engineer is going to promise that their system will meet its requirements without putting enough extra capability—enough margin—into the design to be sure that their requirements will be met even if a lot of things go wrong. Nobody wants to be the one who designed the part that failed. So there's margin tucked into every corner of our rovers, some of it in places that the engineers who built it are reluctant to confess to even today. Margin doesn't come for free; it takes money from the budget and time from the schedule. But when you have that margin, and when things go right instead of wrong, you can cash it in for extra performance and extended life. Today is Sol 227 for *Opportunity* and Sol 248 for *Spirit*, and if we come through superior conjunction okay there's still no end in sight. Even the Mini-TES on *Opportunity* is in good shape, Deep Sleep and all.

With no flight operations to work on these past few days, I've found myself thinking a lot about the rovers, and about what their fate will be. Surely they'll die before long . . . within months, perhaps, or in a year or two at the outside. Even in the most optimistic scenario, it's hard to believe that *Spirit* can survive a second winter on Mars. And *Opportunity*, the good-luck rover in the warm and sunny climate of Meridiani Planum, will succumb, too, one way or another. Maybe it'll be dust buildup on the solar arrays that gets them, with a long, slow decline into hibernation and death. Maybe their mechanical parts will wear out, like the balky wheel on *Spirit*, eventually bringing them to a halt. Or maybe some critical components will fail, like a bullet in the brain that we'll never see coming, and one morning they simply won't wake up and talk to us.

Whatever happens, it'll be a sad day when it comes. But it may be a

strangely satisfying day, too, the inverse of my sadness-amid-the-joy feelings at launch. The rovers will be gone forever, at least to me. But they
will have led rich and successful lives, and their deaths will be honorable
ones. I never could have wished more for them.

And what will become of them then? Mars is a cold, dry place today.
Metal does not corrode there the way it does on Earth, and the ravages of
moisture and vegetation do not take their toll. Even the soft rocks at
Meridiani have withstood hundreds of millions of years of erosion by the
wind, and basalt boulders ejected from craters billions of years ago at Gusev are still there on the surface, unburied by the dust and drifts that swirl
across the surface. Human civilization on Earth is a mere tens of thousands of years old. And yet a million years from now, when JPL, Kennedy
Space Center and Cornell have crumbled away and disappeared, *Spirit* and
Opportunity could still be on Mars, immobile but with their aluminum
surfaces shining.

Twenty years ago, on my first visit to Antarctica, I had the chance to
visit the hut on Ross Island that was built in 1902 by the expedition of
Robert Falcon Scott, on his first unsuccessful attempt to reach the South
Pole. The hut is a palimpsest of Antarctic exploration, used subsequently
by Shackleton in 1908, by Scott again in 1911 and once more by some of
Shackleton's men in 1916 as they waited for the Antarctic crossing by the
Endurance party that never came.

The Discovery Hut, as it is known, was built by Scott's men on a
rocky promontory jutting into the Ross Sea, with the smoking summit of
Mount Erebus on the far horizon. It stands today as it was left almost a
century ago, the wood bleached pale and its grain etched into relief by the
wind, but otherwise in good repair. The carcass of a seal lies, still frozen,
under the eaves. Inside, the hut is much as its final occupants left it, the
primitive technology of late-nineteenth and early-twentieth century polar exploration on vivid display. There are wooden crates, long empty and
converted to shelving, still stenciled in black ink with listings of their
original contents. There are oil-fueled lanterns of green glass on the

shelves, alongside tins of "Hunter's Famed Oatmeal, Edinburgh." The stove is there, with cast-iron skillets atop it containing withered chunks of blubber, and hard wooden bunks with reindeer-skin sleeping bags.

It's a sad scene in some ways, speaking of an era long gone, and of voyages of discovery that were partial successes at best. I carried with me that day polar equipment that outstripped the technology of Scott and Shackleton by a century, and not far from where I stood was the modern research base of McMurdo Station. The hut was a relic of an earlier, more primitive time.

Still, to me, the Discovery Hut was a place with particular significance. When the men of Shackleton's Trans-Antarctic Expedition left it for the last time in 1916, the first age of polar exploration was coming to a close. Some among them, I suspect, must have wondered when and indeed if human eyes would ever see their hut again. And as I think now about the fate of *Spirit* and *Opportunity*, my thoughts are much the same. Will future explorers find our rovers? I hope so. Will they go to them knowing their location well, as I did with the Discovery Hut, or will they come across them unexpectedly, their histories and the people who built them long forgotten? There's no way of telling.

One thing that's certain is that the rovers' technology will appear quaint and primitive to whatever eyes see them next. I wish now that we'd had the time to make the design prettier, to tame the ugly tangle of cables on the front of the WEB, to put all the Mössbauer electronics inside where they belonged instead of taped to the outside of the instrument. I hope whoever sees them next has a forgiving eye, or that they understand because they were once in as much of a hurry to get to Mars as we were.

Above all, I simply hope that *someone* sees them again. A word like *love* is one to be used advisedly, especially when talking about pieces of metal. But I love *Spirit* and *Opportunity*. They were built by a loving family, and I wish more for their fate than eternal abandonment on a distant world.

Don't misunderstand: I wouldn't ever want them to be brought back to Earth. We built them for Mars, and Mars is where they should stay. But *Spirit* and *Opportunity* have become more than just machines to me. The rovers are our surrogates, our robotic precursors to a world that, as humans, we're still not quite ready to visit.

And that's what I really want to see change. There are many things I could wish for our rovers, but in the end, there's only one that matters. What I really want, more than anything else, is boot prints in our wheel tracks at Eagle Crater.

The *MER* Rover

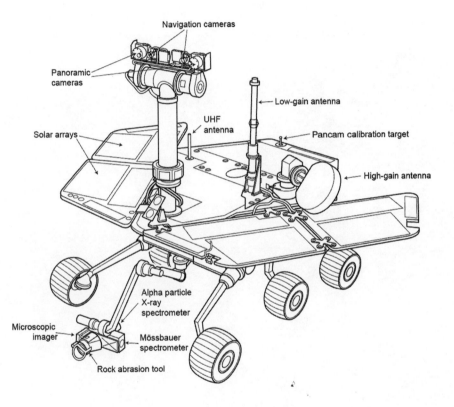

GLOSSARY OF TERMS AND ACRONYMS

ACE: The person who actually transmits commands to a spacecraft.

Aeroshell: Protective shell that surrounds a spacecraft during entry into a planet's atmosphere.

APEX: Athena Precursor Experiment. A set of instruments scheduled to go to Mars on a lander mission in 2001. The mission was cancelled.

APXS: Alpha Particle X-Ray Spectrometer. Instrument mounted on rover's arm that measures the concentrations of chemical elements.

ATLO: Assembly, Test and Launch Operations. The process of building, testing and launching a spacecraft.

Backshell: The rear portion of an aeroshell.

Basalt: The most common kind of volcanic rock.

Beagle 2: Unsuccessful British Mars lander, launched in 2003 and carried to Mars aboard the *Mars Express* spacecraft.

CDR: Critical Design Review. Important design review during the development of a spacecraft.

CMSA: Cruise Mission Support Area. "Mission control" during the flight of *Spirit* and *Opportunity* from Earth to Mars.

Delta II: Boeing-built rocket used to launch *Spirit* and *Opportunity*.

DIMES: Descent Image Motion Estimation System. A camera and software used to determine velocity over the ground during landing.

DSN: Deep Space Network. NASA's network of tracking stations used to communicate with spacecraft that fly beyond Earth orbit. There are three stations: at Goldstone in the Mojave desert, near Madrid, Spain, and in Tidbinbilla, Australia, near Canberra.

EDL: Entry, descent and landing. The process of entering a planet's atmosphere and landing on its surface.

Goethite: An iron mineral that requires water for its formation.

GSE: Ground Support Equipment. Hardware used in testing a space flight instrument. It mimics parts of the spacecraft on which the instrument will fly.

Hazcam: Hazard avoidance camera. Low resolution camera used by *Spirit* and *Opportunity* to avoid obstacles while driving.

HEDS: Human Exploration and Development of Space. A part of NASA concerned with sending astronauts beyond Earth orbit.

Hematite: An iron mineral that sometimes requires water for its formation.

HGA: High-Gain Antenna. Small dish antenna on the rover used to communicate directly with Earth at high data rates.

HRS: Heat Rejection System. The cooling system used by *Spirit* and *Opportunity* on their way to Mars.

IDD: Instrument Deployment Device. The rover's arm.

IRT: Independent Review Team. A team of experts convened by NASA to monitor the progress of a space flight project.

ISIL: In Situ Instruments Laboratory. A facility at the Jet Propulsion Laboratory used for testing rovers indoors in a Mars-like environment.

Jarosite: An iron sulfate mineral that requires water for its formation.

JPL: The Jet Propulsion Laboratory. NASA's lead facility for the exploration of the solar system.

LGA: Low-Gain Antenna. Antenna on the rover used to communicate directly with Earth at low data rates. Unlike the High-Gain Antenna, it does not need to be pointed at Earth to function.

LOX: Liquid oxygen. Used as a rocket propellant.

LST: Local solar time. The local time of day on Mars.

LTP lead: Long-Term Planning lead. Scientist responsible for long-range planning of rover operations.

Marie Curie: Spare model of the *Sojourner* rover used on the *Mars Pathfinder* mission. Was planned for use on the *Mars 2001* lander mission, which was subsequently cancelled.

Mars Express: Successful European Mars orbiter mission, launched in 2003.

Mars Odyssey: Successful NASA Mars orbiter mission, launched in 2001. Helps relay data from *Spirit* and *Opportunity* to Earth.

Mars Pathfinder: Successful NASA Mars lander mission, launched in 1996. Pioneered use of airbags for landing on Mars.

MCO: *Mars Climate Orbiter.* Unsuccessful NASA Mars orbiter mission, launched in 1998.

MER: *Mars Exploration Rover* mission; the mission of *Spirit* and *Opportunity*.

MGS: *Mars Global Surveyor.* Successful NASA Mars orbiter mission, launched in 1996. Helps relay data from *Spirit* and *Opportunity* to Earth.

MI: Microscopic Imager. Close-up camera mounted on rover's arm.

Mini–TES: Miniature Thermal Emission Spectrometer. Infrared spectrometer that can determine rock composition from a distance.

Mössbauer spectrometer: Spectrometer mounted on rover's arm that can identify iron-bearing minerals.

MPL: *Mars Polar Lander.* Unsuccessful NASA Mars lander mission, launched in 1998.

MSO: *Mars Science Orbiter.* Orbiter mission not selected by NASA for launch in 2003. The name was later changed to *Mars Reconnaissance Orbiter.*

Navcam: Navigation camera. Low-resolution black-and-white camera on *Spirit* and *Opportunity.*

Nozomi: Unsuccessful Japanese Mars orbiter mission, launched in 1998.

ORT: Operational Readiness Test. A dress rehearsal for space flight operations.

Pancam: Panoramic camera. High-resolution color camera on *Spirit* and *Opportunity.*

PHSF: Payload Hazardous Servicing Facility. Facility at Cape Canaveral used to prepare spacecraft for flight.

PI: Principal Investigator. The leader of a science team.

PMA: Pancam Mast Assembly. The rover's mast.

QA: Quality Assurance. The people responsible for making sure that a spacecraft is put together properly.

RAD motors: Rocket Assisted Descent motors. Solid rocket motors that slow the lander's descent an instant before impact with the martian surface.

RAT: Rock Abrasion Tool. Grinding tool mounted on rovers' arm, used to expose the interiors of rocks.

REM: Rover Electronics Module. The main set of electronics boards inside the rover... the rover's "brain."

RTV: A kind of glue commonly used in aerospace applications.

SMSA: Surface Mission Support Area. "Mission control" while *Spirit* and *Opportunity* are on Mars.

Sojourner: Miniature rover used on the successful *Mars Pathfinder* mission.

Sol: A martian day. 24 hours and 39 minutes long.

SOWG chair: Science Operations Working Group chair. The daily leader of the science team for a rover on the surface of Mars.

TCM: Trajectory Correction Maneuver. A small rocket motor firing to change the path of a spacecraft in space.

Thermal vac: Thermal vacuum testing. Testing a spacecraft in a vacuum chamber, under space-like conditions.

TSB: Telecom Services Board. An electronics board inside *Spirit* and *Opportunity*.

UHF: Ultra High Frequency. The radio frequency *Spirit* and *Opportunity* use to communicate with the *Mars Odyssey* and *MGS* orbiters.

UTC: Universal Time Coordinated. The time system used for many global applications, including the Deep Space Network.

VOCA: Voice network used for communications during spacecraft testing and operations.

WEB: Warm Electronics Box. The enclosure inside the rover where the electronics are kept.

APPENDIX

The *Mars Exploration Rover* project was a team effort, one involving literally thousands of people. This book tells the stories of a few of those people, but the reality of the effort was far greater. I couldn't fit all the stories into the book, but I thought I should at least try to fit in all the characters.

The list below is my best attempt to name everyone who worked on the *MER* project. There are more than four thousand names here.

Despite my efforts, I'm sure that I've failed badly in putting the list together. Like so many tasks, this one was both impossible and worth trying anyway. To those whom I've missed, and for the names that are misspelled, my sincere apologies!

Lucy Z Abajian
M Abajian
Henry Abakians
Sal Abbate
Derek W Abbott
Randal S Abbott
Mohamed D Abdalla
Jibu Abraham
Leo Abramovitz
Charles H Acton Jr

Josephine Ada
Billy L Adams
Donald E Adams
Douglas S Adams
Jay F Adams
Lynn D Adams
Marc A Adams
Scott J Adams
Amber D Adamsmeiers
Adrian M Adamson

Jerry J Adamson
Mark Adler
Max Adofo
Sanchit Agarwal
Shri G Agarwal
Hrand Aghazarian
D C Agle
Stan Agoot
Rakesh Agrawal
Eric A Aguilar

Ronald J Aguilar
Ynez Aguilar
John Aguirre
Gyan V Ahluwalia
Norman M Ahmad
Mitchell Ai-Chang
Thomas Aiello
Greg Ains
Frederick V Akers
Jacqueline L Akers
Torry L Akins
Elroy M Akioka
Bode A Akisanya
Rafael H Alanis
Jo Anne Alano
Cindy Alarcon-Rivera
Ray Alava
R Albano
J. Luis Alberquilla Garcia
Lee Albers
Phyllis A Albrecht
Alejandro Alcaraz Sanchez
Donnie R Alderman
Dominic Aldi
Wafa S Aldiwan
Douglass A Alexander
James W Alexander
Steven C Alfery
Teresa L Alfery
Nelson B Alhambra
J. Julio Alhambra Garcia
Khaled S Ali
Abdullah S Aljabri
Mark E Alldredge
Bryan L Allen
David Allen
Dean Allen
Gary Allen
Michael J Allen
Scott R Allen
Steven J Allen
Valerie A Allen
William C Allen
William H Allen
Sarah Allgood
Teresa Alonso
Tomas Alonso Hernandez
Raul Alonso Martinez

Juan J. Alonso Queipo
Ricardo J. Alonso Reynolds
Domingo Alonso Santos
Dave Altum
Tracy Altum
N Alvarado
Cristobal Alvarez
Markus Alvarez
J. Ignacio Alvarez Molina
Olegario Alvarez Otero
Arthur V Amador
James Amato
Preston R Amy
Angel Anaya Garcia
Dave Andersen
Jeffrey P Andersen
Alan J Anderson
Alma G Anderson
David A Anderson
Fletcher S Anderson
Gary Anderson
Glenn K Anderson
Graham Anderson
Jeff Anderson
Jeffery A Anderson
John B Anderson
Kalle A Anderson
Keith R Anderson
Mark S Anderson
Paul Anderson
Peter Anderson
Robert C Anderson
Scott R Anderson
Terry Anderson
Wendy J Anderson
Sherry L Anderson-Perry
B Andreichikov
Paul Andres
Francisco B. Andres Farrus
Robert W Andrews
Pam Andrson
Carol Angel
Krisjani S Angkasa
Jeffery A Anglin
Scott Angster
Francisco Angulo Prieto
Peter G Antreasian
Saadat Anwar

Joy K Apgar
Alexander D Appelhans
Jeffrey Appleton
Tarek Arafat
Richard J Aragon
Robert Aragon
Fernando Aragon Calero
Taguhi Arakelian
Brent Archinal
Al Arellano
Albert D Arellano
Tom Arledge
Michael J Armatys
Rafael Armero Hernan-Gomez
Michael Armstrong
Theresa S Armstrong
Heather Arneson
John Arnold
Bob Arrell
Ana Rosa Arreola
David R Arriola
Belinda Arroyo
Manuel Arroyo Sanchez
Betty Arseneault
Dave Arton
Ray Arvidson
Scott C Asbury
Sami Asmar
Robert W Aster
Ahlam Amy Attiyah
Brianna Aubin
James L August
David G Aull
Angela Austen
Charlie Austin
Oscar Avalos
Glenn J Aveni
Arturo Avila
Lawrence W Avril
Henry I Awaya
Mohamed L Ayari
William Ayrey
Ron Baalke
Alan S Baba
Holly Babcock
Jennifer C Baber
Vivek Babtiwale
Debra K Baca

Paul Baca
Alex Bachmann
Paul G Backes
Rich Badalament
James E Badger
Enrique Baez Jr
Curt Baffes
Blaine Baggett
James H Bahder
Candy Bailey
Erik S Bailey
Laurence Bailey
Mark Bailey
Monty Bailey
Christopher E Bain
Graham Baines
Darren T Baird
Brad G Baker
Cheryl A Baker
John D Baker
Kelly Baker
Kym R Baker
Michael J Baker
Raymond S Baker
Alireza Bakhshi
Robert E Bakley
John A Balcerak
Alice Baldridge
Alec T Baldwin
Kenneth L Baldwin
Robert Balkenhol
Steve Ballard
Annie M Balsley
Mark A Balzer
David Bame
Robert Bamerio
Josh Bandfield
Pranab K Banerjee
Ronald S Banes
Don Banfield
Darlene Banks
Sam Bannister
Neil J Barabas
Robert Baranowski
Todd J Barber
Andrea J Barbieri
Leslie A Barboza
Heinz Bareiss

Jack B Barengoltz
Lisa L Barker
Ed Barlow
Phil Barnes-Roberts
Terri L Barnett
Joann Barnette
Sharon L Barr
Janet M Barrett
Peter A Barrett
Pedro Barrio Navarro
Yoaz E Bar-Sever
David E Bartels
Bruce L Barthen
Marda J Barthuli
Paul Bartlett
Manuel Basallote Virues
Cagatay Basdogan
Deborah S Bass
Len Bass
Paul D Bastin
Brian R Bate
David M Bates
Duane Bates
Chris Batting
Richard L Battistelli
Mark R Bauer
James A Baughman
Richard W Baumbach
Eric T Baumgartner
Angel Bautista Castellar
Duane Beach
Scott L Beale
Dolores Beasley
Douglas W Beasley
Robert Beattie
John S Beatty
Mary P Beaver
Lynnette Bechard
John Beck
Terry W Beck
Kris J Becker
Tammy Becker
Wayne T Becker
Matthew H Beckner
Matthias Behrend
Anthony J Beissel
Bryan A Bell
James F Bell

Julia L Bell
Michele D Bell
Paolo Bellutta
J Eric Belz
James N Benardini
Javier Benavente Ruiz
Becky A Benedict
Ray Benesh
Boonsieng Benjauthrit
Darrell V Bennett
Patrick R Bennett
William J Bensler
Charles E Benson
Donald Benson
Keith Benson
Alan Berard
Barry N Berdanier
Charles F Bergh
Sheryl L Bergstrom
Kirk Berhent
Jack Berkery
Ann M Bernath
Bodo Bernhardt
Leslie A Berridge
Scott Berry
Preben Bertelsen
Robert F Bertsch
Laura Berwin
C A Best
Manuel Betanzos Sabao
Nathaniel S Bevins
Ross A Beyer
Shyam Bhaskaran
Ramachand S Bhat
Mark L Bichler
David H Bickel
Donald B Bickler
Bruce W Bieber
Aubyn S Biery
Jeffrey J Biesiadecki
Carl Bigelow
Kenneth W Biggs
Floyd E Biles
John E Biles Jr
Russel D Billing
Charlotte Steffens Binau
Dennis David Binford
Dwight Bird

Morris M Birnbaum

Gajanana C Birur

Leo J Bister

William S Bittman Ii

Gary W Bivins

Jeff G Bixler

Delbert W Black

Mike Black

Derek Blackway

Samaris M Blake

Michael J Blakely

Brian K Blakkolb

Bob Blanchard

R C Blanchard

Alberto E Blanco

Gregoria Blanco

Jaime Blanco Garcia

Luis Blanco Monje

Richard P Bland

Diana L Blaney

Paul J Blasch

Teresa Blaylock

Philip C Blazejak

Fletcher B Bloch

Patricia A Bloomquist

Paul Blume

Jeffrey A Blunck

David J Boatman

Glenn Bobskill

Rudy A Boehmer

Bill Boger

Mary-Hrachoo Boghosian

Pamela C Bogley

David Bolin

Stephen R Bolin

Diane Bollen

Willard E Bollman

Joseph Bomba

Bruce B Bon

Brian Bone

Robert G Bonitz

Uwe Bonnes

George Bonsu

James Booker

Sandra S Boone

Elliott Borden

James S Border

James Borders

Greg Boreham

Mike C Borfitz

Rosana Bisciotti Borgen

Richard V Bork

Kevin Boroczky

Cory Borst

Susan H Bortfeldt

John F Borthwick

Dana Bose

Curtis Boswell

Wayne P Bosze

Joy A Bottenfield

Dick Bourdon

Tony Bourne

John Bousman

Donald V Bousquet

Dhemetrio Boussalis

Paul N Bowerman

Jeffrey G Bowers

Norm Bowler

John Bowlick III

Cassie Bowman

Judd Bowman

Sandy Bowman

Tony Bowman

Ernest Bowman-Cisneros

Tom Bowmaster

Greg Boyd

Larry Boyd

Kobie T Boykins

Gary Boyle

Mark A Boyles

Larry Bracamonte

Richard Brace

Paul Bradford

Catherine A Bradley

N Talbot Brady

Jerome Bragg

Summer Brandt

Tim Brann

James D Brantley

Guillaume Brat

David F Braun

Richard Bray

John Bresina

Deidrea Brewer

Ladonne H Brewer

Casey E Brewer

Scott Brickerhoff

Nathan T Bridges

Steve Bridges

Keith Brierley

Elroy A Briggs

J Brinchmann

Darline Brinegar

Ralph Briones

Phillip E Brisendine

Roy Britten

Charles A Broadwater

Lindsey Brock

Gerald M Brockfield

Larry Broms

Jessica D Brooks

Marc E Broom

Eugene Brower

Laura B Brower

Bill Brown

Chau M Brown

David Brown

Dwayne Brown

Ellen Brown

Jason R Brown

Pamela R Brown

Paula R Brown

Renate Brown

Robert Brown

Robin Brown

Samuel Brown

Steve Brown

Steven Brown

Thomas Brown

Thomas A Brown

Shane W Brown

Dave Browning

Judy Brownsberger

Johannes Brückner

Peter Bruneau

Gary W Bruner

Enrico Bruno

Robin J Bruno

C J Bryan

Larry W Bryant

Belle R Buchholz

Stephanie E Buck

Bryan Buckland

Gerry Buckley

Charles J Budney
Karen A Buerger
Ratnakumar V Bugga
Alan Buis
J Bulharowski
Margaret A Bundschuh
Carol B Bunnell
Terry L Bunting
Noel H Burden
Shala D Burdette
Doug Burg
Vincent Burger
Brent Burgess
Dave Burgess
Bobby Burke
Jennifer E Burke
Kevin A Burke
Paul Burke
Robert Burke
Jon M Burkepile
Peter Burns
Stephen M Burns
Steven P Burns
David E Burow
Devon M Burr
Diana M Burrows
Ekko S Burt
Ron Burt
Matt Burton
Sanford J Bush
Linda F Butler
Robert Butler
Roy L Butler
Karen L Buxbaum
Sharon D Buxman
Stan Byrd
Thea J Byrd
Vaughn P Cable
Velma Cabrera
Francisco Cabrero Botello
Nathalie Cabrol
Cristina Calderon Riaño
Sean E Caldwell
John H Calkins Jr.
Jared Call
John L Callas
John A Callegari
Leslie N Callum

Wendy Calvin
Jesus Calvo H Hernandez
Irene M Camara
Fuencisla Camargo Rodriguez
Leigh Cameron
Lyle Cameron
John Campanella
Jim Campbell
Susan Campbell
Eusebio Campos
George Cannon
Melvin C Cannon
Bruce A Cantor
Christine M Cantrell
Pete Cappadoro
Kris Capraro
Victor Carcamo
John M Cardone
Jerry Carek
Gerard Carlile
Magdi Carlton
Andy Carmain
Michael N Carney
Robert C Carnright
Oscar Caro De Torres
Reiko S Carol
Scott N Carpenter
Frank A Carr
Mike Carr
Pat Carr
Bruce A Carrico
Carmen L Carrillo
Lorena L Carrillo
Todd A Carriveau
Chris Carson
Johnathan W Carson
John M Carson III
Dennis Carter
Gerald R Carter
Kelly Carter
Richard A Carver
James W Carwell
Gregory P Casavant
Warren A Cash
Brent A Caskey
Dick Casper
Richard B Cassell
John Cassidy

Javier Castaño Yagas
Roselia Castanon
Sergio F. Castejon Casado
Mark A Castillo
Dennis P Cate
Andre Caticchio
Michael P Cawley
Michael Cayanan
Helen Caye
Sam Ceballos
J. Luis Celada Gomez
Gloria E Cerda
Roberta Cerda
Michael A Cermak
David Cha
Pamela Chadbourne
Barbara A Chadwick
Brian G Chafin
Joydip Chakravarty
Gene Chalfant
Louise Y Chan
Tak S Chan
Kumar M Chandra
Scott Chaney
Christine Y Chang
George W Chang
Kurng Y Chang
Mark Chang
Chris Chapman
Mary G Chapman
Leonard K Charest
Jeffrey R Charles
Adam Chase
Mark Chatillon
Nigel Chauncy
Cosme M Chavez
Martha M Chavez
Victor M Chavez
Natividad M Chavira
Neil Cheatwood
Albert V Chen
Allen Chen
Amy Chen
Elizabeth A Chen
Emily Chen
Fei Chen
George T Chen
Gun-Shing Chen

Kuan T Chen
Long Y Chen
Siqi Chen
Timothy Chen
Y-K Chen
Jean Y Cheng
Yang Cheng
Kar-Ming Cheung
Regina K Cheung
Steven Chhan
Ben Chidester
Tom K Chikuma
Ed Chin
Keith B Chin
Jack Ching
Gilbert C Chinn
David B Chirdon
Hue M Chiu
Janis L Chodas
'Rich Chomko
Wing-Sang Chong
Natacha G Chough
Phil Christensen
Tony L Christensen
Jeffery J Christensen
Robin Christenson
Herald Christian
Scott Christiansen
Stanley Christopherson
Chengchih Chu
Philip C Chu
Shirley Y Chung
Marjorie A Churchill
Peter Churchill
Alicia Dwyer Cianciolo
Laura Cinco
William J Clancey
Todd Clancy
Alberta L Clapper
Ryan Clapper
Dave Clark
Eileen F Clark
Gerald J Clark
Jerry Clark
John Eric Clark
Ken Clark
Micah H Clark
Steven W Clark

Benton C Clark III
Darren Clarke
Graham Clarke
Kelly A Clarke
Whitney Clavin
Gordon Clee
Gregory C Cleven
Linda J Clifton
Larry Clint
Mark Cmar
James R Coats
Shaun W Coats
Rosemary Cobb
Michael D Cofield
Peter Coiscou
Christe D Cole
Michelle R Coleman
Ron Colier
Laura Colletti
Patrick Collins
Steven M Collins
Stewart A Collins
Jessica A Collisson
Barry S Colman
Isabel Colmenar Del Valle
Farrell J Colmenares
Michael Coluzzi
James E Colvin
David Combridge
Bradley I Compton
Cynthia L Compton
Leslie E Compton
Curtis Conemac
Clara A Conger
Timothy Connors
Keith M Conrad
Brian M Cook
Debbie A Cook
James Cook
Mary Y Cook
Richard A Cook
Gretchen Cook-Anderson
Denise A Cook-Clampert
Brian K Cooper
Margaret S Cooper
Mark S Cooper
Terry L Cooper
Djuna S Copley-Woods

Andrew Coppola
Carolina Coppolo
Thomas F Corbin
Sharon M Corcoran
Harold Corcoran
Christopher Cordell
Ivan D Cordova
Dan Cordray
David J Cormier
Jeff Cornish
Jeffrey J Cornish
Julie A Corpe
Lee Cortelyou
Martha Cortez
Max Cortez
Frank Corujo
Shawn Corwin
Bob G Coss
Joe Costello
Kenneth A Costello
Ed Cowan
William Cowan
Stephanie L Cowans
Nancy Cowardin
Richard T Cowley
Brian Cox
Mike Cox
Tim Cox
Zainab Nagin Cox
Gary G Coyle
Jeffrey W Coyne
Raymond S Cozy
Kenneth Crabtree
Lynn Craig
Kellie A Craven
Beau Crawford
James Crawford
Ken Crawford
Ron Creech
Ken Creel
Harrel B Crenshaw
Katherine A Cresto
Richard Crisa
Josephine Crise
James B Crisp
Joy A Crisp
Robert Crook
J Crosby

George Crose
Michael C Cross
Randall Crouch
Suzan Q Croucher
Susan J Crowe
Larry Crumpler
Eldrin R Cruz
Juan Cruz
Jose A. Cruz Santana
Charles T Cruzan
Gordy Cucullu
Edward F Cuddihy
Julie A Cullen
Russ Cummings
Cynthie Cuno
William A Currie
Kristen Curtis
Dawn M Curtiss
Terry A Curtland
Laura Dabney
Saverio D'Agostino
Mark A Dahl
Timothy Daleo
Sam Dallas
Jerry Dalton
Louis A D'Amario
Jerry Dance
Parviz Danesh
Van T Dang
Alicia R Dangerfield-Benn
Craig Daniel
Heidi M Daniels
Linda A Daniels
Monika J Danos
Mark A Dapore
Michael Dare
Diana G Darus
Peter J Darus
Nick Daskalakis
Nowshirwan S Dastoor
Taher Daud
Hardik Dave
Nimesh Dave
Don Davenport
Evan D Davies
Richard Davies
Stuart Davies
Barry Davis

Carla Davis
Charles Davis
Esker B Davis
Gregory L Davis
Jeffrey C Davis
John T Davis
Kiel Davis
Marilynn Davis
Maryia Davis
Michael C Davis
P Davis
Scott Davis
Jack Dawson
Sandra M Dawson
Travis E Dawson
Darryl Day
Ron Day
Christopher J Dayton
Angel De Benito Hernandez
Richard De Britt
Christian De Jong
Eric De Jong
Roberta De La O
Vicky N De La Trinidad
Greg De Los Santos
Joe K De Mers
David De Nicolas Gonzalez
Carl De Silveira
Paulo Antônio De Souza Jr
Michael De Wildt
Thomas J Dea
Kim Deal
Emily Dean
Karen F Dean
Matthew Deans
David Debevec
Leon Debritt
Dirk H Dedoes
Craig J Deegan
William E Deegan
Robert G Deen
Robert M Deering
Adolfo Delgado
Susan L Deligiannis
Ann M Delmonico
Nicholas J Deluca
Suzanne Delvalle
Michael J Demarco

Fred Demilio
Avo M Demirjian
John S Demmitt
Michael A Denaro
Rosalie Denby
Bill Dengate
Robert W Denise
Elizabeth Dennis
Matthew S Dennis
Richard Densmore
David John Des Marais
Prasun N Desai
Sanjay Desai
Bill Deshler
Govind J Deshpande
Robert Detwiler
Karen Deutsch
Dan Devito
Alan Devoe
Daniel F Devoe
Donn L Devries
Sharon Dew
Elizabeth A Dewell
Kaichang Di
Roy M Dicharry Jr.
Brian Dickerson
Frank Didonna
L Didonna
Roger E Diehl
James Diener
Angela K Difronzo
Robert Dilenno
Todd Dilka
George Diller
Patrick L Dillon
Arsham Dingizian
Shelley A Dinkelkamp
Angelia M Dirks
Michael R Dirr
Salvador Distefano
Michael G Dittman
Paul Dix
Megan Dixon
Liem Do
Dung C Doan
Lake A Dobbins
Nancy S Dobbins
Bill Dobie

Edwin C Dobkowski
Elliot G Dodge
Thomas H Dodge
Deborah E Doganes
David L Dolan
Alice Dominguez
Benjamin J Dominguez
Kareny Dominguez
Henry J Dominik
Wendelin C Donahue
James A Donaldson
Lorie Dongon
Jim Donnelly
Stephen Donohoo
Pat Donovan
Ray Donovan
Linda Doran
Eduardo D Dorantes
Mary K D'Ordine
Matthew D'Ortenzio
Heather A Doty
Scott R Doudrick
Sean Dougherty
Dave Dowen
Mary Downes
Mary B Downes
Pauline J Downes
Berta Doyle
Donald Doyle
Deborah A Drake
Ronald F Draper
Anthony E Dreher
Gerlind Dreibus
Aaron D Driscoll
Naomi Dubarry
Lydia P Dubon
Laura Duchow
Kathe Duckworth
Gary Dudley
Karen Duffy
F C Dumont
Maureen T Dumont
Alexandra J Dunn
Catherine Dunn
Donald E Dunn
Johnny Duong
Ronald V Dupitas
Mark D Duran

Eduardo Durantes
Brian Durgin
David M Durham
Michael O Durham
Claude D'Uston
Elizabeth D Duxbury
Thomas C Duxbury
Carmen R Dycus
Jaime M Dyk
Donald E Eagles
Charles A Eastment
Curtis A Eaton
Ralph A Eaton
Donald H Ebbeler
Bryan E Ebbing
R Ebner
Tom Economou
James W Eddingfield
Jerry L Eden
Will Edgington
Robyn Edlington
Larry D Edmonds
Karl Edquist
Tom Edquist
Helen Edwards
Larry Edwards
Laurence Edwards
Norma Edwards
Thomas E Edwards
Emily K Eelkema
William C Eggemeyer
Joanne Egges
Ronald W Eggleston
Bethany Ehlmann
Stan Eisenbaum
Allan R Eisenman
Charles Elachi
Dina Eldeeb
Nadia Eldeeb
Daniel B Eldred
Jeffrey B Elgar
Larry F Elias
William Ellern
Khanara Ellers
Debbie Ellingsworth
Robert G Elliott
Stythe T Elliott
Kathy H Ellis

Kristan D Ellis
Pam Embree
Walt Engelund
Christopher England
Michael E Engle
Stella J Engles
Alanson T Enos
Douglas J Equils
Franck Ergas
Erdin M Erginsoy
Daniel E Erickson
James K Erickson
Robert C Ernst
Glenda Erskin
Olga Escamilla
David R Escoto
F. Javier Escribano Benito
Gary P Esparza
Victor Espino
Erik Espinoza
Valereen Essandoh
John C Essmiller
Polly Estabrook
Juan R Estergo
Filadelfo (Fila) Estrada
John L Eterno
John S Eterno
Bob Evangelista
Dave Evans
Jon Evans
Kenneth C Evans
Kristan Evans
Howard W Everly III
E Evlanov
Richard C Ewell
Betty R Faber
Loan Fabre
Linda J Facto
Alicia Fallacaro
Chris Farguson
Christine T Farguson
Jack D. Farmer
William Farrand
Kimberly Farrell
Christi M Farris
Perry Fatehi
Lorili Faulkner
Ted Fautz

Jeffrey A Favretto
Jason Fellenstein
Michael E Fellin
Richard Feltman
Guillermo Fenollera Caamaño
Alan J Ferdman
Kerry Fereday
Robin Fergason
Rolf Fergg
Al Ferguson
Patrick F Ferguson
Nayla Fernandes
Connie Fernandez
Jose E Fernandez
Juan P Fernandez
Antonio Fernandez Pedriza
Luis Fernandez Rojo
Daniel Ferre Romero
Fred Ferri
Robert M Ferris
Steve Ferry
Richard Fettig
Royal J Field
Norman Fields
Ernest Fierheller
Donald E Figgins
Orlando Figueroa
David Fike
David K Filbert
Robert E Filman
Craig Fincken
Clifford E Findley
Susan G Finley
Jacqueline L Fiorello
Steve Fischer
Adam Fischman
Dianne M Fisher
Francine Fisher
Linda Fisher
Nova D Fisher
Richard Fisher
Tara Fisher
Terry Fisher
Aaron Fishman
Kathleen M Fisketjon
Hershal L Fitzhugh
John Fitzpatrick
Lori J Flannery

Iris Fleischer
Richard Fleischner
John C Fleming
Kirk Fleming
Timothy S Flick
Shirley A Flickner
Charlayne J Fliege
Gerardo Flores
Richard Flores
C Flores-Helizon
Jeremy E Fobes
Randall H Foehner
Josef Foh
Jeremiah Folds
Christopher J Foley
John K Folke
Bill Folkner
William M Folkner
Gavin R Follenweider
William A Folwell
Allen Fong
Marshall Fong
Antonio Fonseca
Gina Fontes
Eleanor C Foraker
Keven W Forbes
Dennis Ford
Douglas L Ford
Forest L Ford
Virginia G Ford
Xaviant Ford
Bob Forester
John Forgrave
Terri Formico
Scott A Forrest
Gustaf A Forsberg
Sasha W Forsyth
David N Fort
Karl Fortney
Teresa Fortuna
Charles W Foster
Kathleen S Foster
Michelle A Foster
Jason M Fox
Roy Fox
Linda Fracasse
René Fradet
Kathy Fragosa

Judith Fraiko
John Francis
Harvey A Frank
Herbert Franke
Michael Frankhuizen
Brenda Bogan Franklin
Donald A Franzen
George Fraschetti
Tom Fraschetti
Louis D Frausto
Jeff Frazier
Kristen Frazier
Raymond T Frederick
Arthur R Frederickson
Lloyd C French
Robert O French
Joan M Frenzel
Jaunita Fresquez
Irene Friday
Jerry Fridenberg
Alissa Friedman
Louis Friedman
Kenneth Frost
Gregory Fruth
James H Fu
Stephen D Fuerstenau
Kenneth K Fujii
Leonard Funck
Caroline D Furman
Lisa Gaddis
Peter Gagel
Joseph W Galamback
Tsilya L Galin
David A Gall
I Gall
Michael P Gallagher
Juan M. Gallardo Dates
Manuel Gallegos
John Gallon
Donna Galuszka
Edward B Gamble Jr
Luis Gamborino Alvarez
Enrique A Gamez
Nancy Gamez
Gani B Ganapathi
Frank Ganci
Anthony J Ganino
(Melvin) Buddy Ganther

John A Garba
Celina A Garcia
Diana Garcia
Elisa Garcia
Ignacio Garcia Arias
Alberto Garcia Arribas
Luis F. Garcia Centeno
J. Luis Garcia Delgado
Angel Garcia Esteban
Angel L. Garcia Garcia
German Garcia Jimenez
Juan Garcia Lobo
Cristina Garcia Miro
Faustino Garcia Pardo
Raul Garcia Perez
Berj Garibek
Angel Garnica
Henry Garrett
Michael S Garrett
Mike Garrett
Jim Garvin
Robert W Gaskell
Jason D Gates
Gerald S Gaughen
Barrie S Gauthier
William Albert Gavid
Tom Gavin
Stephanie P Gavshon
S Michael Gayle
W Gaylor
William Gaylord
Nola Gear
John Gedde
Mark A Gee
Paul Geissler
Ralf Gellert
Consuelo Gennaro
Bob Gentz
R Gentz
Donald George
Leanne George
Nisha George
Tom Geradi
Dimitrios Gerasimatos
Robert D Gerke
Donald A Germann
Andrew Geselbracht
Reza Ghaffarian

Saina Ghandchi
Vazrik Gharakanian
Eric A Ghesquiere
Stephen E Ghesquiere
Amitabha Ghosh
Mark Gibbel
Charlie Gibbs
Donald E Gibbs
Kathryn H Gibson
Otho Gibson
Ray Gibson
James H Gil
Stephanie Gil
Gary G Gilbert
John B Gilbert
Martin S Gilbert
John J Giles
Christopher J Gillespie
Lamar M Gilliam
John W Gillissen
Greg Gillis-Smith
Jesus Gimeno Avila
Thomas J Giovale
Mac Giovanni
Joseph D Girard
Mark Girard
Bob Gisler
James A Gittens
Roy E Gladden
Gary W Glass
Darwin G Glaze
Dan Glenn
Steve Glenn
Maria Glidewell
Timothy Glotch
Regina D Glover
Renny Glover
David Gockley
Natalie Godwin
Walter Goetz
Brenda L Goforth
Pawan K Gogna
Jay D Goguen
Gregory Golanoski
Mark Golbeck
Steven B Goldberg
Barry G Goldstein
Matt Golombek

Maxine Golub
Ernie Gomez
Fernando Gomez Martin
Lucas Gomez Redondo
Pablo Gomez Ventura
Ike Gonzales
Victor Gonzales
I Gonzalez
Manny Gonzalez
V Gonzalez
Jose C. Gonzalez Barahona
Raquel Gonzalez De La Montaña
Pedro Gonzalez Fernandez
Carlos Gonzalez Niebla
Carlos Gonzalez Pintado
Julio Gonzalez Ugidos
Kirk B Goodall
Charles E Goodhart
Shawn D Goodman
Michael J Gordon
Paul Gordon
Phillip W Gore
Noel Gorelick
Steve Gorevan
Ellissa J Gorham
Zareh Gorjian
Larry R Goss
Joanne M Gosser
Kim P Gostelow
Robert B Gounley
Eric J Graat
Michael J Gradziel
Dustin Graf
Trevor Graff
Janis U Graham
Michael Graham
Sean Graham
Chris Grallelis
Laura A Grammier
Rick Grammier
J. Manuel Grandela Duran
Frederick Douglas Grant
John Grant
Ken Grant
Gary D Gravante
Ricky D Graves
Don Gray

Harmodio Gray
Marisa D Gray
Maureen Gray
Stuart J Gray
George Greanias
Martin E Greco
Ron Greeley
Bette R Green
Chris Green
Daniel J Green
Suzan Q Green
Stephen S Greenberg
Dennis M Greene
Reda Greene
Michael L Greer
Tim Gregor
Freddie O Gregory
John Gregory
Linda L Gregory
Freddie O Gregory Jr.
Tony Greicius
Malcolm Greig
Jennifer Leslie Griffes
James L Griffin
John B Griffin
Paul Griffith
Katherine Grimason
Lorie R Grimes-Ledesma
Edmond Grin
Jonathan F Grinblat
Sharon L Groff
Robert Groman
Konstantin G Gromov
Michael A Gross
John Grotzinger
Shannon J Grover
Myron R Grover III
David C Gruel
Laurie L Guay
Carl S Guernsey
Antonio Guerrero Sanz
Steven M Guertin
Ken Guest
Cecilia N Guiar
Joseph R Guinn
Edward A Guinness
Haraldur Pall Gunnlaugsson
Donald Gurney

Mohanna Gurram
Patrick J Guske
Clemente Gutierrez
Jerry Gutierrez
Marcos A Gutierrez
Ricardo Gutierrez
Jose A Guzman
Jose T Guzman
Natalie E Guzman
Siina I Haapanen
Robert M Haberle
Mike Hacker
Fred Y Hadaegh
Michael L Haddox
Joel Hagen
Edward C Hagerott
Sean J Haggart
Rita Haggerty
Brian Hahn
Kathleen Hahn
R P Haight
Vicki L Hajdu
Albert F Haldemann
Brian Hale
Pete Halewski
Voula Haley
Andy Hall
Chris Hall
Jeff Hall
Robert A Hall
Gerald Halpert
Jerry Halpert
Richard E Halverstadt
Rita A Hamilton
Brian C Hammer
George Hammond
Sherill Hampton
Gregg Hanchett
Gary M Handy
Kristopher D Hanko
Jill Hanna-Prince
Kevin Hanrahan
Candice J Hansen
Mark Hansen
Richard A Hansen
Tom Hansen
David J Happs
Craig Hardgrove

Mary Hardin
Reggie Hardine
Bruce M Hardy
Keith D Hardy
Paul V Hardy
Trent Hare
Stephan A Harman
Maralee Harmon
Ronald M Harmon
Josh Harmony
Josh Harpe
Steve Harrelson
Brian D Harrington
Daniel Harris
Eric S Harris
Jeffery C Harris
Mark Harris
Richard H Harris
Gregory S Harrison
Linda D Hart
Reba J Hart
Wayne Hartford
Frank R Hartman
Cynthia L Hartnett
Hanry Hartounian
Adam C Harvey
Jane H Harvey
Jimmy D Harville
Douglas A Harvison
Michael W Hasbach
Tyler Hase
Larry A Haskin
Craig Haslam
Peter Hatfield
Paul J Hauser
Wallace J Hauswald
Robert J Haw
Chris Hawley
Roderick Hayden
Alex Hayes
Ross Hayes
Thomas Hayes
Les Haynes
James S Hayslett
Kari E Head
Timothy Heaps
Martyn Hearle
Scott Heck

David S Hecox

Dennis Heher

Jan Heinemann

Grant G Helling

Alex Hempstead

Paul W Hendershott

Douglas Henderson

Kelly A Henderson-Nelson

Donald E Hendricks

Teresa A Hendry

Rebecca J Heninger

Adam S Henkle

David A Henriquez

Tonya E Hensley

Kou Her

Maymee S Her

Mayme Her

Benjamin Herbert

David M Herhager

Ken Herkenhoff

David E Herman

Jennifer A Herman

Sandy Herman

Tim Herman

Louie Hermosillo

David C Hermsen

Christopher Hern

Vicente Hernandez Del Castillo

Juan Hernandez Ruiz

Mario Hernandez Santamaria

Mike Herr

Alfonso G Herrera

Jay Herrera

Luis O Herrera

Paul N Herrera

Kelley Hess

Barbara L Hesselgesser

Mark M Hetzel

Matthew C Heverly

Susan Hewett

Geoffrey Hewitt

Gregory S Hickey

Kyle Hickey

Margaret High

Ray Higuera

David Hill

Gay Yee Hill

Heidi Hill

James J Hill

Jennifer R Hill

Leela D Hill

Michael H Hill

Robin L Hill

Thomas C Hill

Steve Hillabrand

Jeffrey E Hilland

Terry W. Himes

Brian D Hinde

Cheryl A Hines

Kenneth G Hissong

Quy Q Ho

Timothy Y Ho

Garry Hobbs

Alan Hoffman

Jack Hoffman

Jim Hoffman

Pamela J Hoffman

Tom L Hoffman

Walter T Hoffman

Carmen M Holder

Denise A Hollert

Wallace H Hollimon

Dean A Holt

Madonna S Holt

Robert A Holt

Rick Honda

Karen E Honohan

Russell O Hope

Ewen Hopkins

Ian Hopkins

William E Hopkins Jr.

Shannon K Hoppe

Larry Horan

Wendy D Horn

Sarah E Hornbeck

Douglas B Horner

Michael B Horner

Christine L Horowitz

William J Horsley

Tony Horton

Victor Horton

Hossein Hosseini

Robert Hostler

Nathaniel Hotchko

D Houser

James S Howard

Rob Howard

William R Howard

John Howell

Annie Howington-Kraus

Cheng-Hsien Hsieh

George H Hsu

Jennifer Hsu

Ming-Ta Hsu

John Huang

Allen B Hubbard

Kim Hubbard

Min Hubbard

Scott Hubbard

Michael W Hubbard

Joseph C Huff

Edwards E Hughes

Michael J Hughes

Robert W Hughes

Scott Hughes

Steve Hughes

Frank Hui

Sarah Hulen

Lawrence A Huley

Howard K Hull

Jeff Hull

David Hulligan

Scott Hulme

Richard A Hund

Bryan L Huneycutt

Vincent Santiago Hung

Thomas Hungerford

Carl N Hunsaker

Beverly S Hunt

John A Hunter

Sheryl R Hunter

Richard E Hunter Jr.

Terrance L Huntsberger

Lorie M Hur

Daniel S Hurley

Joel Hurowitz

Ziad A Hussein

Paul Hutchinson

Kathleen L Hutchinson

David Hutchison

Kenneth T Hutchison

Stubbe Hviid

Ho-Jin Hwang

Pauline P Hwang

Ju Won Hwangbo
Christopher J Hyde
David A Hykes
Al Ibarra
Peter M Illsley
Peter A Ilott
Jack Ingle
Scott W Inlow
Ahmad Iqbal
John W Irving
Christopher E Isbell
Theodore C Iskenderian
Nhan Islam
Denise Island
Keith A Ivanoff
Anton B Ivanov
Yuji Iwai
K Iwanaga
Vicki L Iwata
Gabby Izsak
John Jaatinen
Alfred M Jackson
Ray Jackson
Sheryl A Jackson
Stephanie Jackson
Steve Jackson
Christopher S Jacobs
Michael H Jacobs
Mike Jacobs
Stephen D Jacobs
Richard Jacobsen
Burt Jaffe
Moriba K Jah
Marcos Jaime Alhambra
Geoff James
Louise Jandura
John Jansson
Han Janzen
Maria A Jaquez
Robert W Jarrett
Bruno M Jau
Hamid H Javadi
Barbara A Jefferson
J Jeffryes
Kirk Jellum
Sylvia F Jemison
Jeff Jenkins
Chuck J Jennings

Elsa Jensen
Alice Jerahian
Elmer O Jerez
Doug Jerolmack
Kenneth A Jewett
Chunlei Jerri Ji
Alejandro Jimenez
Lisa J Jimenez
Juan Jimenez Carrion
Sandra Johan
Kim Johansen
Bonnie John
Greg L Johns
Roger L Johns
Alan F Johnson
Albert S (Sid) Johnson
Andrew E Johnson
Bill Johnson
Christine Johnson
David W Johnson
Debra K Johnson
Gregory B Johnson
Guadalupe M Johnson
Jackie Johnson
Jason B Johnson
Jeffrey R. Johnson
John Johnson
Margaret F Johnson
Mark E Johnson
Mary L Johnson
Matt Johnson
Michael R Johnson
Michele A Johnson
Miles Johnson
Norman Johnson
Peter B Johnson
Sarah Johnson
Sid A Johnson
Steve Johnson
Theodore R Johnson
Allan H Johnston
Ashley Johnston
Doug V Johnston
Jean Johnston
Paul J Johnston
Sharon F Johnston
Brad Jolliff
Byron B Jones

Chris Jones
Dan Jones
Donna Nigh Jones
Harrison Jones
Jason Jones
Loren E Jones
Margarete E Jones
Mark A Jones
Steve Jones
Steven M Jones
Vernon Jones
Wade Jones
Ralph Jones
Andre P Jongeling
Ari Jonsson
David Jordan
Ed Joscelyn
Jonathan Joseph
Lawrence Juare
Insoo Jun
Amy J Jurewicz
Charles J Kaczinski
Michael W Kahler
Dan Kalcic
Theodore Kaliher Jr.
Lewis Kalin
S Kamine
Richard Kaminsky
Edwin P Kan
Mike Kandis
Bob Kanefsky
Julie A Kangas
Egbert Kankeleit
Michael Kapareliotis
Kathryn Kapelke
Andrea S Kapitanoff
Paul F Kaptchen
Raymond Kariger
Paul B Karlmann
Rodney J Karre
Suniti Karunatillake
David Kass
S Katanyoutanant
Jonathan Katz
Steven G Katz
Guy Kauffman
Kristy Kawasaki
Kevin Kay

Selahattin Kayalar
Sammy Kayali
Gayane Kazarians
Dale B Kechter
Rolph Keehn
Gary Keel
Leslie Keely
Terry Kehoe
Sotiris Kellas
Don Keller
Wayne D Keller
Mark W Kellner
Keith Kelly
Kenneth C Kelly
Michael Kelly
Stephen Kelly
Beth H Kelsic
Shirley T Kemper
Michael Kempf
Richard Kemski
Annabel R Kennedy
Brett A Kennedy
Brian M Kennedy
John A Kennedy
Mark Kennedy
Kevin E Kerk
Albert Kern
Dennis L Kern
Roger Kern
Tom Kern
William D Kert
Robert W Keskinen
Laszlo Keszthelyi
Kenneth Kettering
Matthew S Keuneke
Abdur R Khan
Teerapat "Tk" Khanampornpan
Garen Khanoyan
M Khorrami
Derek L Kiang
Edward H Kieckhefer
William C Kiehl Jr.
Aaron B Kiely
Thanh C Kieu
Danny J Kiewicz Jr
Teresa J Killmon
James W Killough
Andrew Y Kim

Richard Kim
James C Kimberling
Kjartan Kinch
Andrew Kindler
David D King
Michael J Kinnel
Gary M Kinsella
James A Kipstuhl
Kathleen M Kipstuhl
Randy Kirk
Mike Kirsch
Larry E Kirschner
Andrew Kissil
Robert F Kitz
Roger S Klammer
Marlin F Klatte
Bruce Klein
John W Klein
Tyler Klein
Roger W Klemm
Matthew A Klimesh
Henry Kline
Rob Kline
Göstar Klingelhöfer
Vadim S Klochko
John Klohoker
Gerhard J Klose
Gregory L Klotz
Sheri Klug
Allan R Klumpp
Don Knaepple
Jason B Knapp
Anthony Knight
Jennifer M Knight
Kevin Knights
Philip C Knocke
Andy Knoll
Carl Knoll
James Knoll
Carl Knoll Jr.
Michael D Knopp
Jens Martin Knudsen
Amy Knudson
Adans Y Ko
Patrick Ko
Robert Wai Ko
Michael C Kobold
Peter A Kobzeff

John S Koehler
Mark Koehler
Jason L Koelliker
John S Koenig
Lorraine Koger
Paul Kohorst
Steven G Kondos
Edward H Konefat
Hans A Konstabel
Edward H Kopf Jr
Nicholas K Koranda
Mark A Kordon
Nick Kormanik
Richard P Kornfeld
David Korsmeyer
Paul Kostura
Robert C Koukol
Louie T Koumoutsos
Anthony S Kouri
Hrysi Kourkoumelis
Robert M Kovac
Theresa D Kowalkowski
James E Kowalski
Marge Kozaki
Nobby Kozlowski
Joel A Krajewski
Bradley S Kramer
Michael Kramer
Paul Kramer
Sanford M Krasner
Terry E Krayenhagen
Ernest J Kreiner
Satish A Krishnan
Peter M Kroger
Sheerae F Kruck
Duane Krumweide
Ed Kruzins
Robert J Krylo
James E Kubitschek
Albert M Kuchler
Joanne C Kuenning
Teddy S Kuenning
Gregory Kuhlman
Eric A Kulczycki
Stephen Kulczycki
James G Kulleck
Clay Kunz
Frank Kunz

Chin-Po Kuo
Neal R Kuo
C Eric Kurzweil
Alastair Kusack
Garrett F Kusakabe
Eug-Yun Kwack
Richard Kwan
Tony Kwok
Helenann H Kwong-Fu
Charles Kyriacou
Myron La Duc
Jason La Rosa
Emily M La Rose
Gregory R Laborde
Daniel Lacanilao
Veronica Lacayo
Deidre Laclair
Jennifer L Lafkas
Martha C Lageschulte
Diana I Laguna
Erin M Lahr
Thomas A Laing
Mike Laird
Dan Lake
Geoffrey T Lake
Margaret W Lam
Mary M Lam
Andrew U Lamborn
Richard G Lamborn
Matt Landano
Geoff Landis
Robby R Landis
James M Landolt
Christopher P Landry
Pete Landry
Preston D Landry
Arthur L Lane
Robert W Lane
Minh Lang
Leticia Langman
Kurt Lankford
Gabor E Lanyi
Jacques Laramee
Calvin N Larson
Tim Larson
Sandi Lasako
Jeffrey Lasco
Joseph W Latiolais

Chi-Man E Lau
Chi-Wung Lau
Sharon L Laubach
Paul E Laufer
Geoff Laugen
Dave Lavery
Sue Lavoie
Honey Law
Jennifer T Law
Russell A Lawton
William Layman
Chuck Lazansky
Scott M Lazaroff
Gwen Lazzeri
Hoang M Le
Thu Le
Tram Le
Renee Leake
Adolfo Leal
Alex Leal
Stephanie Lear
Zoe Learner
Alan W Lee
B Gentry Lee
Brett Lee
Charles H Lee
Chern-Jiin Lee
Darlene S Lee
Debbie Lee
Ella Mae Lee
Huyn Lee
Hyun H Lee
Jane Lee
Kathy Lee
Lawrence W Lee
Linda M Lee
Peter Lee
Shu-Chen Susan Lee
Siend Lee
Siu-Chun Lee
Wayne J Lee
Yee N Lee
Young H Lee
Mai Lee
Kristoffer Leer
Dave Lees
Craig E Leff
Patrick C Leger

Victor N Legerton
Michael E Lehmann
Donald T Lehr
Jian Y Lei
John Leimgruber III
Cynthia A Leinweber
Charles J Leising
Timothy W Leisy
Nelson Leiva
Tim Lemesurier
Mark Lemmon
Bruce A Lemoine
Darrel D Lemon
Christopher Leng Jr
Terry Leonhart
Darryl R Leopold
Jerry Leslie
Jeff Lesovsky
Anita L Lettieri
Chris Leung
Mark Leung
Anthony Levay
Scott G Lever
Michael E Levesque
Joe Levine
David Levitt
Steve Levoe
Christopher A Lewicki
Bruce Lewis
Donald F Lewis
James Lewis
Joseph C Lewis
Juan M Lewis
Steve Lewis
Terry L Lewis
Karan L L'Heureaux
Jennie Chin-Yee Li
Rongxing Li
Samuel S Li
Karen G Liao
Steve M Lichten
Brad Liddy
Carl C Liebe
Robert R Liebersbach
Paul J Lien
Gregory P Lievense
Kimberly A Lievense
Stephenie H Lievense

Damian Lilley
Timothy Lilley
Annie Lim
Tenny W Lim
Daniel Limonadi
Susan L Linch
Erik M Lind
Charlotte Linde
Randel A Lindemann
Brett A Lindenfeld
Susan H Linick
J Linke
Slava Linkin
Matt Linton
Michael E Lisano
Lane Liston
Carson Little
Edward C Litty
Todd E Litwin
Dankai Liu
Brynn Llewellyn
James W Lloyd
Michael W Lloyd
Carmela C Lloyd
C John Lo
Francisco Loaiza
Frank Locatell
Lois E Loche
Robert E Lock
Clayton Locke
Frank L Lockwood
Charles Lodestro
Phyllis A Lofland
Ron Logan
Mike Logothetis
Thomas H Lohaus
David J Lokshin
Winthrop C Lombard
Jon Lomberg
Hartwell Long
James L Long
Mark Loomis
Genevieve Lopez
Kathery E Lopez
Mario Lopez
Patricia Lopez
Carlos W Lopez
Vicente Lopez Busto

Manuel Lopez De Ayala
Javier Lopez Perea
Enrique Lopez Rubio
Alfonso Lopez Villa
Mark Lorah
David Lorimer
George A Lorman
Jean Lorre
Susan E Lotocky
Robert Lotter
Nathalie Louge
John J Louie
Nancy Lovato
Linda Love
Mark E Lovely
Corey Lovers
Eugene Lovett
Stephen T Lowe
Michael Lowry
Ti Luangpraseuth
Don F Luczai
Roger Ludwig
Jan M Ludwinski
Günter Lugmair
Vince Luh
Debra L Lujan
David Lukachko
Matthew T Lum
John J Lumia
James Lumsden
Cathy Luna
Jaime Luna
Karen M Ly
Daniel T Lyons, Dr
Jacqueline C Lyra
David N Lysohir
Robert K Lytle
James Lyttleton
Dmitriy Lyubimov
John Lyver
Betty Macdonald
Minh (Eric) Mach
Jose M Macias
John Maciejewski
Greg Mackert
David H Mackey
Paul Macneal
Duncan Macpherson

Daniel Esmarch Madsen
Morten Bo Madsen
Iñigo Madurga Escribano
Viswanath S Magapu
Lisa De Lange Maginnis
Lee Magnone
Todd G Magor
Donna Magowan
Melody Mahirka
Gary Mahonchak
Mark W Maimone
Merrell W Maine
Earl H Maize
Walid A Majid
Gary Major
Joanne A Major
Ronald Mak
Justin N Maki
Andrew Makovsky
Kevin P Makowski
Mike Maksymuk
Pierre F Maldague
Carolina Maldonado
Juanita Maldonado
Shawn N Malik
Mike Malin
Melissa Mallis
George J Malone
Brian Malott
Kin Fung Man
Al Mancilla
Michael J Mangano
Ken Mankoff
Diane M Mann
Dennis Manning
Eleanor R Manning
Mark Manning
Robert M Manning
Roland R Manning
Miguel Manresa III
Joe A Manriquez
Jaime C Mantel
Kamesh Mantha
Miguel A. Manzanares Parra
Francisco Manzanares Velasco
Benigno J. Manzano Vazquez
Bruce W Maples
Steven March

Brian D Marchbank
Carmen Marco
Margaret Rose Marco
Michael Marcucci
Nick Mardesich
Amelia Marez
Lucia M Marino
Scott D Markham
Masis Markosyan
Manuel Marmol Concejal
Eric D Marquardt
Alfred Marquez
Dana L Marrs
Patricia V Marsh
Janis F Marshall
Sarah L Marshall
Vern Marston
Anthony C Martin
Bill Martin
Keith E Martin
Kyle W Martin
Lisa Martin
Patrick Martin
Terry Z Martin
Paul Martin
M. Helga Martin Diaz
M. Carmen Martin Duran
Angel Martin Garcia
Heliodoro Martin Gomez
Harry Martin Iv
Fernando Martin Parra
Federico Martin Rodriguez
Luis D. Martin Ruiz
Carolina Martinez
Eduardo B Martinez
Elmain Martinez
Elsa Martinez
Gilbert Martinez
Henrietta L Martinez
Louis G Martinez
Maria J Martinez
Patricia A Martinez
Michael D Martinez
Jose M. Martinez Colina
Enrique Martinez Fajardo
Angel Martinez Marcos
Ignacio Martinez Rubio
Jorge Masdeu Rodriguez

Robert Mase
John Mason Jr.
Kim Massey
Mark W Massey
Armond Matevosian
David M Mathews
Valerie Mathews
Jacob R Matijevic
Jim Matlock
Mark D Matonic
Harout Matossian
Bruce Mattarocci
Jaret B Matthews
Larry H Matthies
Scott A Maxwell
Brian E Maxwell
John T Mayer
Linda M Mayo-Cupples
Alan Mazer
Jason Mazotta
Joseph M Mazzotta
Matthew A Mazzotta
Jeffrey McAllister
William J McAlpine
Myche Mcauley
James W McBarron III
Larry McCain
Elaina McCartney
René McClendon
David R McClure
Gary McClure
James R McClure
Linda S McClure
Louise U McClure
Timothy W McClure
Randy McConaughey
Darrell L Mcconnell
Jeff N McCowen
Fred L McCown
David K McCoy
Michael E McCracken
Dennis L McCreary
Michael McCurdy
Mame McCutchin
Clifford C McDaniel
Val W McDannel
Kevin McDevitt
Ian McDonald

Mark McDonald
Timothy P McElrath
Connor McGann
Paul L McGrath
Sally McGrath
Gail T McGraw
Kenneth E McGraw
Veronica McgrGegor
Barbara A McGuffie
Ernest C McGuire
John McGuire
Helen McIlroy
Tamsin McInerney
Robert P McIntosh Jr.
Kieran McKay
Richard J McKenzie
Kathy McLeon
Tom McLeown
Kristen S McLallen
Alan R McLaughlin
Scott McLennan
Jeremy McMartin
Robert L McMaster
Gordon McMenemy
Robert McMillan
Thedra A McMillian
Dennis E McMurray
Vernon B McNeill
Des McNicholas
Angus McRonald
Hap McSween
Bruce McWilliam
Paul McWilliams
David O Mead
Gerald L Mead
Jack Meadows
Peter Meakin
Walter A Meares
Cheryl A Mechalke
Theodore W Mecum
Alison A Medbery
Alan C Medzoyan
Kerry Meek
Shelby J Meek
Vincent Meglio
Greg Mehall
Laura Mehall
Gregory Mehle

Raka A Mehra
Atul C Mehta
Jitendra S Mehta
Pamela L Meintzer
Cindy A Meisner
Johnny Melendez Jr
Jose C Melendez Jr.
Joseph P Melko
David M Meller
Jeffrey A Mellstrom
Barbara V Mendoza
Ricardo Mendoza
Stephen K Menikheim
Robert J Menke
Mark S Menke
Jerami M Mennella
Marian Meridieth
Susan J Merrill
Jonathan Merrison
Christine L Merwin
Douglas M Merz
Richard A Messersmith
Blair A Messervy
Amy Z Meyer
Donald Meyer
June Meyer
Steven D Meyer
Virginia D Meyer
Timothy L Meyers
John H Meysenburg
Rudy Meza
Mike Miccio
Tim Michaels
Joe Michalski
Ian Middleton
Tim Middleton
David T Mih
Steven C Mikes
Igor Mikhaylov
Keith Milam
Troy Miles
Ben R Miller
Bobby Miller
Bonnie L Miller
Christopher R Miller
John E Miller
Mark Miller
Steve Miller

Chase Million
Perry L Millman
Fred Milstein
Trena M Mims
Jennifer A Mindock
Ray Mineck
Doug Ming
Rufino Minguez Arribas
Zara Miralek
Adam A Miranda
Koorosh Mirfakhraie
Zara Mirmalek
Richard L Miseroy
Andrew H Mishkin
Philip J Mislinski
Robert T Mitchell
Stephen Mitchell
Robert A Mitcheltree
Anthony P Mittskus
Tetsuo F (Tets) Miyahira
Masashi Mizukami
Robert Mockler
Karen J Modafferi
Jeff Moersch
James B Mohl
Merle Mohring
V. Esther Moll Moreno
Hiwot A Molla
John A Mollura
Martha A Molodowitch
Bob Moncada
Jack F Mondt
Doug Monick
Curtis Montano
Mike Montgomery
Ryan H Montgomery
Diane M Montini
Paul V Moomjean
Charles J Moore
Charles W Moore
Craig Moore
Kirk G Moore
Peggy N Moore
Percy Moore
Randall J Moore
Vickie L Moore
William V Moore
Alejandro Mora

Alex Mora
Joseph L Mora
Mario L Mora
Victor D Mora
Juan G Mora Fernandez
Fabian Morales
Marvin Morales
Misrahim Morales
Pedro Morales
Kelly C Moran
Nancy Moran
F. Ramon Morcillo Masero
Harvey Morehouse
Pedro E Moreira
Daniel W Morgan
Feiming C Morgan
Justin K Morgan
Gary E Morgan Jr.
Michael L Morian
Pedro Moriera
Ronald Morillo
Charles Morris
Chuck Morris
Dick Morris
J Richard Morris
Jerry A Morris
Joe Morris
Mark A Morris
Paul Morris
William B Morris Jr.
Andrew D Morrison
Jack Morrison
Helen Mortensen
Uri Moszkowics
Amin H Mottiwala
Edward L Motts
Samih Mouneimne
Alex G Mourad
Daniel J Moyers
Mike Mueller
Robert D Mueller
Suparna Mukherjee
John J Mullin
Jonathan Mullins
Kevin F Mullins
Mary Mulvanerton
Erik Mumm
John A Munnerlyn

Clem W Munninghoff
Rosina Munoz
Javier Muñoz Garcia
David Muñoz Mochon
Marilyn Murakami
Jeff Murch
Edgar M Murillo
Charlie Murphy
Doug Murphy
James Murphy
James R Murphy
Juliana A Murphy
Mary Murphy
Patrick Murphy
Patrick J Murphy
Annie Murray
Cory Murray
Demarcus Murray
James F Murray
John Murray
Maralee Murray
Ross Murray
Scott W Murray
Mary B Murrill
Richard Murth
Nicola Muscettola
Jenica Muse
Walt Mushagian
Michael E Mussler
Dave Musson
Tom Myrick
Firouz Naderi
Glen Nagle
Adrian F Nagle IV
Sara Najjar-Wilson
David I Nakamoto
Lori L Nakamura
Michael A Nakashima
Albert Y Nakata
George H Nakatsukasa
Barry Nakazono
Ron Naldrett
Sumita Nandi
Pablo Narvaez
Annette K Nasif
Sal Nasser
Pamela S Nation
Maher C Natour

Surasak Natvipada
Charles J Naudet
Robert Navarro
Keith Naviaux
Diane M Naylor
Joe P Necas
Joe P Necas Jr.
Laura Needels
Gary Neel
Peggy S Neff
William C Neiderheiser
Tracy A Neilson
Luis Neira Martinez
Karl T Neis
Mark A Neitenbach
Barry Nelson
Lloyd Nelson
Marlynn L Nelson
Nils E Nelson
Robert W Nelson
Stephen E Nelson
Victor Nelson
Linda H Nenadovic-Cantuna
David Neri
Bill J Nesmith
John O Nesseth
Linda S Nesseth
Lloyd J Nessler
Terry Neuendorf
Peter F Neuroth
Orval E Neuschwanger
Ray A Nevarez
Oly Neverida
Timothy A Newby
James M Newell
Paul S Newer
Mark Newfield
Laura E Newlin
Kelly Newman
Russell T Newnes
Calvin Ngo
Chau M Ngo
Tran Ngo-Luu
Alexander G Nguyen
Danny H Nguyen
Dave Nguyen
Duc N Nguyen
Kien Nguyen

Paul R Nguyen
Quit B Nguyen
Tam T Nguyen
Tien T Nguyen
Tuyet T Nguyen
Vivian Nguyen
Vu H Nguyen
Thuy C Nguyen-Onstott
Vincent Niave
Fabian Nicaise
Bud Nicholas
Mark J Nicholas
Curt Niebur
Sandra S Nielsen
Donald G Nieraeth
Troy Nilson
David Nimish
Steven L Nissen
Guy M Niswender
Ashly Nix
Sophia M No
David Noble
Sam H Noble
Eldar Noe
Manuel Nogales Arroyo
Steve Noland
Don E Noon
Kate Noonan
Tyler Nordgren
Garth O Nordstrom
Richard Norman
Richard A Norman
Shelley R Norman
Thomas E Norman
Per Nornberg
Jeffrey S Norris
Bradford E Northup
Mylan Nosar
George Nossaman
Keith S Novak
Carl Nuckolls
Jeffery A Nunes
Rick Nybakken
Bill Nye
Eric R Nysather
Eric Oakes
Dave Oates
Markus F X J Oberhumer

Patrick O'Brien
Robin A O'Brien
Thomas J O'Brien
Michael R O'Connell
Michael D O'Connor
Cindy H Oda
Edwin H Odell
Deborah D O'Dell
Bill O'Donnell
Frank O'Donnell
Carol A Oelker
Daniel L Oenes
Steve T Ogawa
Ken Ogden
Dean C Okamoto
R Okamoto
Brian K Okerlund
James O Okuno
Pat Olagues
Eric Olds
Robert Olds
Patrick M Olguin
Oskar A Olivares
Randii Oliver
Ayoola K Olorunsola
Malte Olsen
Jane O'Neill
Brad M Oney
Larry Opper
Clyde F O'Quinn
Rafael P Ordonez
Taifun O'Reiley
Joe Orman
Basilio Ormeno
Juan F. Ortega Martin
Gary M Ortiz
Dennis R Orton
Tee Osborne
Teresa Osminer
Thomas O'Toole Jr
Frank M Ott
Tom L Otto
Lorenz Ou
Kamal Oudrhiri
Ricky Overcash
Kyran J Owen-Mankovich
Amalaye Oyake
Suzanne M Oyama

Tetsuo Ozawa
Jon Pabst
J Pacheco
Richard G Pacheco
Marc M Pack
Doug Packard
Newton R Packard
Andrew Paddison
Fred Padgett
Lisa Padgett
Richard S Padilla
Dennis N Page
James P Page
John R Page
Christopher M Paiz
Ricardo Palavecino Rodriguez
Jeremy M Pallotta
Mark A Palm
Naomi K Palmer
Timothy Palmer
Paul Pangburn
Jose M Pantaleon
Gonzalo Parada Crespo
Rajendra B Parikh
Anthony D Paris
Oleg Pariser
Eric Park
M Park
Peggy Park
Sung M Park
Bonny Parke
Judy Parker
Timothy J Parker
Michael D Parks
Patricia L Parrett
Kristine A Parrish
Mary E Parrish
Priscilla P Parrish
Heather A Parsons
Rino Passaniti
Mark E Pastor
Gregorio Pastor San Roman
Susan Pastrano
Scott E Paswaters
Kirti Patel
Tarang Patel
Robert Path
Santo Patinella

Vlad P Patrangenaru
Ian M Patrick
Jean E Patterson
Margarete Patterson
Pat Patterson
Karen Patton
Timothy Patton
Jack D Patzold
Michael T Pauken
Larry Paul
David Paule
Rich Pavlosky
Thomas Pawlowski
Ricardo Payan
Dan A Payne
Meredith Payne
Victoria D Payne
Jon M Payne
Richard L Paynter
Mike Paz
Harry Peacock
Brian Peale
James Pearson
Scott Pearson
William R Pearson
H B Pedersen
Shannon Pelkey
Tracy Pellegrino
Robert P Pelletier
Tony Pelling
Laura Pelton
Joseph Pem
Chia-Yen P Peng
John Penix
George Pennybaker
Rick Peralta
Alex Perez
Haydee Perez
Marisol Perez
Ramiro Perez
Raul M Perez
Reinaldo J Perez
Fernando Perez Montes
Pablo Perez Zapardiel
Scott Perl
Edward Pernicka
Jonathan D Perret
Douglas Perry

Minnie V Perry
Doug Petercsak
Barbara Peters
Gregory H Peters
Stephen F Peters
Gene C Peterson
Todd D Peterson
Richard D Petras
Steven M Petree
Robert L Pewthers
Henry Pfanner
Samuel T Pfister
James Pflieger
Lien Pham
Phuonghanh M Pham
Timothy T Pham
Stephen W Phelps
Charles J Phillips
J Mark Phillips
Jessica L Phillips
Keith Phillips
Kim Phillips
Michael J Phillips
Robert Phipps
Frank Q Picha
Craig D Pickett
Kevin Piechowski
Thomas Pierce
John J Pierok
Wayne F Pierre
Stephanie Pieruccini
Muthango Pillai
Paul A Pinarretta
Daniel J Pinion
David R Pinkley
Michael S Pirkey
Stephen Pitts
Don Pitzer
Gregory G Pixler
Thusitha D Piyasena
Jane Platt
Martin Platupe
Glenn Pleasant
Susan E Pleasanton
Carl W Plehaty
Gary A Plett
Jill Pocock
Harry C Poehlmann Jr.

Brian Poggiali
Ronald J Pogorzelski
Dave Polak
Lee C Polet
Robert L Poley
Bill Polk
Robert Polleyfeyt
Christine Pollock
Grant M Poly
Elly Ponce
L Ponce
Gary Ponikvar
Peter T Poon
Phillip Pope
Sheasen G Porrine
Charles D Porter
Christopher C Porter
Margaret Porter
Jesus Portero Alcaraz
Brian M Portock
Betty R Potter
Don T Potter
Tom Potter
Christopher L Potts
Ali M Pourangi
James Powderly
Betty D Powell
Dawn H Powell
Edward W Powell
Kenneth W Powell
Mark Powell
Eugene Poyorena
Mariano Pozas Herranz
Kevin Pracon
L. Alberto Pradilla Yunta
Lo Pranolo
David Pravlik
Betty L Preece
Christine S Preheim
Scott Premo
Elizabeth A Preston
Humphrey W Price
Jim Price
Guy D Prichard
Oleg Priluyskiy
Robert C Proctor
Roger K Proctor
David S Proffitt

Walter K Proniewicz
Jon Proton
Robert A Provine
Edward Prstec
Mikel Pruett
John M Prunk Jr.
Gilmer Puccini
Maria Pulsonetti
Matthew Purdie
Kip Puri
Gregory L Purvis
Marco B Quadrelli
David Quarles
Eric Queen
Bob Quick
John J Quicksall
Megan S Quigley
Phillip C Quigley
Juan Quinones
James A Quintal
Elizabeth Quintanilla
Bill Raabe
Ronald D Race
Caroline S Racho
Viktor Radchenko
Jim Rade
Joel D Rademacher
Andrew Rader
Michael R Radovich
Mark A Radovich Jr.
Guy Raede
Shaheed Rafeek
Scot C R Rafkin
Fred Rahrig
Richard A Rainen
Ben Raiszadeh
Kanna Rajan
Rajeshuni Ramesham
Rudy Ramierez
Alfred J Ramil
Brenda J Ramirez
Frank E Ramirez
Richard Ramirez
Wes Ramm
Linsee L Ramsay
Paul Ramsay
Steve A Ramsay
Thomas L Ramsay

Vincent L Randolph
Socorro M Rangel
Selena Elizabeth Ransom
Behzad Raofi
Paul M Rapacz
Anthon J Rasmussen
Robert D Rasmussen
Ludivina L Ratledge
Ruth A Ratledge
Martin Ratliff
Serene Rawlings
J Richard Rawls
Richard Rawls
Bernard G Rax
Darrell J Ray
Pamela Ray
Benno Rayhrer
Francisco Razo
Gustavo Razo Jr.
Paul Reager
Stephen A Rearden
M J Reardon
Mary E Reaves
Kevin Recker
Bonnie Redding
Fred B Redell
Evelyn Reed
Russell Reed
Timothy Reed
William F Reed
Kate Reedy
Timothy J Reeve
Glenn E Reeves
Lorraine A Reeves
Robert Reeves
Fernando Regalado Espino
Sandra L Reichel
Laurence E Reinhart
Julie M Reiz
Jane F Reller
Kathleen F Remner
Michael Renbarger
Robert D Renken
Nilton Renno
Franz Renz
John Repar
Ben Retan
James Reuter

Juanita Reyes
Timothy R Reyes
Walt Reyle
Ivan L Reznor
E Rhinehart
Brian E Rhodes
Len Ricardo
Carlos D Rice
D Rice
Eric B Rice
James William Rice Jr.
Reva D Rich
Stephen J Rich
Matthew Richards
Randy Richards
Tamara J Richards
Mark I Richardson
Bernie Richter
Lutz Richter
Manfred A Richter
Paul Richter
Thomas Ridgway
Rudi Rieder
Benjamin D Riggs
Mark Righetti
Elaine L Riley
Elizabeth A Riley
Emilio Rioja Juarez
Audra K Riskus
Kevin J Ritchie
Tommaso P Rivellini
Duveen Rivera
Gerardo Rivera
Ronald J Rizzi
Judy D Roach
Curt Roberts
James A Roberts
Jean Roberts
Rachel Roberts
Stanley Roberts
Susan M Roberts
Teresa Robinette
James M Robinson
Kyoko Robledo
Edna Robles
Erik Roche
Daniel Rodionov
Lela J Rodriguez

Maria A Rodriguez
Patricia Rodriguez
Theresa L Rodriguez
Tina Rodriguez
Zenovia A Rodriguez
Susana Rodriguez Blanco
Jose Rodriguez Botello
Gregorio Rodriguez Pasero
Jonathan S Roehrs
Richard L Roelecke
Voleak Roeum
Cheryn M Roff
Deanne Rogers
Greg Rogers
Katherine L Rogers
David H Rogstad
Stephen P Rogstad
Steve Rogstad
John A Rohr
Gil Roldan
Walter F Roll
Paul A Rollings
Robert Roman
Debi Romero
Joe Romero
Raul A Romero
Rafael Romero Juarez
Diego Romero Sabio
Elizabeth J Romo
Ralph B Roncoli
David Rooney
Ramin Roosta
Gregory T Rosalia
Debra S Rose
James Rose
Mark Rosekind
Frederic H Rosenblatt
Mark Rosiek
Ana Ross
Ron Ross
Ronald G Ross Jr.
Tony Ross
Dino J Rossetti
Diane R Rossi
Sara Rosso Hernandez
John C Roth
Heinz Rott
Eugene D Rouse

William J Rousey
Cary G Rowe
Sherri Rowe-Lopez
Chris L Rowland
Johnny Roy
Patricia Roy
Eric B Roybal
Jeffrey Royse
John Rozycki
Dale Rudolph
Betty A Ruff
Steve Ruff
Ronald P Ruiz
Jose M. Ruiz Guinea
Angel Ruiz Tovar
Elizabeth A Rulla
Allan J Runkle
Edward R Ruskus
Dennis Russell
Robin L Russell
Ross H Russell
Karen C Russo
Richard S Ruth
James W Ryan
Marty L Ryan
Rich Rynders
Dara Sabahi
Mike Sachse
Shazad Sadick
Lauri L Sager-Devirian
Raymond Sajka
Andrea R Salazar
Ron Salazar
Mark A Sallee
Wesley Sallee
Phil M Salomon
Wade Saltzgiver
Alberto Salvador Galan
Carlos Salvador Vallejo
Christopher G Salvo
Anthony Sampere
Yvonne Samuel
Leroy L Samuels
Alejandro M San Martin
Javier Sanchez
Oscar Sanchez
Juan A. Sanchez Colmenarejo
Alfonso A. Sanchez Pascual

Antonio L Sanders
C Dennis Sanders
Jane-Ann Sanders
Mike R Sanders
Jasmine Sandhu
Michael A Sandor
Carlos Sandoval
Roberta A Sandoval
Sergio Sanjuan Valor
Ada Sann
Joseph T Sanok
Andrew Santa
Maria Santellan
Manuel A Santi
Abraham A Santiago
Richard Santiago
Luis C. Sanz Luna
Randy Sargent
Felix Sarkissian
Edgar H Satorius
Ana Saucedo
Don Savage
Joseph L Savino
Gary Savona
Dmitry Savransky
Khamphou Saycocie
Rocco P Scaramastra
Bruce W Scardina
Mark M Schaefer
Raimond Schaffer
David Scharpenberg
Terry D Scharton
Donald V Schatzel
Leif Z Scheick
Craig R Scheir
Lawrence M Scherr
Lynn W Schick
John J Schilling
Helmut Schiretz
Philip A Schluckebier
Brian Schmid
Philip M Schmidt
Ronald L Schmidt
Duane H Schmitz
Michele F Schneider
Paul Schneider
Mark Schoenenberger
John G Schofield

John T Schofield
Eric Scholten
David B Scholz
Marcel J Schoppers
John Schreiner
Christian Schröder
Curtis L Schroeder
Jeff Schroeder
Todd Schroeder
Wayne W Schubert
Charli Schuler
Frank Schwager
Werner Schwarz
Kurt Schwehr
Jeff A Schweinsberg
Doug Schweitzer
Mark A Schwochert
Andres Scola Ezquerro
Chris Scolese
Aaron J Sconzert
Bradley A Scott
Bryan Scott
Carol J Scott
Dale D Scott
Earl W Scott
Sheldon S Scott
Terri A Scribner
Larry Scrivens
Roy L Scrivner
Susan G Scrivner
Victor Scuderi
Nelson Seabolt
Chin Seah
Lori Sears
Dana J Seaver
Marian Seder
Frank Seelos IV
Ron Segura
Ruben C Segura
Boris V Semenov
Frank Semerano
Frank A Semerano
Caesar Sepulveda
Linda S Sepulveda
Charles Serafy
Homayoun Seraji
Frederick Serricchio
Frank Sesko

Yong Set
Paul J Setty
Donald R Sevilla
Faith A Sexson-Rayl
Ted Shab
James Shafer
Donald Shaffer
Scott J Shaffer
Michael G Shafto
Thomas W Shain
Michael Shakar
Partha Shakkottai
Valerie Shalin
Kurt Shalkhauser
Arshaluys Shamilian
Phil Shampinato
Alex Shapero
Andrea Sharff
Boz Sharif
Colleen Sharkey
Neelam Sharma
Khaled Sharmit
Jacquelyn J Sharp
Stephanie L Sharp
William O Sharp
George Sharpe
George D Sharpe Jr.
Janet Sharpless
Tom Sharpless
Edward E Shaw
Mark J Shay
Dennis J Shebel
Kim R Shelley
Glenn Shephard
Laurence M Sheridan
Conrad D Sherman
Statley Sherman
Katalin M Sherwood
David Shiels
Alex Shih
Choon-Foo Shih
Jac Shinaman
Fred Shipp
Lori R Shiraishi
Brad A Shogrin
Alan B Short
I-Hsiang Shu
Chris Shunn

John Shuping
Herbert L Siegel
Matt Siegler
Jon Siegrist
Keith Siemsen
Maarten Sierhuis
John F Sietz
David L Sieving
Marlene Siewert
Burton C Sigal
Deborah A Sigel
Elliott H Sigman
Baltasar Siguero Pascual
Art Silva
Leroy N Silva
Steve Silverman
David Simmonds
Jerry Simmons
Marvin Simmons
Aaron L Simo
Roger Simond
James K Simons
Dick Simpson
Michael Sims
Jeff Sincell
Elias Sinderson
Sase Singh
Mitra S Singhal
Frank L Singleton
Samuel W Sirlin
Allen R Sirota
John V Sisino
Donald Sitterly
Norm Skiby
Dawn J Skinner
John Skok
Richard Slaght
John S Slater
Robert M Slatkin
Susie Slavney
Glenda L Small
Marshall C Smart
Garry Smee
Daniel E Smidt
Chad Smith
Christopher Smith
Dane Smith
Ella Smith

G H Smith
Gary L Smith
Gene Smith
Giana Smith
Glenn Smith
Harman Smith
James A Smith
James R Smith
John Smith
Ken Smith
Kevin S Smith
Larry D Smith
Marnell Smith
Matthew W Smith
Michael Smith
Michael D Smith
Nelson W Smith
Pamela R Smith
Patrick J Smith
Peter Smith
Philip S Smith
Richard P Smith
Sung S Smith
Sye Smith
Todd A Smith
Trevor Smith
Trey Smith
Bradley G Smock
Roberta Smolik
David Smyth
Nathan Snider
Karen J Sniff
Roosevelt Snipes
Clark Snowdall
Daniel Snyder
Gerald C Snyder
Joseph F Snyder
Michele E Snyder
Jason Soderblom
Larry Soderblom
Jascha Sohl-Dickstein
Jeffrey T Sokol
Stan Soll
Al Solla
Raphael Solomon
Deborah Soltesz
Mark Somerville
Kenneth C Sommerfield

Gregory B Sorber
Peter L Sorci
George F Soule
David M Soules
Kathy Sovereign
Dan Spadoni
Philip M Spampinato
John T Spanos
Nicole Spanovich
Barry L Spath
Roberta Spaulding
Ted D Specht
A B Spector
Kathleen P Spellman
Cherise Spencer
Amelia A Spendlow
Richard B Spitler
Rose Spizzirri
Keith Splawn
Williams K Splawn
Stephen M Spohn
Terry Sprague
Ed Springer
Mark L Springer
Richard J Springer
Jim Sprunck
Kathryn A Spurrier
Steve Squyres
Beverly C St Ange
James R Staats
Jonathan Stabb
John R Stagner
Matt Staid
Alice Stanboli
Chuck Standen
Kelly W Stanford
Carol L Stanley
Thomas W Starbird
Michael D Starkey
Richard Starznyshi
Lynn R Stavert
Kimberly B Steadman
Adam Steege
Mark Steele
Andre T Stefanovich
Ivan Stefek
Richard A Steffke
Daniel R Steidle

Christopher A Stein
David Stein
James Stein
Tom Stein
Mj Steinbacher
Carl N Steiner
Robert C Steinke
Christopher B Stell
Paul M Stella
Adam D Steltzner
Edward Stenger
Jane A Stephens
John Stephens
Richard Stephenson
Robert L Stephenson Jr.
Thomas Stepleton
James A Steppe
Kirsten S Sterrett
Pryce Steve
Philip P Stevens
Shirley J Stevenson
Cynthia K Stewart
Harry Stewart
John Stewart
Roger Stewart
Richard J Stiebel
Ralph B Stocker
Derrick L Stockhausen
Karen Stockstill
Barry Stokker
Andy Stone
Benjamin E Stone
Dusty J Stone
Ernest W Stone
Henry W Stone
James A Stone
Frank R Stott
Erik F Stover
Kennis Stowers
Sheryl Strahs
William D Strauss
Scott Striepe
Sergiu Stroescu
William D Stromberg
Dellon R Strommen
Ashley W Stroupe
Frederick V Stuhr
J Stultz

Kathryn F Sturdevant
Laura P Su
Ling Su
Aranzazu Suarez Perez
Olga Suarez Rodriguez
Bob Sucharski
Tracie Sucharski
Jay M Sucher
Jason J Suchman
John B Suchsland
M Sucy
Larry D Suedmeier
Mike Sullivan
Rob Sullivan
Woody Sullivan
Susan L Summerhays
Kim Sun
Nien-Tung W Sun
Eric T Sunada
Richard F Sunseri, Dr
Karl-Heinz Suphan
Subbarao Surampudi
Bob Surratt
Kathleen A Sutton
Shigeru Suzuki
Chris Swan
Jerry T Swanson
Brian Swayze
George A Sweeney
Robert L Sweet
Edward R Swenka
Bradford L Swenson
Gary M Swift
John A Swift
Matt Switzer
Betty J Sword
Christine E Szalai
Thach Q Ta
Yaro Taeger
Miranda Tafoya
Hiro Takeda
Kevin P Talley
John M Tallon
Kevin Tan
Grace H Tan-Wang
Suzanne Tapia
Patricia A Tate-Belmont
Mike Tauber

Eric P Tauer
Tim Tawney
Peter Tay
David J Taylor
F H Jim Taylor
Frances M Taylor
Keith Taylor
Michael A Taylor
Randall L Taylor
Russell Taylor
Wayne Taylor
Zachary Taylor
James Tedesco
Farinaz Tehrani
Andres Teijo Barbeito
Cristina Temprano De La Pe̅na
Vicente Tendero Romero
Imelda Terrazas-Salinas
Dee Terry
Vickie E Terry
Avo Terzian
William N Tesch
John Teter
Dennis A Teusch
Timothy Thames
Bonnie C Theberge
Rick Theis
Peter C Theisinger
Michael P Thelen
Demitri G Theodoropoulos
Larry Theriault
Daniel Theroux
M Jose Thiery Manrique
David L Thiessen
Shannon Thiessen
Benjamin L Thoma
Amanda Thomas
Andrea Gail Thomas
Guy Thomas
Marlene Thomas
Nick Thomas
Reid C Thomas
Maruk Thomasian
Arthur D Thompson
Charles Thompson
James Thompson
Mark Thompson
Patrick M Thompson

Patrina G Thompson
Randolph W Thompson
Shane Thompson
Stanley K Thompson
Toni Thompson
Craig Thomsaian
Patricia L Thornley
D Thornton
Janine M Thornton
Marla S Thornton
Scott Tibbitts
Michael J Tickner
Darrin Tidwell
Jo E Tillis
Jason T Tillman
R Frank Tillman
Hubert Garard Timmerman
A Timpe
Aurelio Tinio
Robert E Tio
Christopher S Tippit
Daniel A Titus
Claudia Tobar
Ruth Ann Tobler
John W Tobler Jr.
Joseph F Toczylowski
Anthony D Toigo
John T Tolbert
Irene Tollinger
Preston Tollinger
Fred A Tomey
Alan Tomotsugu
Thara Tongvanit
Jerry Toogood
Nicholas Toole
Jeffrey Tooley
Teresa F Toops
Carlos G. Topham Reguera
Jordan L Torgerson
Donald H Tormohlen
Livio Tornabene
R Jay Torres
Ricardo G Torres
Jim Torson
Nick Tosca
Daniel C Tostevin
Tom Tourville
Arno J Toutounjian

Julie A Townsend
Lisa Townsend
Richard W Townsend
Ben Toyoshima
Albert H Tran
Hung Tran
Huy Tran
Mau-Huu Tran
May H Tran
Phong Tran
Steven T Tran
Tuan C Tran
Ashitey Trebi-Ollennu
David C Tresch
Sarah R Trester
Virginia A Trester
James V Tribbett
Jeff Tricarico
Lawrence H Trilling
Jay Trimble
Robert Trimble
Hung Trinh
Joseph T Trinh
Rosary Troia
Jennifer H Trosper
Ung T Troung
Dennis Troxell
Robert F Troy
David True
Paul C True II
Pablo Trueba Manzano
Letricia Trujillo
Marc C Trummel
Brian Truong
Chi Truong
Hung Truong
Phuong Truong
Dean C Tsai
Jack Tse
David S Tseng
Stephen W Tseng
Wan B Tsoi
Walter S Tsuha
Glenn T Tsuyuki
Eldred F Tubbs
Howard Tucker
James L Tucker
Ramona H Tung

Edward W Tunstel
Todd Turnbull
Cheryl J Turner
Eric Turner
Peter Turner
Ron Turner
Bert E Turney
Perica Turudija
Andrew Jack Tuszinski
Mark Tyson
Santi Udomkesmalee
Hal Uffelman
Larry D Uhlenkott
Antonio Ulloa-Severino
David B Ulrich
Jeffrey W Umland
Muhammad Ali Ummy
Jesse Umsted
Mark L Underwood
S Ung
Jose Uribe
Danelle S Uyeda
Robert T Uyeda
Joseph D Vacchione
Arvydas Vaisnys
Alex Valdez
Michael Valdez
Paul Valdez
Larry L Valenta
Gerald L Valentine
Lorraine V Valenzuela
Juan A. Valera Ramirez
Adolfo A Valerin
Steven V Vallejo
Mark Vallejos
Paul Van Damme
John Van Dommelen
Charles N Van Houten
Lynn Van Meter
Merrell L Van Sickle
Paul Van Velzer
Marc Vancampen
Scott Vanderzyl
Hawk Vanek
Joseph V Vanella Jr.
Charles A Vanelli
Charles N Vanhouten
Lynn Vanmeter

Philip Varghese
Ozan Varol
Raymond Vasas
John E Vasbinder
Michael L Vaughan
Thomas E Vaughan
Ashton G Vaughs
Manuel F. Vazquez Cortes
Jesus Vazquez Garrido
Carlos L. Vazquez Sala
Felix E Veal
Mark J Veile
Klaus-Dieter Veit
Anthony R Velasquez
Vincent Venegas
Arnaud Venet
Alonso Vera
James M Vercammen
Jerry Verlinde
Christopher Versfelt
Cheryl L Verstraete
William Vialpando
Jason H Vierra
Kenneth J Vigil
Sonny P Villa
Chad M Villacorta
Susana Villalba Merino
Victor Villanueva Morales
Enrique B Villegas
Ambrosio Villeta Garcia
Gustavo Villeta Sanchez
Michelle A Viotti
Manny Virata
Vincent M Visco
William G Vlahos
Hung D Vo
Quince K Vo
Kent Volkmer
Richard A Volpe
William Von Kamp
Vincent P Von Ruden
Marsette A Vona
Somsak S Vongphasouk
Cami Vongsouthy
Christopher J Voorhees
Vatche Vorperian
Richard M Vose
Sarah D Vose

Pim W Vosse
Duc T Vu
Thien-Ly Vu
Hemali Vyas
Samuel S Waddles
Debra A Wade
Julie A Wade
Mark Wadsworth
Bruce C Waggoner
Michael Wagner
Herman Wagner Jr.
Amy M Wakefield
Roxana Wales
Nancy J Walizer
Jamie Walker
Tony Walker
Walter Walker
Stuart D Walker
John M Wall
Mark S Wallace
Matthew T Wallace
Brian Wallace
Bill Wallach
Jack Wallick
Francine D Walls
Dennis Walstrum
Alfred Walters
Brad Walters
Bruce F Walters
Yolanda Walters
Joan D Walton
Alian Wang
Gordon Wang
June Wang
K Charles Wang
Wei-Min Wang
Heinrich Wänke
Donald O Ward
Jennifer Ward
Leonard Ward
Valerie A Ward
Robert M Warden
Joanne M Ware
Keith R Warfield
Zachary G Warfield
Michael J Warner
Ian Warren
Richard M Warrington

Robert R Warzynski
Lance Watanabe
Susan Watanabe
Alfred Waters
Elsa G Waters
John J Waters
David R Watkins
Michael M Watkins
Aubrey Watson
Deborah W Watson
Wesley Watters
John B Watterson
Jerry Watts
Geoffrey G Wawrzyniak
Allan H Wax
David Way
Peter J Waydo
Stephen J Waydo
Leonard R Wayne
Tom Wdowiak
Marilyn Weaver
Paul D Weaver Jr.
Rick Webb
Guy W Webster
Neil Webster
Richard Grayson Webster
Kevin E Weed
John Weenink
John W Wehner
Zongying Wei
Colleen A Weil
Ed Weiler
James Weiler
James C Weiler Jr
Craig Weinstein
Stacy S Weinstein
Ralph Weis
Jerry Weisbaum
Gerald Weisbaum Jr.
Freia R Weisner
Michael A Weiss
Catherine Weitz
Richard V Welch
Ronald T Welch
Roger Welker
Aaron V Wells
Edward E Wells
Kendra Wells

Kevin Wells
Samuel D Wells
Tim Welton
John J Wenckus
Robert G Wendlandt
Brian Wenham
Tim Werner
Mike Wert
Randii R Wessen
Stephen West
Sylvia West
Timothy J Westegard
Jenna M Westrick
Richard E Whalen
Patrick Whelley
Charles W Whetsel
Larry D Whitcanack
Christopher V White
Ed White
Leslie A White
Mark White
Mary L White
Richard White
Ronnie White
Vernie White
James L Whitman
Barbara Whitney
Albert Whittlesey
Justin Wick
Joe Wieclawek III
Michael C Wiedeman
Douglas J Wiemer
Brandi Wilcox
Brian H Wilcox
Jamila A Wilcox
Reed E Wilcox
Katherine D Wilde
Bruce Wiley
James Wilia
Rebecca Wilkins
Elizabeth A Wilkinson
Mark W Wilkinson
Colette M Wilklow
William H Willcockson
Becky Williams
Brian D Williams
Brian M Williams
Gregory L Williams

Joseph Williams
Randolph Williams
Robert Williams
Robert F Williams
Robert L Williams
Will Williams
Lonnie R Williamson
Richard E Williamson
Christine Willis
Jason R Willis
Kirby Willis
Paul B Willis
Reg G Willson
Micah Wiloth
Darrel "Skip" Wilson
Darrell L Wilson
David A Wilson
Jack Wilson
James R Wilson
Michael G Wilson
James E Wincentsen
Robert Wing
Robert E Wingren
John W Wirth
Susan M Wirth
Roy Wiseman
Sandra Wiseman
Paul Withers
Terry Withers
Eddie Witherspoon
Al Witkowski
Allen Witkowski
Mona M Witkowski
Joseph W Witt
Steven M Wolery
Jay Wolf
Michael J Wolff
Peter J Wolff
Pamela J Woncik
Conway Wong
Li Ann Wong
Bobbie W Woo
Erika B Woo
Steve B Woo
Eric G Wood
Mark D Wood
Catherine A Woodall
Russell W Woodall

Jack Wooddell

Paul R Woodmansee

Thomas J Woodruff

Bruce Woodward

David Woodward

Scott A Woodward Sr.

Scott M Woolaway

John T Woolhouse

Ted Woolhouse

Deborah A Work

Timothy Work

Mark Worley

Edward J Worner Jr.

Stephen Wragg

Anne Wright

Gerald T Wright

Jesse J Wright

John R Wright

Shawn Wright

Shonte J Wright

Steve Wright

Michael Wu

Yi-Chien Wu

Richard Wujek

Mike Wyatt

Nelson Wyatt

Yme Wyminga

Terry R Wysocky

Jonathan Wysong

Fengliang Xu

Gary M Yagi

Flora Yang

George Yankura

Jill M Yarger

Andre H Yavrouian

Rosendo Ybarra

Ross Ybarra

Scott A Yeats

James Yee

Albert Yen

Jeng Yen

Connie P Yencer

Robert A Yenne

Sergey Yentus

Jack E Yerigan

Greg Yeseta

Byron G Yetter

Jeffrey Yglesias

Shyuan-Ju Yin

Homer D Yocum

Chuck Yoder

Lauren York

Lewis G York

Ron York

Alan F Young

Anthony A Young

Bill Young

Eric P Young

Jeffrey Young

Maria A Young

Michael L Young

Ron Young

Simeon Young

Tammy L Young

Thomas E Young II

Dan Yucht

Peter A Yuritich

Serjik Zadourian

Payam Zamani

Stephen M Zapp

Steve Zapp

Tony Zavala

Annie Zeiger

Daryl Zeiger

Pete Zell

Ron Zellitti

Mathew A Zemel

Kathleen J Zetune

Liwei Dennis Zhang

John Ziats

Robert Ziemke

Samuel H Zingales

Lyle Zink

Jutta Zipfel

Jack Zivic

Phillip J Zulueta

James F Zumberge

Chris Zuniga

Richard W Zurek

Pedro Zurita San Andres

Sam Zwick

John M Zynsky

INDEX

accretionary lapilli, 301, 302, 305

Adirondack, 260, 261, 263, 265, 303

Adler, Mark, 74–75, 76, 82–83, 90, 91, 101–2,
   126, 239, 241, 243–44, 269, 280, 286, 288

airbags, 229
   Athena-in-Bags, 69–70, 71, 73, 74
   on *Mars Pathfinder*, 69, 71, 74, 79–80, 108, 124–26
   MER, 112, 124–27, 138, 141, 144, 201
   on *Opportunity*, 297–98, 304
   on *Spirit*, 243, 246, 248, 255, 258–59

alpha detectors, 169, 170–71

American Geophysical Union (AGU), 29, 30

Ames Research Center, 9, 10, 38
   wind tunnel at, 136–40, 141

Amundsen, Roald, 237–38, 329

Antarctica, 376–77

AO (Announcement of Opportunity), 10–11, 12
   for "Discovery" program, 31–37, 44
   for integrated payload, 22–28
   for *Mars 2001*, 37–41
   for *MESUR Pathfinder* camera, 17–20, 27

APEX (Athena Precursor Experiment), 50–51,
   66–71

Apollo, 104, 320

APXS (Alpha Proton X-ray Spectrometer), 25,
   32, 33, 50, 69, 75, 81, 165, 168–78, 185
   on *Opportunity*, 295, 305–7, 312, 314–16, 333,
   368
   on *Spirit*, 253, 264, 364, 369, 370

Arizona, University of, 18, 19, 20

Arvidson, Ray, 26, 29–30, 32, 88, 262, 288, 308,
   350, 351, 364–67, 369

Athena, 39, 45, 50–51, 66–71, 78–79, 104, 126
   as "Discovery" mission, 33–37, 38, 44
   Mars Geologist Pathfinder and, 75–76
   Mars Mobile Pathfinder and, 76–83

Athena-in-Bags, 69–70, 71, 73, 74

Athena-Pathfinder, 75

ATLO (Assembly, Test and Launch Operations),
   142–63, 166–78, 179, 179, 181, 183–85,
   187, 210, 233

Baez, Omar, 210–11, 216

Ball Aerospace, 14, 15, 18, 26, 39, 161

basalt, 308, 322, 323, 324, 338, 347, 352, 363,
   364, 372, 373

Beagle 2, 221–22, 224, 228, 229, 250

beamsplitter, 333–34

Beisadecki, Jeff, 349

Bell, Jim, 251, 254

Bernhardt, Bodo, 259

Berry Bowl, 318, 319, 320

Bister, Leo, 156, 158–62, 332

Blake, Dave, 38

Bogdanovski, Oleg, 171, 172

Bollen, Diane, 39–40, 41

Bonneville Crater, 264–65, 322–24, 325, 330,
   331, 351, 352, 375

Boyce, Joe, 19–20, 27, 40
Boyles, Mark, 195
bromine, 312–13, 315–16, 317
Burke, Kevin, 246
Burns, Roger, 338
Burns Cliff, 338, 339
Burt, Ron, 175–77

Callas, John, 233
cameras, 12, 14, 15–16, 23, 38, 39, 56,153
    DIMES, *see* DIMES
    Hazcam, *see* Hazcam
    MESUR Pathfinder, 17–30, 27
    MI, *see* Microscopic Imager
    Navcam, *see* Navcam
    Pancam, *see* Pancam
    Suncam, 109, 128
Cape Canaveral, 52–53, 158, 164
    range safety at, 211–13
Cape Canaveral Air Force Station (CCAFS), 164
    PHSF at, 164–67, 171
carbonates, 323
Carr, Mike, 288, 353
Casani, John, 75
Cassini, 74
CDR (critical design review), 110–12, 113,
    117–18
Challenger, 62, 200
chlorine, 316, 364, 370, 372
Christensen, Phil, 24, 26, 122, 298, 342
Clinton, Bill, 34
CMSA (Cruise Mission Support Area), 238–45,
    288
Collis, Sofi, 201
Columbia, 163, 180, 256, 352
Columbia Hills, 256, 265–66, 322, 324, 325,
    327–29, 350, 352, 354, 357, 363, 370–73,
    375
concretions, 301, 311, 312, 320, 363
Cook, Richard, 77, 105–6, 108, 112, 129, 135, 162,
    174, 177, 193, 228, 239, 240, 269, 277, 278,
    282, 288, 290, 337–39, 348
Cornell University, 1–2, 9–10
    Mars Room at, 2–3
craters, 253, 299–300
    Bonneville, 264–65, 322–24, 325, 330, 331,
        351, 352, 375
    Eagle, 294–321, 323–25, 329, 331, 336, 338,
        344, 362, 378

Endurance, 321, 322, 324, 329, 330, 334–39,
    350–52, 357, 371
    Fram, 329
    Gusev, *see* Gusev Crater
    naming of, 256–57
Crisp, Joy, 75, 111–12
crystal molds, 310–11, 312
Cunningham, Glenn, 145, 146

Dallas, Saterios "Sam," 33, 37, 40
Deep Sleep, 332–34, 340, 341, 342, 343, 346, 357,
    375
Delamere, Alan, 14–15, 18, 19, 26, 39
Dells, 317, 318, 319, 374
Delta rockets, 198, 200–203, 205, 206, 212,
    217
Des Marais, Dave, 327–29
DIMES (Descent Image Motion Estimation
    System), 129–30, 135, 138, 141, 157, 227
    on *Spirit,* 250, 256, 264–65
Discovery Hut, 376–77
"Discovery" program, 31
    AO for, 31–37
    Athena proposal for, 33–37, 38, 44

Eagle Crater, 294–321, 323–25, 329, 331, 336,
    338, 344, 362, 378
    naming of, 320
EDL (Entry, Descent, and Landing), 120–41,
    227–28
    of *Opportunity,* 264, 280, 290–91, 299
    of *Spirit,* 240–44
Eisen, Howard, 75
Elachi, Charles, 32, 71–72, 150, 187–89, 192–93,
    195, 241–42, 288–90, 338, 346
El Capitan, 308, 310–17
Elysium Planitia, 124, 125, 141
Endurance, 321, 376
Endurance Crater, 321, 322, 324, 329, 330,
    334–39, 350–52, 357, 371
engineers, 375
    scientists and, 11–12, 98, 152
Estabrook, Polly, 240–44, 289
evaporites, 316, 338, 374

flight software, 154–56, 270–71,
    277–87
Fram, 237, 329
Fram Crater, 329

Galileo, 21
Garvin, Jim, 88–89, 90
Gavin, Tom, 73
Gavsat (*Mars Science Orbiter*), 73, 74, 76, 78, 79, 83–84, 86–92, 101
Gellert, Ralf, 169–71, 172–78, 185, 261, 306, 316
geology, 1, 4–5, 81, 82
Germany, 25, 48–49
glass, 301
Giuliani, Rudy, 116
goethite, 372
Goldin, Dan, 21–22, 49, 64, 87–92
Goldstein, Barry, 37, 40–41, 51, 66–70, 72, 73, 74, 77–78, 90–92, 162, 185, 186, 243
Gooding, Jim, 24, 26
Gorevan, Steve, 34–35, 92, 113–14, 259, 303, 313, 366
    World Trade Center and, 113–16
Grant, John, 288
Greeley, Ron, 288
Grotzinger, John, 260, 303, 308–11, 314–15, 317–20, 336, 337, 339, 340, 374
GSE (ground support equipment), 172–73, 174, 176
Gusev Crater, 152, 225, 230, 288, 308
    Columbia Hills and, 256, 265–66, 322, 324, 325, 327–29, 350, 352, 354, 357, 363, 370–73, 375
    consideration of site, 121–25, 128, 129, 141
    Pot of Gold at, 353–56, 358–72
    rocks in, 254, 258–61, 263–65, 303, 352–56, 358–73
    Spirit on, 245–66, 267–87, 288, 303, 322–29, 337, 350–70
    Spirit's approach to, 225–29
    Spirit's landing on, 237–45, 265
    see also Spirit
Guzman, Jose, 178

Hansen, Candy, 38
Harri, Ari-Matti, 25
Haskin, Larry, 30, 82
Haynes, Norm, 34
Hazcam (hazard avoidance camera), 128, 157
    on *Opportunity,* 292, 299, 349
    on *Spirit,* 247, 252, 258, 259, 263
HEDS (Human Exploration and Development of Space) instruments, 44–49

hematite, 122–23, 224, 298, 300, 301, 304, 305, 307, 311, 312, 317, 318, 320, 361, 362, 363, 369, 372
HGA (high-gain antenna), 109, 195–96, 263, 271, 276, 287
hill-climbing tests, 343–46
Hinners, Noel, 31–32
Honeybee Robotics, 34–35, 92, 113, 303, 313, 365–67, 369
    RAT and, 82
    World Trade Center attack and, 113–15
horizontal velocity sensors, 126–27
HRS (Heat Rejection System), 240
Hubbard, Scott, 86–92
Hubble Space Telescope, 66
Huck, Fred, 15
Huntress, Wes, 20, 40, 45, 46
Husband, Rick, 352
Husband Hill, 352–53, 360
    West Spur of, 352–53, 360, 367, 368, 370, 371

IDD (Instrument Development Device), 310, 347
    on *Opportunity,* heater for, 295–96, 298, 315, 332–34
    on *Spirit,* 257, 259, 261, 263–65, 359, 360
IRT (Independent Review Team), 110, 117–18, 143–46, 150, 180
ISIL (In-Situ Instruments Laboratory), 233, 235, 258

jarosite, 306, 307, 311, 314, 315, 317, 338, 373
JAXA, 221
JPL (Jet Propulsion Laboratory), 17, 21, 31–34, 36–40, 42–47, 51, 68, 69, 71, 73, 74, 104, 146, 150
    Athena-in-Bags and, 69–70
    building 264, 228, 230–36, 249, 263
    Flight Operations Facility, 239
    Mars Climate Orbiter failure and, 55, 56
    Mars Geologist Pathfinder and, 75–76
    Mars Mobile Pathfinder and, 76–78
    Mars Polar Lander failure and, 62–63, 64, 65, 67–68, 70, 77
    Mars program meeting of, 70–72
    Mars time schedules and, 326
    MER project and, 97, 99, 105, 110, 117, 129, 130, 135, 143, 145, 151–53, 158; *see also Mars Exploration Rover*

JPL (*continued*)
  NASA restrictions and, 64–65
  QA engineers at, 175, 178
  Spacecraft Assembly Facility, 147–48, 151, 152, 160–61, 162

Karatepe, 337, 339, 340, 343, 344, 346–47, 357–59, 367
  hill-climbing tests and, 343–46
Kennedy Space Center (KSC), 164
Klingelhöfer, Göstar, 25, 171, 174, 222, 224, 252, 253, 260–61, 305, 306, 361, 362
Knoll, Andy, 305, 311, 318
Kondos, Steve, 115
Krajewski, Joel, 299

Langley Research Center, 15
lapilli, 301, 302, 305
Larry's Leap, 336, 337, 339
Last Chance, 317, 318, 319, 374
launch locks, 182
lava, 254–55, 258, 266, 297, 301
Lawler, Andrew, 91, 92
Lee, Gentry, 112–13, 269, 275
Lee, Wayne, 124, 125, 126, 137, 228, 242–44, 289, 290, 363
Level One Requirements, 102
  for *MER*, 102, 103–4, 112
Lewicki, Chris, 245, 246, 248–49, 291, 292, 293
Limonadi, Daniel, 276, 278, 292
Lindemann, Randy, 104–5, 106–7, 110, 296
Lockheed Martin, 26, 31–32, 50, 70
  Mars Climate Orbiter and, 55–56
Lookout Point, 359, 360
Lowry, Lynn, 36
Lunakhod, 25
Lunar and Planetary Institute, 86

McKay, Chris, 38
McKay, Dave, 33–34, 35
McKittrick MiddleRAT, 313
McSween, Hap, 305
magnetite, 123
Maki, Justin, 156, 222, 245–47, 291–94, 304, 353
Malin, Mike, 17–21, 23–26, 29, 42, 56, 83, 250, 297, 311, 330
Malin Space Science Systems, 17, 57
Manning, Rob, 75, 125–30, 161, 162, 164, 227–28, 240, 243, 244, 264, 278, 289, 290

Marie Curie, 50, 81
Mariner, 21
Mars:
  days on, *see* sols
  dust on, 3, 225, 226–29, 249, 323, 324
  fossils in meteorite from, 33–34, 35
  rocks on, *see* rocks
  soil on, *see* soil
  temperature of, 3, 376
  water on, evidence of, 3, 83, 88, 123, 254, 264, 306, 307, 308–21, 323, 355, 362, 364, 367, 370–74
  winds on, 122, 124, 125, 126, 128, 129, 130, 265
Mars Climate Orbiter (MCO), 54–56, 62–63, 64, 65, 86, 91, 99
Mars Environmental Survey (MESUR), 16–17, 20–21, 30, 71, 73
  Pathfinder, see Mars Pathfinder
Mars Exploration Rover (MER), 92–93, 97–113, 117–19
  airbags for, 112, 124–27, 138, 141, 144, 201
  ATLO for, 142–63, 166–78, 179, 181, 183–85, 187, 210, 233
  budget for, 90–92, 102–3, 112, 118–19, 130, 143, 144, 150
  cameras in, *see* cameras
  CDR for, 110–12, 113, 117–18
  EDL in, *see* EDL
  egress ramps for, 110, 111, 112, 255, 257–58
  Elysium Planitia considered as site for, 124, 125, 141
  Gusev Crater site for, *see* Gusev Crater
  independent review of, 110, 117–18, 143–46, 150, 180
  landing system for, 74, 79–80, 83, 108, 112, 124–27, 144
  launch window for, 98–99
  length of mission, 100–104
  Level One Requirements for, 102, 103–4, 112
  as *Mars Geological Rover (MGR),* 83–85, 86–92
  as *Mars Geologist Pathfinder (MGP),* 75–76
  as *Mars Mobile Pathfinder (MMP),* 76–83
  Melas Chasma considered as site for, 121–22, 124
  MER-1 rover, *see* Opportunity
  MER-2 rover, *see* Spirit
  Meridiani Planum site for, *see* Meridiani Planum
  names of workers on, 379–407

parachutes for, 112, 130–41, 144, 228, 242, 265
pyros in, *see* pyrotechnic devices
readiness reviews for, 189, 190, 210
REM in, 147–48, 151
schedule for, 93, 98–99, 104, 112, 130, 143, 148–51, 157–58, 161, 162, 167, 177
size and weight issues in, 108–12, 124–25, 127, 128, 130, 138
solar arrays in, *see* solar panels
sols and, 99–103
spectrometers in, *see* spectrometers
thermal vacuum tests for, 152–58, 170, 171, 187, 246, 279, 284
two rovers in plan for, 89–93, 118–19, 143–46, 150–52
World Trade Center pieces in, 115–17
Mars Express, 221, 222
Mars Global Surveyor (MGS), 22, 24, 26, 54, 61, 67, 83, 109, 122, 225, 250, 251, 270, 272, 273, 275–78, 330, 333, 368
Mars Observer, 18, 20–22, 24, 26, 57
Mars Odyssey, 109, 225, 230, 244–46, 249, 250, 263, 268–70, 273, 277, 333, 334, 360
Mars Pathfinder, 20–21, 22, 31, 62, 75, 77, 99, 103, 228
camera for, 17–20, 27
landing system of, 69, 71, 74, 79–80, 83, 108, 124–26
Sojourner, 20, 28, 31, 42, 50, 80, 81, 146
solar arrays of, 100
Mars Polar Lander (MPL), 26, 27, 28, 37, 38, 56–71, 73, 77, 86, 91, 99, 126–27, 221–22, 228
Mars program, 21–22, 30, 31, 35, 52, 56
changes made in, 66
costs of, 64, 65, 90–92
funding for, 35, 36, 38
meeting on, 70–72
2003 and 2005 plans for, 30, 44–49, 51, 65, 66, 69, 71, 72, 73, 75, 86, 187
2003 mission choice, 78–85, 86–93, 101
restrictions on, 64–65
Mars Room (Cornell University), 2–3
Mars Sample Return, 65, 66, 68
Mars Science Orbiter (MSO; Gavsat), 73, 74, 76, 78, 79, 83–84, 86–92, 101
Mars 2001 project, 40, 42–51, 66–71, 76, 77
AO for, 37–41

budget for, 45, 46–47, 49
schedule for, 46
Martin Marietta, 18
Matijevic, Jake, 42, 46, 75
MAV (Mars Ascent Vehicle), 51–52
Mazatzal, 324
Melas Chasma, 121–22, 124
Melko, Joe, 291
MER, see Mars Exploration Rover
Meridiani Planum, 152, 225, 230, 288
consideration of site, 121, 122–24, 125, 141
Endurance Crater at, 321, 322, 324, 329, 330, 334–39, 350–52, 357, 371
Fram Crater at, 329
Karatepe site at, 337, 339, 340, 343–47, 357–59, 367
Opportunity on, 291–321, 322–24, 327, 329, 330, 334–49, 350–51, 355, 357–59, 364, 367, 368, 371
Opportunity's approach to, 224, 225, 271
Opportunity's landing on, 273, 280, 288–91
rocks at, 293–95, 297–300, 303–21, 335–36, 347, 349, 368, 371, 376
water at, 311, 313, 315–17, 319–21, 373–74
see also Opportunity
MESUR, see Mars Environmental Survey
meteorology, 23, 25, 31
MI (Microscopic Imager), 32, 33, 75, 81, 109, 128, 153, 157
on *Opportunity,* 295, 300–301, 302, 304, 305, 310, 312, 313, 317, 319, 368
on *Spirit,* 253, 264, 359, 360, 362, 363, 369–70
Mini-Corer, 35, 81, 82
Mini-MAV, 51–52
Mini-TES, 24, 26, 32, 39, 42–43, 46–48, 50–51, 69, 75, 78, 81, 153–54, 328
Deep Sleep and, 333–34, 340, 341, 342, 343, 346, 357, 375
on *Opportunity,* 298, 299, 304, 305, 333–34, 347, 375
on *Spirit,* 245, 255, 258, 271, 323, 352
Morris, Dick, 88, 306, 360–62
Mössbauer spectrometer, 25, 32, 33, 48–51, 69, 75, 81, 165, 171, 224, 377
on *Opportunity,* 222, 223–24, 253, 295, 298, 302, 304–7, 314, 315, 318–20, 333, 368
on *Spirit,* 222, 223, 224, 251–53, 257, 259–61, 264, 360–63

Mulvanerton, Mary, 213–14
Myrick, Tom, 116–17

Naderi, Firouz, 87, 90, 92, 144, 145, 272
Nanokhod, 25, 26, 28
NASA:
    Ames Research Center, *see* Ames Research
        Center
    Announcement of Opportunity, *see* AO
    failed missions and, 62
    fever charts of, 193
    foreign instruments and, 25, 48–49
    IRT, 110, 117–18, 143–46, 150, 180
    Jet Propulsion Laboratory, *see* JPL
    Langley Research Center, 15
    Level One Requirements of, 102, 103–4
    Mars program, *see* Mars program
Navcam (navigation camera), 128, 157
    on *Opportunity*, 292–95, 334–36
    on *Spirit*, 246, 247, 250, 252, 253
Nozomi, 221

O'Keefe, Sean, 194, 241–42, 288–90
olivine, 264, 361, 362, 364
Opportunity, 5–6, 222, 374–78
    building 264 facilities for, 230–36
    Endurance Crater and, 321, 322, 324, 329, 330,
        334–39, 350–52, 357, 371
    fate of, 376, 377
    first data from, 291–98
    freeing from lander, 298–99
    heater problem of, 295–96, 298, 315, 332–34
    launch of, 213–17
    launch window for, 203, 205–13
    as *MER-1*, 151–52, 155, 177, 178
    on Meridiani, 291–321, 322–24, 327, 329, 330,
        334–49, 350–51, 355, 357–59, 364, 367,
        368, 371
    Meridiani approach of, 224, 225, 271
    Meridiani landing of, 273, 280, 288–91
    mobility of, 324–25, 331
    Mössbauer on, 222, 223–24, 253, 295, 298,
        302, 304–7, 314, 315, 318–20, 333, 368
    pyros on, 191
    solar panels on, 296, 332, 339–42
    solar storms and, 224–25
    superior conjunction and, 370–71
    trajectory corrections for, 271
ORTs (operations readiness tests), 233–36, 247

Paige, Dave, 26–28, 30, 31, 37, 38, 56–57, 60, 71
Pancam (panoramic camera), 16, 17, 18, 23, 26,
    32, 39, 46–48, 50, 69, 75, 78, 80, 81, 109,
    128, 155, 234
    on *Opportunity*, 295, 300, 303–5, 309, 310,
        317, 330, 337, 338, 343, 344, 346, 349, 368
    speckling problem with, 156–63, 246
    on *Spirit*, 245, 246, 251, 253, 254, 259, 323,
        352, 353, 358, 362, 369
parachutes, 112, 130–41, 144, 228, 229
    on *Spirit*, 242, 265
    squidding in, 138–39, 140, 228
Parker, Tim, 250, 335
Pathfinder, see Mars Pathfinder
phosphorus, 370, 372
PHSF (Payload Hazardous Servicing Facility),
    164–67, 171
Pilcher, Carl, 86–90
Pillinger, Colin, 222, 250
PMA (Pancam Mast Assembly), 155, 245, 247,
    250, 263, 292
Pot of Gold, 353–56, 358–72
Powell, Mark, 267
Proton, Jon Beans, 213–14
pyrotechnic devices, 182–83, 185–86, 190–91
    on *Opportunity*, 191
    on *Spirit*, 178, 180–84, 188, 190, 191–92,
        195–96, 240, 245, 257

QA (quality assurance) engineers, 175, 178

RAD (rocket assisted deceleration) motors, 122,
    126, 164–65, 227, 229
Rademacher, Joel, 172, 175, 178
Raman spectrometer, 30, 31, 32, 33, 38–39, 40,
    45, 81, 82
Ranger, 21, 241
RAT (rock abrasion tool), 82, 97, 109, 191
    on *Opportunity*, 307, 313–14, 315, 317, 347,
        368, 371
    on *Spirit*, 255, 259, 303, 324, 364–66, 369,
        370, 372
    World Trade Center pieces in, 116–17
Reeves, Glenn, 271, 273, 277, 278, 281, 282
REM (Rover Electronics Module), 147–48, 151,
    158–59, 162
resistors, ballast, 185–86, 188–92, 195
Rice, Jim, 257, 264, 328
Rieder, Rudi, 25, 81, 169–71, 174, 178, 316

Rio Tinto, 373
rockets:
    cork on, 203–5, 207–11, 217
    Delta, 198, 200–203, 205, 206, 212, 217
    launches of, 197–200
    naming of, 256
    range safety and, 211–13
rocks, 4–6, 24, 27–28, 30, 328
    Adirondack, 260, 261, 263, 265, 303
    Dells, 317, 318, 319, 374
    El Capitan, 308, 310–17
    at Gusev, 254, 258–61, 263–65, 303, 352–56,
        358–73
    instruments for studying, 24–25, 31, 35, 38; see
        also RAT; spectrometers
    Last Chance, 317, 318, 319, 374
    at Meridiani, 293–95, 297–300, 303–21,
        335–36, 347, 349, 368, 371, 376
    naming of, 260
    Pot of Gold, 353–56, 358–72
    ripple cross-beds, 308–10, 312, 314–15,
        317–20, 338, 374
    Stone Mountain, 303, 305–7, 311–13
    vesicles in, 254, 259
    volcaniclastic, 372
    weathering of, 81–82, 355–56
    white, 323–24
Rodionov, Daniel, 252, 260, 261, 361, 362
Ruff, Steve, 355

SAF (Spacecraft Assembly Facility), 147–48, 151,
    152, 160–61, 162
Salvo, Chris, 278, 341
sand, 337, 338, 343, 344, 347
San Martín, Miguel, 127, 128, 141
Science, 91
scientist-engineer relationship, 11–12, 98,
    152
Scolese, Chris, 194
Scott, Robert Falcon, 376–77
September 11 terrorist attacks, 113–17, 133
Sexton, John, 116
Shackleton, Ernest, 321, 376–77
Sheirer, Richard, 116
Shiraishi, Lori, 259
Shirley, Donna, 34, 43–44, 45
Sleepy Hollow, 253, 255, 260, 323
Smith, Peter, 20, 21
SMSA (Surface Mission Support Area), 230–31,

239, 244, 245, 252, 258, 263, 267–68, 272,
    274, 275, 283, 286, 291, 293, 300
Soderblom, Larry, 32, 262, 288, 308, 336, 339,
    353, 368, 370, 371
soil, 24, 26, 31, 33, 255, 257–58, 259, 264, 292,
    297–98, 302, 304–6, 319, 328, 370
    spherules ("blueberries") in, 300–302, 304,
        305, 307, 311–13, 317–20, 329, 344, 354,
        371
    see also rocks
Sojourner, 20, 28, 31, 42, 50, 80, 81, 146
solar panels, 99–108, 110–12, 121, 153, 225
    on *Opportunity,* 296, 332, 339–42
    on *Spirit,* 245, 246, 249, 269, 325, 351,
        356–57, 367
    on *Spirit,* pyros for, 178, 180–84, 188, 190
solar storms, 224–25
sol quartet, 328–29, 350, 351
sols (martian days), 99–103, 249, 262, 326–27,
    350
    planning process and, 100
    restricted, 327
Southard, John, 319–20
Soviet Union, 25
    symposium in, 13–15
SOWG (science operations working group),
    232–33, 262, 288
Space Station program, 49
spectrometers, 23, 24, 38, 45
    Alpha Proton X-ray, *see* APXS
    Mössbauer, *see* Mössbauer spectrometer
    Raman, *see* Raman spectrometer
    Thermal Emission, *see* TES
spherules ("blueberries"), 300–302, 304, 305,
    307, 311–13, 317–20, 329, 344, 354, 371
Spirit, 5–6, 222, 374–78
    building 264 facilities for, 230–36, 249, 263
    Columbia Hills and, 256, 265–66, 322, 324,
        325, 327–29, 350, 352, 354, 357, 363,
        370–73, 375
    communication problem with, 267–87, 288
    Deep Sleep and, 357
    fate of, 376, 377
    first data from, 245–49
    first in-flight health check of, 222–23
    freeing from lander, 257–59
    fuse blown in, 179–85, 188–90, 192
    on Gusev, 245–66, 267–87, 288, 303, 322–29,
        337, 350–70

Spirit (continued)
Gusev approach of, 225–29
Gusev landing of, 237–45, 265
launch of, 197, 201–3
launch window for, 187–94, 195–96
as *MER-2,* 151–556, 158–63, 164, 174, 178
mobility of, 324–25, 331, 371
Mössbauer on, 222, 223, 224, 251–53, 257, 259–61, 264, 360–63
Pot of Gold and, 353–56, 358–72
pyros on, 178, 180–84, 188, 190, 191–92, 195–96, 240, 245, 257
solar panels on, 245, 246, 249, 269, 325, 351, 356–57, 367
solar storms and, 224–25
superior conjunction and, 370–71
trajectory corrections for, 225–26, 229
wheel problem of, 351, 371
Sputnik, 13
squidding, 138–39, 140, 228
Squyres, Katy, 18, 19, 33, 41, 151, 206, 214, 221, 237, 307, 352
Squyres, Mary, 9–10, 18, 19, 33, 41, 151, 186–87, 206, 221, 222, 237, 272, 307, 352
Squyres, Nicky, 18, 19, 33, 41, 151, 206, 214, 221, 222, 237, 307, 352
Steltzner, Adam, 130–41, 290, 363
STGs (science theme groups), 231–32
Stone, Ed, 75, 77, 87
Stone Mountain, 303, 305–7, 311–13
sulfur, 306, 307, 311, 312, 313, 315, 317, 329, 338, 347, 364, 367, 370, 371, 372
Suncam, 109, 128
Surveyor, 21

Tattini, Gene, 187
TCM (trajectory correction maneuver), 225–26, 229
TDR (time domain reflectometer), 160–61
TES (Thermal Emission Spectrometer), 24, 122
Mini-TES, *see* Mini-TES
Theisinger, Pete, 77, 84, 90, 97–98, 103, 104, 111, 112, 129, 130, 143–45, 162, 179, 194–96, 203, 204, 207, 228, 229, 239, 242–45, 269, 272, 275–76, 283, 288, 290
thermal vacuum tests, 152–58, 170, 171, 187, 246, 279, 284
Thompson, Art, 231, 278

TIRS (Transverse Impulse Rocket System), 126, 128, 264, 265
transponders, 192–94
Trosper, Jennifer, 267–68, 269, 282, 324, 325, 326
TSB (Telecom Services Board), 178
Spirit pyros fired in repair of, 178, 180–84, 188

UHF antenna, 109, 245, 271, 272

Valles Marineris, 121
Venus, 33, 71, 72, 73
Viking, 1–2, 15, 21, 29, 79, 99, 127, 132, 228–29
volcanic ash, 297, 298, 308
lapilli from, 301, 302, 305
Voorhees, Chris, 194, 349
Voyager, 21, 230
cameras of, 12

Wallace, Matt, 146–52, 157, 161, 162, 168, 173–74, 177–78, 190–91, 193, 196, 228, 269, 276, 289, 299, 324, 337–39, 343–47
Wänke, Heinrich, 25
water, 4
bromide and chloride salts in, 316
crystal molds and, 310–11, 312
on Mars, evidence of, 3, 83, 88, 123, 254, 264, 306, 307, 308–21, 323, 355, 362, 364, 367, 370–74
ripple cross-beds and, 308–10, 312, 314–15, 317–20, 374
WEB (Warm Electronics Box), 101, 159–60, 270, 272–73, 277, 377
Weiler, Ed, 65–66, 68, 76–77, 87, 88, 118, 119, 141, 144–46, 150, 338, 346, 348
Opportunity landing and, 288–89, 290
Spirit landing and, 241–42
Spirit launch decision and, 187, 188–89, 192–94, 195
Welch, Rick, 341–42
West Spur, 352–53, 360, 367, 368, 370, 371
Willis, Jason, 239, 240, 241, 244
Wirth, John, 171
World Trade Center attack, 113–15, 133

X-band fault, 277, 280–81, 282, 287

Zurek, Rich, 84